# Biological Flow Modelling

*Editors*

**Suvash C. Saha**
School of Mechanical and Mechatronic Engineering
University of Technology Sydney (UTS)
Sydney, New South Wales
Australia

**Akshoy Ranjan Paul**
Department of Applied Mechanics
Motilal Nehru National Institute of Technology Allahabad
Prayagraj, Uttar Pradesh
India

CRC Press
Taylor & Francis Group
Boca Raton London New York

CRC Press is an imprint of the
Taylor & Francis Group, an **informa** business
A SCIENCE PUBLISHERS BOOK

*Cover credit*: Designed by Adrita Saha.

First edition published 2026
by CRC Press
2385 NW Executive Center Drive, Suite 320, Boca Raton FL 33431

and by CRC Press
4 Park Square, Milton Park, Abingdon, Oxon, OX14 4RN



*CRC Press is an imprint of Taylor & Francis Group, LLC*



*Library of Congress Cataloging-in-Publication Data (applied for)*

ISBN: 978-1-032-26311-3 (hbk)
ISBN: 978-1-032-26312-0 (pbk)
ISBN: 978-1-003-28763-6 (ebk)

DOI: 10.1201/9781003287636

Typeset in Times New Roman
by Prime Publishing Services

# Preface

In the dynamic domain of biomedical engineering and Computational Fluid Dynamics (CFD), the exploration of bio-fluid flow phenomenon stands at the forefront of scientific inquiry and innovation. This book serves as a guiding beacon, illuminating the path toward a profound understanding of the complex interplay between fluid dynamics and biological systems. Within these pages, readers will embark on a captivating voyage through the vast expanse of bio-fluid flow modeling, traversing a diverse landscape of research endeavors and technological advancements.

Chapter 1 opens the gateway to the region of airway structure modeling, where the complex architecture of the respiratory system is precisely dissected and reconstructed through advanced computational techniques. Here, researchers investigate the complexities of airway geometry and flow dynamics, unraveling the mysteries of respiratory mechanics with unprecedented clarity and precision. Chapters 2 and 4 transport readers into the pulsating heart of fluid flow and particle transport within the upper airway model, offering profound insights into the mechanisms underlying normal physiological function and the aberrations that manifest in conditions such as stenosis. Through meticulous simulations and analytical scrutiny, researchers illuminate the complex movement of molecules and particles, painting a vivid portrait of bio-fluid dynamics in health and disease.

The symbiotic relationship between computational modeling and drug delivery is the focal point of Chapters 3 and 10, where the transformative potential of CFD in enhancing pulmonary drug delivery and nanoparticle-based therapeutics is laid bare. From inhalers to intricate nanoparticle formulations, researchers navigate the complex terrain of drug transport and distribution, forging new pathways toward more efficacious and targeted treatment modalities. Chapter 5 ventures into the domain of cardiovascular biomechanics, offering a penetrating analysis of physiological blood flow through an abdominal aortic aneurysm. Here, the intersection of computational modeling and clinical insight converges, shedding light on the complex hemodynamic forces that govern vascular pathophysiology and guiding the development of novel diagnostic and therapeutic strategies.

The narrative unfolds further with Chapters 6, 7, and 8, which investigate the simulation of biomagnetic fluid flow in arterial aneurysms, cerebrospinal fluid dynamics, and ocular flow modeling, respectively. Within these chapters, researchers peer into the depths of physiological complexity, unraveling the mysteries of fluid dynamics within the context of neurological and ophthalmological disorders. Chapter 9 expands the horizons of bio-fluid flow modeling, exploring a diverse array of miscellaneous flows that span the spectrum of biological systems. From cerebrovascular dynamics to lymphatic transport, researchers illuminate the multifaceted nature of bio-fluid phenomenon, offering fresh insights and innovative methodologies for understanding and harnessing the power of biological fluids.

Finally, Chapters 11 and 12 cast a gaze toward the future, where cardiovascular assistive devices and emerging trends in bio-fluid flow modeling converge to shape the landscape of healthcare and biomedical innovation. Here, readers are invited to envision a world where computational prowess and scientific ingenuity converge, ushering in a new era of precision medicine and transformative healthcare solutions.

In summary, this anthology stands as a testament to the boundless potential of bio-fluid flow modeling, offering a panoramic vista of discovery and innovation. As readers embark on this journey of exploration and enlightenment, may they be inspired to push the boundaries of knowledge and forge new pathways toward a healthier, more vibrant future.

Suvash C. Saha
Akshoy Ranjan Paul

# Contents

*Preface* **iii**

1. **Techniques of Airway-Structure Modeling** **1**
*Xinlei Huang, Isabella Francis, Akshoy Ranjan Paul* and *Suvash C. Saha*

2. **Fluid Flow and Particle Transport in Upper Airway Model: With and Without Stenosis** **29**
*Bao Vuong Duong, Adnan Munir, Shahid Husain, Hamidreza Mortazavy Beni, Mohammad S. Islam* and *Puchanee Larpruenrudee*

3. **Role of CFD in Pulmonary Drug Delivery using Inhalers** **50**
*Vivek Kumar Srivastav, Akshoy Ranjan Paul, Anuj Jain* and *Suvash C. Saha*

4. **Drug Delivery Through Naso-pharyngeal Routes** **74**
*Nazrul Islam* and *Shafiqur Rahman*

5. **Analysis of Newtonian Physiological Blood Flow through an Abdominal Aortic Aneurysm** **87**
*Md. Anwar Hossain, Mohammad Matiur Rahman, Most. Nasrin Akhter, Md. Mamun Molla* and *Suvash C. Saha*

6. **Finite Volume Simulation of Biomagnatic Fluid Flow in a Model Arterial Aneurysm** **100**
*Md. Mahadul Islam, Md. Mamun Molla, Most. Nasrin Akhter* and *Suvash C. Saha*

7. **Computational Fluid Dynamics of Cerebrospinal Fluid Flow** **118**
*Isabella Francis, Akshoy Ranjan Paul, Md. Mamun Molla* and *Suvash C. Saha*

8. **Ocular Flow Modelling** **140**
*Vivek Kumar Srivastav, Aman Raj Anand* and *Akshoy Ranjan Paul*

9. **Modeling of Miscellaneous Bio-Fluid Flows** **158**
*Akshoy Ranjan Paul, Anuj Jain* and *Suvash C. Saha*

10. **Nanoparticle-Based Drug Delivery** **189**
*Keshav S. Moharir*

**11. Advancements in Cardiovascular Assistive Devices and Future Innovations for Post-COVID Care** **214**
*Yamini Singh, Tej Pratap, Ambak Kumar Rai* and *Abhishek Kumar Tiwari*

**12. New Trends in Bio-Fluid Flow Modelling** **231**
*Akshoy Ranjan Paul* and *Suvash C. Saha*

***Index*** **247**

# Chapter 1

# Techniques of Airway-Structure Modeling

*Xinlei Huang,*[1] *Isabella Francis,*[1] *Akshoy Ranjan Paul*[2] and *Suvash C. Saha*[1,*]

## 1. Introduction

Lung cancer, cystic fibrosis, tuberculosis (TB), asthma, Chronic Obstructive Pulmonary Disease (COPD), and pulmonary hypertension. These are a few of the most prevalent lung diseases that have brought great suffering to our society (Das et al., 2021). COPD and lung cancer are the third and sixth leading causes of death worldwide, according to the World Health Organization (WHO). Lower respiratory infections, which continue to be the most lethal communicable illness globally and claimed 2.6 million lives in 2019, are the fourth leading cause of mortality (WHO, 2022b). According to the Centers for Disease Control and Prevention (CDC), the severe acute respiratory syndrome coronavirus 2 (SARS-CoV-2)-caused coronavirus disease 2019 (COVID-19) was the third leading cause of mortality in the United States in 2020 (CDC, 2022). In addition, WHO reports that as of April 2022, there were more than 500 million confirmed cases of COVID-19 worldwide (WHO, 2022a).

Applying medical therapies directly to patients' respiratory systems is the most direct method of treating respiratory illnesses. When compared to other non-invasive techniques of administering medicine, pulmonary drug administration provides the highest possible level of bioavailability (Fahad et al., 2021). A multitude of factors, including the vast surface area, the thin epithelial layer, the faster blood flow, the

[1] School of Mechanical and Mechatronic Engineering, University of Technology Sydney (UTS), 15 Broadway, Ultimo, NSW-2007, Australia.

[2] Department of Applied Mechanics, Motilal Nehru National Institute of Technology Allahabad, Prayagraj-211004, Uttar Pradesh, India.

* Corresponding author: Suvash.Saha@uts.edu.au

DOI: 10.1201/9781003287636-1

high solute permeability, and the avoidance of first-pass metabolism, led to an environment that was ideal for the administration of drugs (Gill et al., 2007; Ruge et al., 2013; Willis et al., 2012).

The history of humans inventing medical therapies that focus on the pulmonary pathway is a long and illustrious one. According to research that was published in 2007, the oldest archeological evidence for the inhalation of therapeutic substances dates back to the year 1554 B.C. (Sanders, 2007). Christopher Bennet's inhaler in the seventeenth century, vapors device in the eighteenth, metered dose inhaler in the twentieth, etc., all contributed to the advancement of therapy. And today, at the beginning of the twenty first century, researchers from numerous sectors are adopting cutting-edge technologies to support respiratory medical treatments, such as nanotechnology and deep learning networks (Pathak and Islam, 2021; Shen et al., 2017).

Two of the most significant advances in technology that have been made in the field of medical imaging are Computed Tomography (CT) and Magnetic Resonance Imaging (MRI). For his invention of X-rays in 1895, Wilhelm Roentgen received the Nobel Prize in Physiology or Medicine in 1901. Through the creation of computerized axial tomography by Allan MacLeod Cormack, X-ray imaging was eventually brought into three dimensions. Between 1963 and 1964, he published two studies that would later serve as the basis for CT theories. Later on, the scientific community acknowledged his contributions, and as a result, he was awarded the Nobel Prize in Physiology or Medicine in 1979 (Tan and Poole, 2020). The same 1979 Nobel Prize was also given to Sir Godfrey Hounsfield, who prototyped the first few CT scanners from 1967 to 1971 (Hounsfield, 1980). The announcement of the first set of CT images in 1972 caught the public's attention, which led to the development of subsequent generations of CT scanners that had faster image generation rates and higher resolutions (Ahookhosh et al., 2020). In 1974, Paul C. Lauterbur and Peter Mansfield laid the groundwork for the development of MRI technology when they suggested the use of magnetic field gradients to assist in the spatial localization of nuclear magnetic resonance signals (Geva, 2006). In recognition of their work, the Nobel Prize in Physiology or Medicine was bestowed on them in 2003. And in the following ten years, the research fields witnessed many firsts, such as the first cross-sectional MR image through a finger by Peter Mansfield and Andrew Maudsley (Geva, 2006); the first MR image of the wrist by Hinshaw et al. (1977); the first MR image of the chest by Damadian et al. (1977); the first transverse MR image of the head by Hugh Clow and Ian R. Young (1978), etc. The development of MRI is accelerating at a rate that has never been seen before in modern times, with novel techniques such as kinetic MRI being suggested to assist clinical diagnosis (Michelini et al., 2018).

Many researchers turn to computer simulations to gain a better understanding of respiratory diseases and their associated therapies. Over the course of the last two decades, there has been a rise in the application of Computational Fluid Dynamics (CFD) in several fields of study, including biomedical engineering. As a result of the advancement of computer technology, computers are becoming more powerful than ever before. With the newly acquired computational resources, the application of numerical methods in the simulation of airflow in the airways is within our capability (van Ertbruggen et al., 2005). With the use of CFD, tasks such as the modeling of processes involving the distribution of particles, aerosols, and gaseous medication formulations, as well as the assessment of drug toxicity and permeability, may all be accomplished (Larpruenrudee et al., 2021; Patton and Byron, 2007; van Ertbruggen et al., 2005). However, before initiating any kind of CFD simulation, it should go without saying that one must first obtain the airway geometry. This chapter aims to provide a list of airway-structure modeling strategies utilized by researchers in recent years as well as some of our own. Four modeling strategies are covered in the chapter. In summary, they are: construction of idealized geometry, derivation of digital model from cast model, derivation of digital model from medical imaging, and neural network-assisted modeling.

## 2. Airway Geometry & Terminology

The human airway is divided into two regions determined by their respective functions: the conduction zone, which is responsible for the transmission of airflow, and the respiratory zone, which is the location where gas exchange takes place.

Figure 1 provides a diagrammatic representation of a complete airway. Each component's name is labeled, which can be used as a navigational aid throughout the remainder of the chapter. The green circle, as shown in the illustration, highlights the respiratory zone. It includes the bronchioles, the alveolar ducts, and the alveolar sacs as components. The conducting zone encompasses the remainder of the airway,

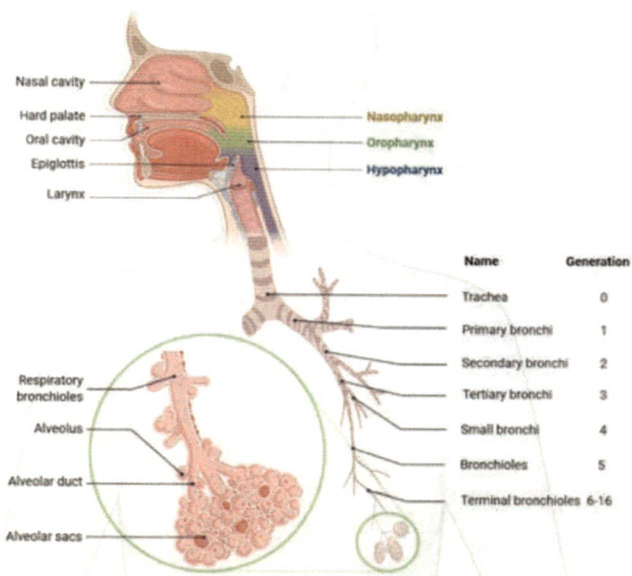


**Figure 1.** The entire airway is depicted with the names of each component indicated.

from the nasal and mouth cavities to the terminal bronchioles. Three terminologies that are frequently discussed are bronchi, bronchioles, and respiratory bronchioles. These names correspond to three organs that constitute a significant part of the pulmonary airway. Bronchi are surrounded by circular or plate-like cartilage, whereas a significant portion of smooth muscle creates bronchioles without cartilage, and respiratory bronchioles are just regular bronchioles with the inclusion of alveoli to distinguish them from a structural standpoint (Weibel, 1963). In this chapter, the generations of the tracheobronchial tree will be counted using Weibel's approach. For more information, one may refer to Morphometry of the Human Lung (Weibel, 1963).

## 3. Name and Abbreviation of Software and Hardware

Table 1 is a collection of the names listed in this chapter and their explanations. The list includes software, hardware, and facilities.

**Table 1.** Software and hardware names in this chapter.

| Name | Purpose | Notes |
|---|---|---|
| Alya | CFD simulation | A high-performance computational mechanics code designed for massively parallel supercomputers (Alya, 2022). |
| AVL FIRE | CFD simulation | Software package that integrates CFD pre-processing, standard simulation, and post-processing (AVL, 2022). |
| CaMEL | CFD simulation | Stands for "Computation and Modeling Engineering Laboratory", a hybrid flow solver (Akbar et al., 2017). |
| CFX | CFD simulation | A flow solver that excels in turbomachinery simulation (CFX, 2022). |
| FIDAP | CFD simulation | A general-purpose finite element fluid dynamics analysis program (Engelman, 1982). |
| FLUENT | CFD simulation | A general-purpose finite volume fluid dynamics analysis program (FLUENT, 2021). |
| GAMBIT | Grid generation | Generate the grid or mesh for the CFD solver (GAMBIT, 2004). |
| Geomagic Wrap | Geometric design & processing | Software that excels in reverse engineering 3D data (Wrap, 2022). |
| ICEM | Grid generation | Software that provides geometry modeling, mesh generation, and mesh optimization (ICEM, 2012). |
| Mimics Medical | Image reconstruction | Software that can visualize and segment medical images and render 3D geometry (Medical, 2021). |
| OpenFOAM | Grid generation & CFD simulation | A collection of applications that includes CFD meshing, simulation, and post-processing (OpenFOAM, 2022). |
| PyCharm | Coding | Integrated development environment (IDE) dedicated to Python. |
| PyTorch | Coding | An open-source machine learning Python library. |
| SGI Octane workstations | Geometric design & processing | High-performance computer, Industry standard hardware for 3D modeling, video editing, etc. (Octane, 1999). |
| Siemens NX 12 | Geometric design & processing | Software package that integrates geometric design, simulation, and manufacturing solutions (Siemens, 2017). |
| SOLIDWORKS | Geometric design & processing | Software package that integrates geometric design, simulation, etc. (SolidWorks, 2022). |

# 4. Modeling Techniques

## *4.1 Ideal Geometry*

Undoubtedly, one of the first methods of airway computational modeling was the creation of ideal models. In spite of this, we continue to use it and do research on it today due to the clear advantages of controllability and simplicity that it offers. When it comes to microscopic or internally intricate structures, reconstructing Three-Dimensional (3D) models from medical imaging such as CT or MRI is a significant challenge. Constructing an ideal geometry presents fewer obstacles and can be a starting point for further investigation. With the help of commercial engineering software such as SOLIDWORKS or Siemens NX, ideal geometry can be easily constructed. Here are three examples of ideal geometry models from the trachea to the alveoli. Manual construction (as opposed to algorithm generation) of idealized respiratory airway models can be done with Computer-Aided Design (CAD) software. Here are several examples of how to create trachea, tracheobronchial trees, and alveoli models.

### *4.1.1 Trachea*

Russo et al. (Russo et al., 2008) constructed a trachea and primary bronchial model with cartilage rings in SOLIDWORKS. The purpose of the research was to ascertain the effect that cartilage rings have on the flow of air and the deposition of particles. In the beginning, the parameters of Weibel dichotomous morphometry were employed to create a smooth-wall model from the trachea all the way to the primary bronchi. After that, scientists examined the geometric qualities of cartilage rings using a whole lung specimen taken from a deceased person. In the end, the ring structure was added to the model based on the mean value of their measurements. It is clear from looking at their work that both the shape of the airway and the ring pattern of the cartilage have been idealized. This degree of idealization is required because there is a risk that increasing the degree of realism in geometry may result in the introduction of additional factors that will undermine the purpose of the study.

### *4.1.2 Tracheobronchial Tree*

There are many models of the bronchial tree. One example is the one used in the study by Larpruenrudee et al. (2021). To explore how pharmaceutical aerosols are carried through a narrowing in the airway, they constructed six 5-generation 3D airway models based on Weibel's lung model. As can be seen from their work, changes were made to the original model in order to construct five models with stenosis parts that represented blocked airways in various locations. The models used in this research were created using SOLIDWORKS 2019, much like the example given before. Controllability is shown to be one of the benefits that ideal geometry modeling offers via the use of this example. Producing a number of different models, each with its own unique modifications is possible via the use of the parametric design strategy. Users are able to edit the model's settings by utilizing an Excel spreadsheet, for instance, thanks to the "design table" function included in SOLIDWORKS. Other similar works can be found in Table 2. Here is an example of the utilization of parametric design in the modeling of an obstructed airway:

(a)

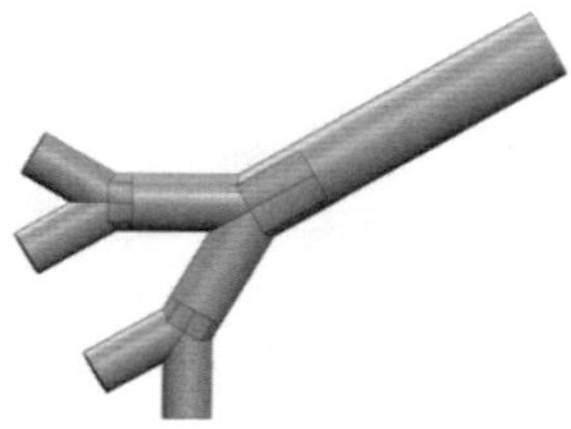

(b)

| | A | B | C | D |
|---|---|---|---|---|
| 1 | Design Table for: parametric design example | | | |
| 2 | | Distance to Bifurcation | Stenosis Region Length | Stenosis Region Width |
| 3 | Modification-1 | 9.00 | 12.00 | 6.00 |
| 4 | Modification-2 | 9.00 | 12.00 | 9.00 |
| 5 | Modification-3 | 9.00 | 5.00 | 9.00 |
| 6 | Modification-4 | 15.00 | 5.00 | 9.00 |
| 7 | Modification-5 | 9.00 | 15.00 | 9.00 |
| 8 | Modification-6 | 9.00 | 15.00 | 6.00 |

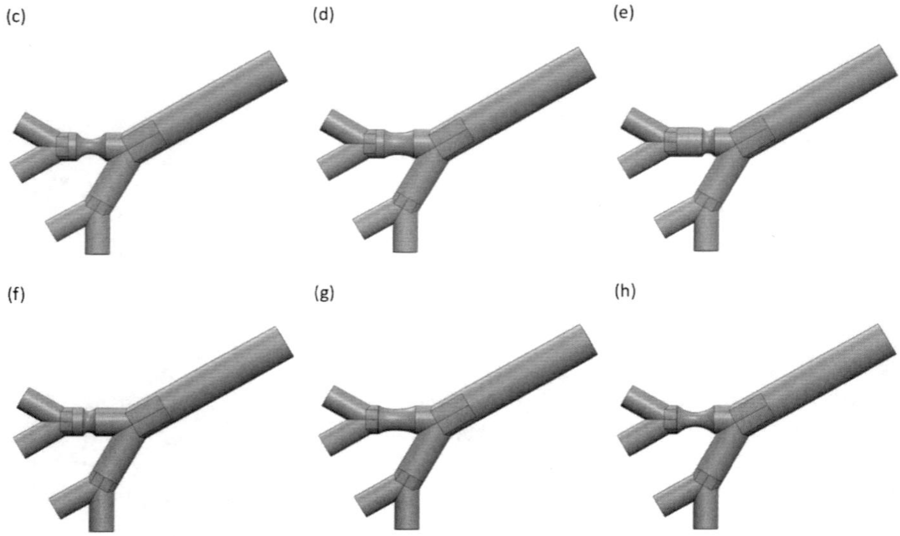


**Figure 2.** Application of the design table function in SOLIDWORKS 2020. (a) is a 3-generation bifurcation model similar to the one used in the study of Larpruenrudee et al. (2021). (b) is the design table interface, where a user can input parameters to a spreadsheet. (c) to (h) are the models generated by the design table in (b) and correspond to Modification-1 to Modification-6.

A 3-generation bifurcation model may be constructed with the help of SOLIDWORKS, as shown in Fig. 2a. This model is analogous to the 4-generation model that was used in the research carried out by Larpruenrudee et al. (2021) is constructed in SOLIDWORKS 2020. The parameters are as follows: (1) the distance to the bifurcation (measured from the beginning of the stenosis area); (2) the length of the stenosis region; and (3) the width of the stenosis region (diameter of the narrowest point). One has the ability to create models with a variety of values for the aforementioned parameters by using the design table. As can be seen in Fig. 2c to h, six different modifications with random values for the parameters are created.

**Table 2.** Some selected literature on ideal geometry modeling of human tracheobronchial airway.

| Author | Topic | Simulation Software | Generation Count | Shape |
|---|---|---|---|---|
| (Comer et al., 2001) | airflow | CFX | G3 - G5 | symmetric |
| (Liu et al., 2002) | airflow | FLUENT | G5 - G7 | symmetric |
| (Balashazy et al., 2003) | particle deposition | AVL FIRE | G3 - G4 | symmetric |
| (Liu et al., 2003) | airflow | FLUENT | G5 - G11 | asymmetric |
| (Farkas et al., 2006) | aerosol drug delivery | FLUENT | G1 - G5 | asymmetric |
| (Russo et al., 2008) | particle deposition | CFX & FLUENT | G0 - G1 | symmetric |
| (Zhang and Kleinstreuer, 2004) | particle deposition | CFX | GO - G3 | symmetric |
| (Zhang et al., 2009) | particle deposition | CFX | G0 - G15 | symmetric |
| (Soni et al., 2011) | particle deposition & airflow | CaMEL | G0 - G9 | symmetric |
| (Mead-Hunter et al., 2013) | particle deposition | OpenFOAM | G0 - G3 | symmetric |
| (Tian and Ahmadi, 2013) | fiber transport and deposition | FLUENT | G0 - G3 | asymmetric |
| (Soni and Aliabadi, 2013) | airflow & particle deposition | CaMEL | G0 - G10 | symmetric |
| (Larpruenrudee et al., 2021) | particle deposition | FLUENT | G0 - G4 | asymmetric |

### *4.1.3 Alveoli*

Due to the intricate nature of the alveoli's internal structure, modeling them is undoubtedly difficult. Xi et al. (2018) and their following work Talaat et al. (2019) presented three types of alveolar models in their investigation of nanoparticle transport and deposition in terminal alveoli. They modeled alveoli using the findings of Ochs et al. (Ochs et al., 2004), which stated that the diameter of a single alveolus is 0.2 millimeters. According to their work, the first model consists of a single alveolus that is connected to an alveolar catheter. The second model depicts four individual alveoli, each of which has a center that is on the same plane. The final one has 45 alveoli banded together, representing an alveolar sac. In their model, geometric specifics such as the wall and tunnel that are found between each alveolus were included. If, on the other hand, we are able to give the model some degree of unpredictability, such as by allowing the alveolus diameter and shape to change, then the model will be much closer to reality.

## *4.2 Algorithm Generation*

A number of different computer algorithms and methodologies have been created to construct the bronchial tree structure on a larger scale. Here are two instances, one dealing with symmetrical airway generation and the other dealing with asymmetrical airway generation.

### *4.2.1 Symmetrical*

A data-driven surface modeling technique was employed by Spencer et al. (2001) to develop a 3D symmetric Weibel model of the human lung from the trachea to the 23rd generation. They generated the graphical display with the help of SGI Octane workstations, with the goal of creating airway geometries that can be used in the medical field and for CFD analyses. The fundamental concept is that they divide the surface of a bifurcation with one parent branch and two daughter branches into three surface patches. The patch arrangement is shown in Fig. 3. Each patch is constructed from control points generated by algorithms using parameters that include length, diameter, and orientation. Finally, the surface patches are stitched together to form a complete airway model. Obviously, there is room for improvement to make the model more realistic. However, the modeling idea behind it is worth learning for future research.

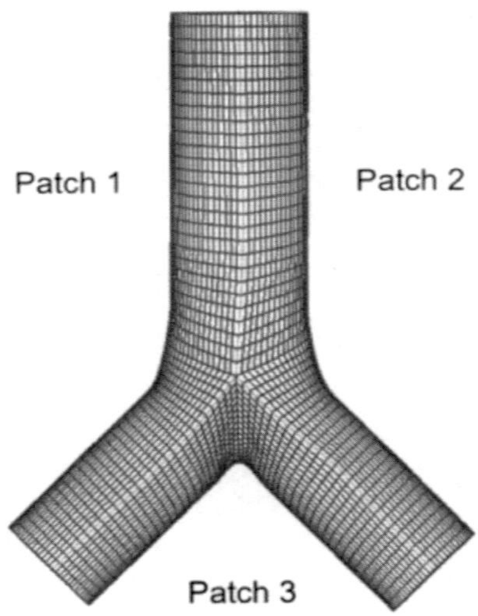


**Figure 3.** We created an illustration to show the surface patch arrangement in the work of Spencer et al. (2001).

### *4.2.2 Asymmetrical*

Using a deterministic algorithm, Kitaoka et al. (1999) constructed a 3D airway branching duct system with approximately 54,000 branches. The method makes certain that (A) the amount of fluid supplied by a branch is proportionate to the volume of the area that is supplied by the branch, and (B) the terminal branches are uniformly distributed among the lobes. They began by constructing the fundamental structure of the algorithm, which is comprised of nine basic rules. After that, they presented four variable rules, and then they experimented with different combinations of the fundamental and supplemental rules in order to identify the most suitable combination. The resulting lung geometry is nearly indistinguishable from the real airway tree, with only minor shortcomings such as branching stop too early around concave surfaces. Tawhai et al. (2000) took on the challenge of modeling asymmetrical bronchial tree models using the volume-filling algorithm. They expanded the Monte Carlo approach from two-dimensional space to three-dimensional space in an effort to develop the computational mesh of airway models. To put it another way, the initial step is to build a uniform host mesh to fill the host volume and act as the foundation for the structure. The host volume is then subdivided into two smaller volumes by a splitting plane that is determined by the location of the host volume's center of mass as well as the vector that represents the parent branch. Connecting the parent branch to the mass centers of the two smaller volumes results in the formation of a bifurcation with two daughter branches. Repeat the procedure in order to raise the total number of bifurcations and completely fill up the available area with airway branches. In addition, measures were taken to ensure that the models created by this algorithm were anatomically realistic. These steps included defining bifurcation angle limitations and branching stop criteria, among other things.

## 5. From Cast Model to Digital Model

The production of *in vitro* cast models may be a challenging endeavor at times. For the casting, a cadaver sample is required, but unfortunately, the technique will render the sample unusable. In addition, it should go without saying that it is not feasible for patient-specific modeling. This approach, on the other hand, has the potential to produce the most realistic and comprehensive 3D airway model of any other method.

### *5.1 From Oral Cavity to Upper Trachea*

Based on the measurements of a cast model, Cheng et al. (1997) created a model of the airway that extends from the mouth cavity to the upper trachea. First, the oral cavity was cast from an alginate dental impression of a living male adult. The rest of the model (from the oropharynx to the upper trachea) was cast from a human cadaver. The next step was to assemble the cast models so that a whole negative wax model could be made. Finally, the coupled model was recreated from the negative wax model. They also made another model with polyester resin and sliced it. The slices were used to measure the dimensions of the airway. Later, Farkas et al. (2006) and Zhang and Kleinstreuer (Zhang and Kleinstreuer, 2004) adapted those measurements to construct their digital models for CFD simulation.

### *5.2 Bronchial Tree*

Numerous researchers have performed lung casts in the past (Esteban et al., 2017; Phalen et al., 1973; Yeh and Schum, 1980). However, with the help of modern CT technology, Schmidt et al. (2004) were able to convert the cast model into a digital model with incredible detail and high precision. They started by preparing a human lung sample, after which they injected it with a solution of liquid rubber. The lung tissue was removed when the rubber had cured and became firm. The final step is to acquire volumetric data by segmenting the High-Resolution Computed Tomography

(HRCT) of the rubber lung cast. However, they did not specify the casting process. One can read the work of Phalen et al. (Phalen et al., 1973) for detailed information. In addition, the next section provides more details on the picture reconstruction process. The HRCT scan gives us an accurate portrayal of the lung anatomy by allowing us to see even minute details of the cast model. The surface representation of the lung was also displayed in their paper, and from this, we can see that even the cartilage ring pattern is mirrored in the model. This surface data, according to the author, may also be transformed into the stereolithographic (STL) format, allowing for post-processing and grid generation for CFD simulations.

## 6. From Image to Digital Model

The reconstruction of realistic respiratory airway models is made easier with the advancement of medical imaging technology and segmentation software. The improved spatial resolution of scan pictures enables us to observe microscopic characteristics such as cartilage rings and terminal bronchioles, and the automated segmentation of the airway saves us a tremendous amount of time and effort. The semi-automatic airway segmentation capability that is provided by commercial software such as Mimics Medical, which we have used in the past and can attest to its effectiveness, can cut the segmentation time down to one minute.

### *6.1 Nasal Cavity*

Recently, modeling the nasal cavity and nasopharynx has been a major topic of discussion, as shown in Table 3. Due to the fact that it is the primary entrance to the human respiratory system, it is an extremely important aspect to consider when researching topics such as airborne virus transmission, nasal medication release, inhalation of pollutants, and so on. The early modeling method documented in the work of Inthavong et al. (Inthavong et al., 2006) provided us with some insight into nasal modeling techniques. To develop the model in GAMBIT, they sliced the 3D CT scans from the nasal cavity to the anterior larynx at different intervals. During reconstruction, the CT scans serve as the cross-section, and the minimum spacing between these cross-sections was set to 1 mm for areas with complex geometry. A surface mesh was subsequently generated on the solid model prior to the generation of the body mesh. Researchers continue to use the same modeling approach, with minor modifications to software and algorithms. Dr. Xi and his colleagues, for instance, have published several studies (Si et al., 2021; Xi and Si, 2017) on the modeling of Computational Fluid-Particle Dynamics (CFPD) in the nasal cavity. They explained that multi-slice MRI images are segmented by Mimics Medical and converged into a set of cross-sectional contours, which are then connected to form a surface geometry. Table 4 contains a number of other comparable works.

**Table 3.** Some selected literature on image reconstruction of human nasal cavity.

| Author | Topic | Meshing Software | Simulation Software | Image |
|---|---|---|---|---|
| (Inthavong et al., 2006) | spray particle deposition | GAMBIT | FLUENT | CT |
| (Kimbell et al., 2007) | spray particle deposition | FIDAP | FIDAP | MRI |
| (Shi et al., 2007) | particle deposition | ICEM | CFX | MRI |
| (Inthavong et al., 2008) | spray parameter optimization | GAMBIT | FLUENT | CT |
| (Shang et al., 2015) | particle deposition: influence of breathing zone | ICEM | FLUENT | CT |
| (Tong et al., 2016) | spray parameter optimization: device & its orientation | GAMBIT | FLUENT | CT |
| (Calmet et al., 2018) | particle deposition: subject-variability | ICEM | Alya | MRI & CT |
| (Calmet et al., 2019) | spray parameter optimization: inhalation conditions | ICEM | Alya | CT |
| (Vachhani and Kleinstreuer, 2021) | particle deposition: micron- vs. nano-particle | ICEM | OpenFOAM | MRI |
| (Si et al., 2021) | spray particle deposition: liquid film translocation | ICEM | FLUENT | MRI |

**Table 4.** Some selected literature on image reconstruction of human tracheobronchial tree.

| Author | Topic | Generation | Image |
|---|---|---|---|
| (Garrity et al., 2002) | development of four-dimensional NURBS-based cardiac-torso phantom | G0 - G5 | CT |
| (Tawhai et al., 2004) | geometry analysis | G0 - G8 | CT |
| (Luo and Liu, 2008) | airflow simulation | G0 - G4 | CT |
| (Miyawaki et al., 2016) | particle transport | G0 - G5 | CT |
| (Pourmehran et al., 2016) | magnetic drug delivery | mouth to G6 | CT |
| (Rahimi-Gorji et al., 2016) | particle deposition & airflow simulation | mouth to G6 | CT |
| (Kannan et al., 2017) | airflow simulation | G0 - G8 | CT |
| (Miyawaki et al., 2017) | aerosol transport | G0 - G9 | CT |
| (Islam et al., 2018) | aerosol transport | G0 - G15 | CT |
| (Qi et al., 2018) | particle deposition & airflow simulation | 9 models with max G0 - G8 | CT |
| (Van de Moortele et al., 2018) | geometry analysis | G0 - G7 | CT |
| (Zhang et al., 2018) | particle deposition & airflow simulation | 6 models with max G0 - G8 | CT |

## 6.2 Tracheobronchial Tree

When compared to modeling the nasal cavity, the task of modeling the tracheobronchial tree using clinical CT/MRI scans presents a more difficult challenge. A limited number of investigations were successful in creating airways that extended from the trachea to the terminal bronchioles (6–16th generations). There is no indication, based on the available records, that there was any research that reached the respiratory bronchioles (17–19th generation). The only exception was Schmidt et al. (2004), whose model has 18 generations (0–17th generation). However, they generated their HRCT from an *in-vivo* cast model rather than a live patient. Cast models allowed for the elimination of interference from biological materials, which resulted in extraordinarily high-quality CT scans. A threshold-based technique was then applied to the HRCT volumetric data, which resulted in the production of a surface geometry and a branching topology geometry of the tracheobronchial tree. The latter was then subjected to further processing using a smoothing technique in order to eliminate noise and discretization errors. The resulting geometry may be exported as an STL file for use in CAD or meshing applications.

This model has been the subject of a number of different CFD studies. Using the Large Eddy Simulation (LES) approach, Gemci et al. (2008) simulated the flow of air through the top 17 generations of the airway using a mesh that had roughly 6.7 million mixed cells. Based on the findings, it seems that the secondary vorticial flows are affected differently depending on the anatomic characteristics. Islam et al. (2017a) generated a mesh with 6.74 million tetrahedral cells and adopted the LES and Euler-Lagrange-based Discrete Phase Model (DPM) method to simulate aerosol transport and deposition in the respiratory tract. They discovered that the majority of particles with a diameter of 10 microns were deposited in the first 17 generations. However, independent of parameters such as breathing pattern, particles of a smaller diameter (one micron) may still pass through the respiratory system in significant quantities. They continued the CFD research in their second paper (Islam et al., 2017b), investigating the transport and deposition of ultrafine particles (with diameters ranging from 1 to 1000 nanometers). They discovered that the breathing pattern as well as the diameter of the particles each had an influence on the location of the disposition. Multiple disposition hot spots have been found for the purpose of designing a targeted medication release mechanism. Novel research regarding the simulation of SARS CoV-2 (which caused the pandemic of the acute respiratory disease called COVID-19 (Hu et al., 2021)) aerosol transport and deposition can be seen in the work of Islam et al. (2021). The numerical findings demonstrated to us how far SARS CoV-2 aerosols may travel inside human lungs, as well as their disposition pattern under a variety of breathing situations. And finally, Gemci et al. (2022) used the same model to study the deposition of dry pulmonary surfactant aerosols. They came up with a recommendation for an acceptable dose of dry surfactant aerosols for individuals suffering from acute respiratory distress syndrome with the aid of the results of their simulations.

The work of Zhang et al. (2018) and Qi et al. (2018) documented a reliable approach for reconstructing the tracheobronchial airway from clinical CT scans. They used the same methodology to rebuild the patients' airways, but this time for six and nine patients, respectively. In the first phase of the process, CT image segmentation was done using an algorithm embedded in the software Mimics Medical. The outputs, which were STL-formatted surface geometry files, were then processed in Geomagic Studio before being sent to SOLIDWORKS in Parasolid format for the final processing phase.

Recently, we were able to generate tracheobronchial airway models of up to 14 generations by using the approach described above with a few adjustments. In the subsequent sub-sections, the approach will be presented in a detailed manner.

### *6.2.1 Image Segmentation*

Importing the CT scan into Mimics Medical is the first step. After that, one may make use of the Advanced Segment toolbar to access the Segment Airway tool. To start the algorithm, one must first click on the picture to designate the beginning point and the direction in which the segmentation will proceed. The segmentation is fully automatic from this point on. Nonetheless, if mistakes such as leaking occur, the user may directly intervene by halting the algorithm to stop the segmentation area's expansion and remove the incorrectly segmented area. Finally, choose the "calculate part" option to generate the geometry. The outcome is shown in Fig. 4a. The lump-shaped irregularity shown in Fig. 4b is the leakage error, which is a common phenomenon at this stage. The corresponding segmentation label on the axial CT image is shown in Fig. 4c.

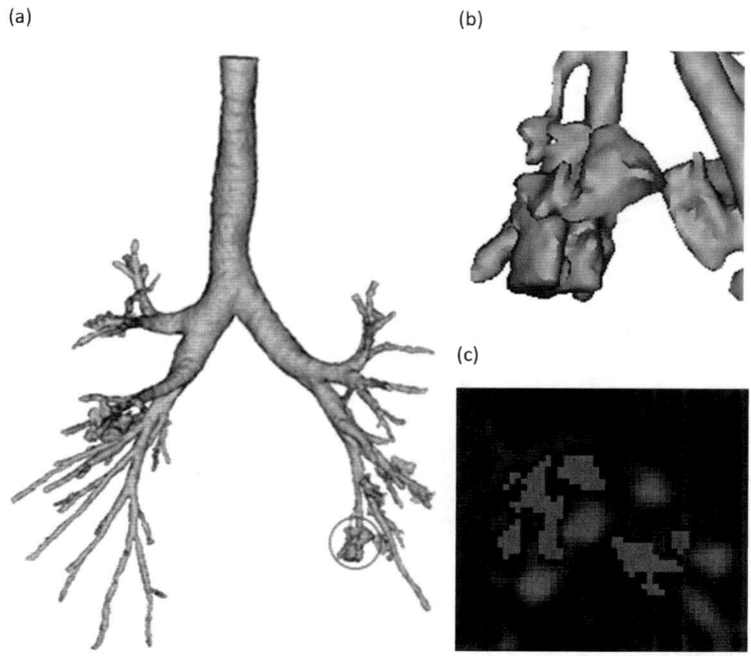


**Figure 4.** (a) The geometry created using the Segment Airway tool in Mimics Medical. (b) An example of leakage error. (c) The leakage segment in b is shown on CT.

Similar to the work of Zhang et al. (2018) and Qi et al. (2018), we could eliminate the leakages using Mimics Medical or another CAD program, and then we could accept this geometry with a restricted number of airway generations (generations 0 to 12) and bronchial lengths. Another option is to extend the bronchi by manually labeling the CT images. When the data from the anterior and posterior slices are taken into consideration, it is easy to determine that the white circle that is present in the axial CT picture (Fig. 5a) indicates the bronchus's cross-sectional image. The "Thin Structure" segmentation function provided by Mimics Medical is a useful tool to extend the bronchi. As the user clicks on the center of the white circle in successive CT slices, a tube-shaped geometry is generated. As illustrated in Fig. 5b, this geometry has its center line traveling through all of the points set by clicking.

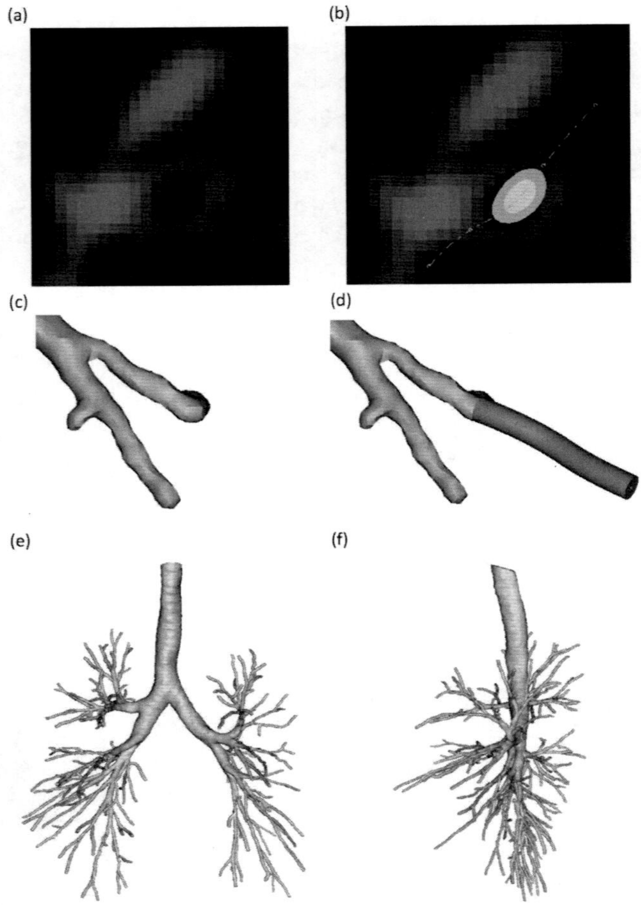


**Figure 5.** (a) The CT image with an airway that the algorithm failed to identify and segment. (b) Manually labeling the CT images using the "Thin Structure" function in Mimics Medical. (c) The 3D geometry created by the algorithm at the location is shown in (a). (d) The manually labeled cylindrical mask corresponds to (b). (e) The anterior view of the manually extended airway model. (f) The left lateral view of the manually extended airway model.

The 3D visual effect of the addition of the thin structure is shown in Fig. 5c and d. Repeat this process to ensure that all of the branches of the airway are covered by tubes to the extent that the human eye can determine from the CT image. The final thing that has to be done is to take these cylindrical constructions and utilize them as masks (also known as controlled borders) so that another segmentation may be carried out and then add the newly segmented geometry to the existing one. These cylindrical structures are, in essence, regions of interest. They restricted the segmentation region to a small area, thereby removing any prospect of leaks being produced by segmentation algorithms, whether those techniques were threshold-based or seeded region growth methods. As a direct consequence of this, the model progresses to generation 14 with longer bronchi, as displayed in Fig. 5e and f.

### *6.2.2 Geometry Processing*

It is possible to export the model in stereolithographic (STL) format so that it may be processed further. The surface geometry may be refined with the assistance of Geomagic Wrap, and the surface can be segmented into orderly patches with its use. Unnatural surface characteristics are created as a result of picture noise, restricted image resolution, and format compatibility concerns. These features include tiny components (Fig. 6a), self-intersections (Fig. 6b), sharp points (Fig. 6c), and small tunnels (Fig. 6d). The STL processing software Geomagic Wrap or another program

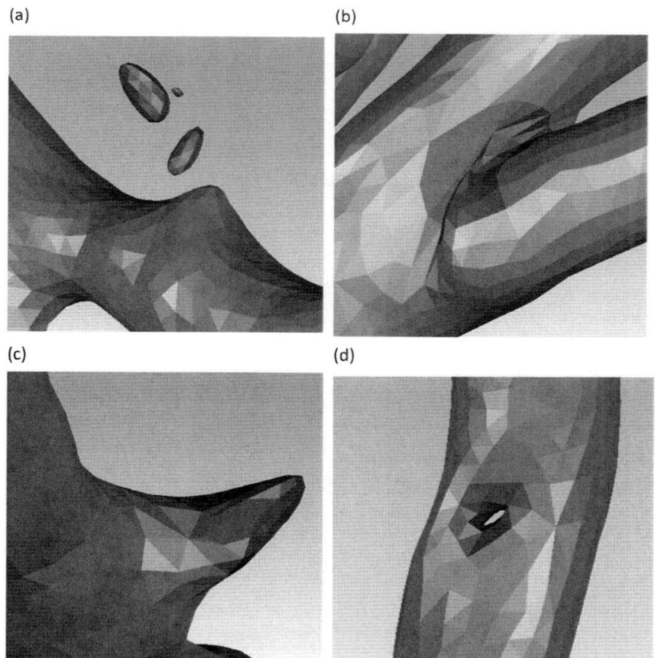


**Figure 6.** Examples of abnormal surface features: (a) small components, (b) self-intersections, (c) sharp tip, (d) small tunnel.

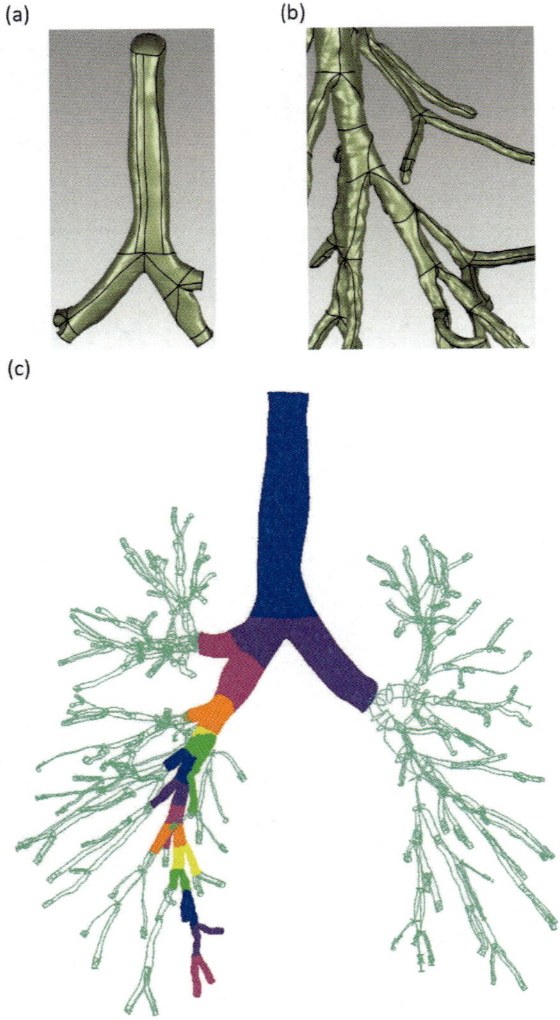


**Figure 7.** Surface patch arrangement. (a) Surface division of generation 0 to 2. (b) Surface division of airways in the right lower lobe. (c) Boundary condition setup in grid generation software Pointwise.

with a comparable function may eliminate these irregularities. Reconstructing surface patches in the appropriate manner is yet another significant role that Geomagic Wrap plays in the whole process. The surface is manually segmented in accordance with the formation of the tracheobronchial tree, as shown in Fig. 7a, b. When using CFD software, one has the ability to choose surface patches from each generation and then either build up generation-specific boundary conditions or evaluate generation-specific metrics, such as the local heat flux, as shown in Fig. 7c.

The surface geometry may be exported from Geomagic Wrap in either the Initial Graphics Exchange Standard (IGES) or a comparable format. This allows the surface

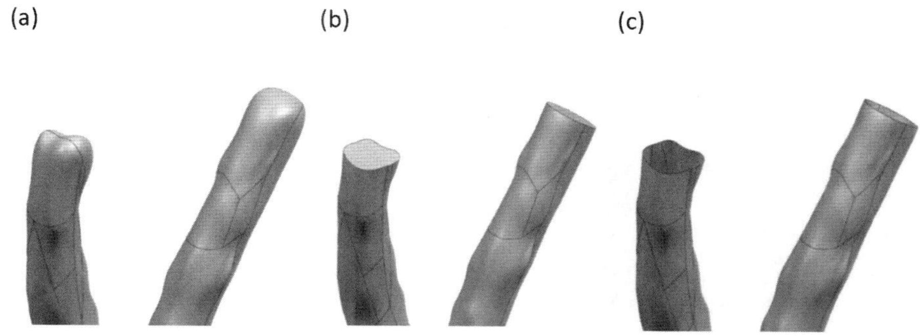


**Figure 8.** Body geometry processing. (a) A close view of the end of the airway after the sewing operation. (b) Flat surfaces are created at the end of the airway to serve as outlets in CFD simulation. (c) Flat surfaces cannot be created if the model is hollow.

geometry to be further processed in engineering CAD software. The following is a rundown of the stages in Siemens NX 12: Import the IGES file in the first step. Check to see whether there are any surface patches that are substantially deformed or absent in the second step. The next step is to sew the patches together by using the "Sew" feature, which can be found in the "Surface" toolbar. The stitching procedure will fill the volume that is contained by the surface, which will result in the formation of a solid body if the surface geometry is watertight. Finally, given that the geometry is no longer hollow, the user is now able to define inlet and outlet borders for further simulation by making use of the "Extrude" feature located on the "Home" toolbar, as shown in Fig. 8a and 8b, respectively. In addition, if one fails to form a solid body when doing the stitching operation, as shown in Fig. 8c, then one will not be able to complete the last step.

## *6.3 Image Reconstruction Plus Ideal Algorithm Generation*

Even though high-resolution thin-layer CT was used as the foundation and a significant amount of manual work was used to manually expand the airway, the finished product still only comprises 15 generations. The model of the lung airways is not yet fully developed. However, it is feasible to utilize computer techniques to artificially produce the remaining bronchioles that are too tiny to be captured by clinical CT or MRI scans. Garrity et al. (Garrity et al., 2002) incorporated the 3D tree structure generation algorithm proposed by Kitaoka et al. (Kitaoka et al., 1999) to extend their manually segmented 5-generation CT-based model. Because of this, the model has a significant number of additional generations. However, their airway tree did not assume the form of the lung lobes, which resulted in the presence of significant voids in which there were no bronchioles. Tawhai et al. (Tawhai et al., 2004) overcame this problem as they extracted the shapes of lobes from CT scans to use them as the virtual boundary to grow the bronchiole tree. First, they segmented the lobes and airways with approximately 11 generations. After converging the CT-segmented model to a centerline model, they applied the volume-filling algorithm (Tawhai et al., 2000) (which we mentioned in the Algorithm Generation section) to create up to 23 generations of airways. When the bronchioles reach the spot condition, which is the border between lobes, the algorithm is programmed to halt the development of the bronchioles. Consequently, their airway model fits perfectly into the lobes and, therefore, is more realistic than that of Garrity et al. (Garrity et al., 2002).

Featuring numerous anatomical details, the completed geometry has been employed in several studies to analyze heat and mass transfer within the respiratory tract (Huang et al., 2024a; Saha et al., 2022a), investigate particle dynamics (Huang et al., 2024b; Huang et al., 2024c; Saha et al., 2022b), and aid in the design of pharmaceutical devices (Saha et al., 2024).

## 7. Convolutional Neural Networks in Airway Modeling

The modeling technique described in the preceding section is labor- and time-intensive. We spent nearly two days semi-automatically segmenting the airway branches that the algorithm had not found. Although patient-specific modeling is obtained, the effectiveness of the technology hampers its practical uses. The algorithm for deep learning can automatically execute medical image synthesis, allowing modelers and physicians to evaluate data quantitatively. Clearly, deep learning is not the sole solution. However, in light of its recent advancements and accomplishments, it is one of the most promising answers. This section will introduce Convolutional Neural Network (CNN) techniques for medical image synthesis.

### *7.1 U-Nets*

The most frequently used CNN structure in medical image segmentation is U-Net, which was first proposed by Ronneberger et al. (2015). The architecture of U-Net consists of a series of down-sampling steps and up-sampling steps. Every down-sampling step performs a maximum pooling operation in addition to two convolutions, two rectified linear units (ReLU), and a maximum pooling step. As a consequence, the number of feature channels will increase by a factor of two at each stage. Every iteration of the up-sampling process consists of an up-convolution, a concatenation, two convolutions, and two ReLU operations. At each stage, the number of feature channels will be reduced by half. There is also a bottleneck step and an output step. The former connects the last down-sampling step and the first up-sampling step, while the latter is a 1 × 1 convolution layer. Since that time, a multitude of U-Net-based algorithms have been developed, all with the intention of enhancing the original algorithm's capabilities in a variety of areas, including its capacity for segmentation. For other instances, please see Table 5.

Using the U-Net framework, Erattakulangara and Lingala (2020) developed an automated segmentation code to segment the vocal tract airway from a speech MRI data set. As a consequence of the examination, it was determined that U-Net had attained a higher mean dice score than the seed-growing method that was already in use. Based on 3D U-Net segmentation carried out over Cone-Beam CT (CBCT) data sets, Wu et al. (2021) created a 3D model that extended from the nasal cavity all the way up to the larynx. A high-precision three-dimensional model was produced as a

**Table 5.** Examples of U-Net-based algorithms published in recent years.

| Algorithm | Author | Notes |
|---|---|---|
| Res-UNet | (Xiao et al., 2018) | Adding a weighted attention mechanism to U-Net. |
| Res-UNet++ | (Jha et al., 2019) | Adding squeeze and excitation blocks, attention blocks, and ASPP to Res-UNet. |
| UNet++ | (Zhou et al., 2018) | Replacing the skip connection in U-Net with nested skip pathways. |
| UNet 3+ | (Huang et al., 2020) | Replacing the nested skip pathways in UNet++ with full-scale skip connections. |
| Dense-UNet | (Guan et al., 2020) | Adding dense connectivity to U-Net. |
| MultiResUNet | (Ibtehaz and Rahman, 2020) | Replacing the convolution bock and skip connection in U-Net with MutiRes block and ResPath. |
| nnU-Net | (Isensee et al., 2021) | Can automatically configure itself. |
| Double U-Net | (Jha et al., 2020) | It has two U-Nets with skip connections connecting the first U-Net's encoder and the second U-Net's decoder. |
| KiU-Net | (Valanarasu et al., 2022) | Combine Ki-Net with U-Net. |
| DC-UNet | (Lou et al., 2020) | Replacing the MutiRes block in MultiResNet with a DC block. |
| TransUNet | (Chen et al., 2021) | Adding a transformer layer to U-Net. |

consequence of the use of an attention mechanism during the segmentation process and an algorithm known as the marching cube during the post-processing of the model. Shujaat et al. (2021) segmented the pharyngeal airway in CBCT using 3D U-Net, achieving a dice score of $0.97 \pm 0.02$, which indicates a high resemblance of the segmentation prediction to the ground truth. Park et al. (2020) performed the segmentation of lung lobes in CT data sets using the 3D U-Net CNN algorithm. The trained algorithm demonstrated high accuracy and reasonable speed over multiple data sets, segmenting lobes in the left and right lungs at a maximum of 17.37 seconds. The problem that is caused by the pulmonary airway branch size change was addressed by Meng et al. (2017) when they introduced an adaptive volume of interest analysis to the 3D U-Net program. The researchers found that their approach had a greater airway identification rate and fewer instances of false positives in comparison to other cutting-edge segmentation techniques. A completely autonomous network that can handle massive 3D patches in a single run was constructed by Garcia-Uceda et al. (2021). This network was developed on top of the basic U-Net architecture. When the segmentation results of two different public data sets are examined, it becomes clear that their technique is capable of achieving matching performance on par with more complicated networks like U-Net-FRAD. Recently, we conducted an experiment that utilized a simple 2D U-Net code to perform pulmonary airway segmentation. An introduction to the U-Net segmentation method may be found in the following subsections, which discuss the procedure in further depth.

### *7.1.1 Data Preparation*

The objective of this experiment was to train a neural network based on the U-Net architecture that can segment small airways that the algorithm in Mimics Medical

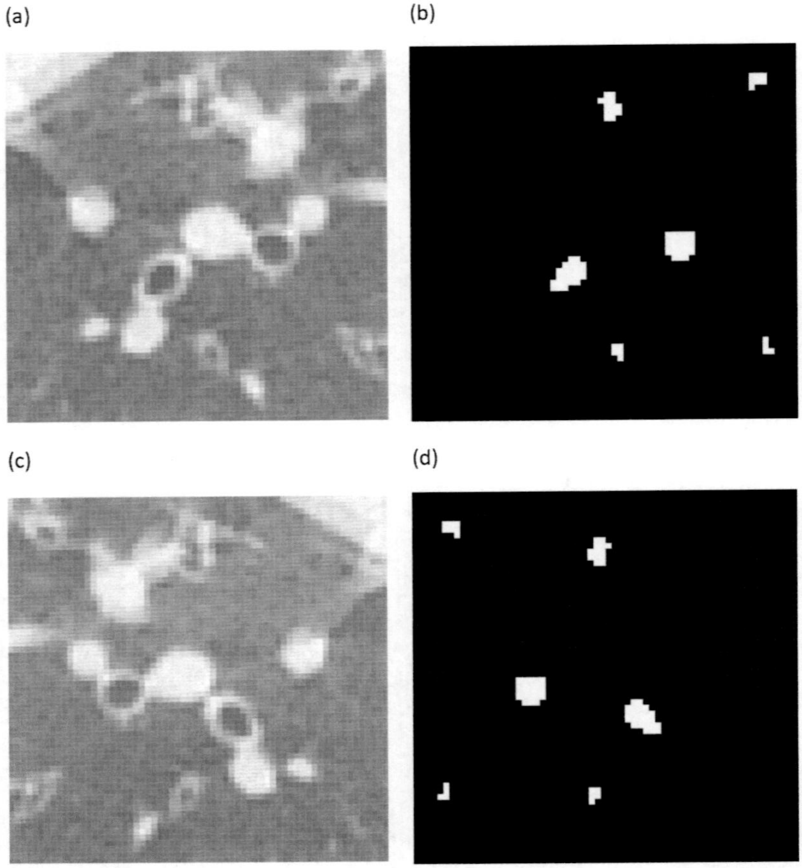


**Figure 9.** Data augmentation for the training data set. (a) Eighth image that fed to the data augmentation code. (b) The manually segmented mask of the eighth image, in which the white pixels represent the cross-sectional area of the airway. (c) Eighth image after the vertical flip operation. (d) Eighth image's mask after the vertical flip operation.

failed to segment (as documented in the previous section 3.2.1). To begin, the CT data set (which had a pixel size of 512 by 512) that was used in the previous part is cropped to produce 13 pictures with a pixel size of 64 by 64. The cropping position is somewhere in the center of the left lower lobe, and the axial CT scans were cropped from head to toe in an intermittent fashion (one in every three consecutive slices). The sixth through thirteenth photos were unsuccessful for the algorithm used by Mimics Medical in identifying the airway architecture. The first 10 images are supplied to the data augmentation algorithm (Fig. 9a), together with their matching manually segmented mask (Fig. 9b), to provide extra samples for neural network training. The final three images, which have even narrower airways, are kept for validation. The data augmentation code can perform one of the following operations each time: a vertical flip (Fig. 9c and d), a horizontal flip, or a 45-degree rotation. Consequently, the training data set is expanded fourfold.

### *7.1.2 Implementation of the U-Net Structure*

PyCharm was used as the IDE, while the algorithm was constructed in Python version 3.9 with assistance from the Pytorch module. The U-Net is built with the following structure in mind: for each encoder block, three-by-three convolution, batch norm, and ReLU operations are looped twice before a 2 × 2 max pooling operation; and for each decoder block, two-by-two convolution transpose and concatenation operations are carried out prior to looping three-by-three convolution, batch norm, and ReLU operations twice. The bottleneck and output processes are exactly the same as they were in the original U-Net. Utility codes such as dataloader were also added to complete the network. After that, the network was trained for a total of 300 epochs.

### *7.1.3 Results*

The trained network will be evaluated based on the 11th through 13th photos. Figure 10 shows both the network inputs and the predictions made from them. As can be seen, the U-Net was successful in segmenting three out of five airways to a respectable degree, both in terms of their form and their location. The algorithm in Mimics Medical did not identify these airways. Having said that, given that this is simply an introductory experiment to show the working progress of the U-Net, there is certainly an opportunity for further improvement. A more extensive training set, an increased amount of time spent in training, the incorporation of an image preprocessing method, and several other adjustments may all help the network produce more accurate predictions.

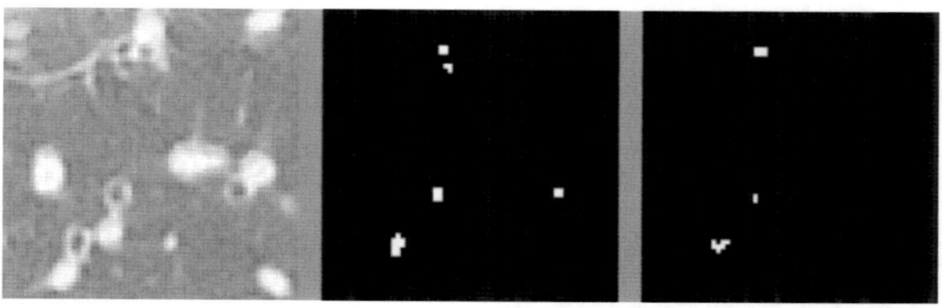

**Figure 10.** The inputs and outputs of the U-Net have been concatenated. From left to right: the thirteenth image of the CT data set; manually segmented mask (ground truth); U-Net prediction.

## *7.2 Generative Adversarial Networks*

Generative Adversarial Networks (GANs) also play a vital role in medical image processing. A fundamental GAN consists of two neural networks: the generator and the discriminator. Its modifications and extensions encompass a vast array of functionalities, including image enhancement and segmentation, which are of great assistance to the reconstruction of medical images (Kazeminia et al., 2020).

### 7.2.1 Image Enhancement

The use of contrast may improve the visibility of body components in CT images. On the other hand, having a higher contrast is not always ideal since it exposes patients to an unnecessary amount of radiation. GAN is one of the numerous deep learning algorithms, and it is widely recognized as a potentially useful solution for noise and artifact filtering as well as high-contrast image synthesis. Wolterink et al. (2017) trained a GAN that successfully generated regular-dose CT scan images from low-dose CT scan images. They concluded that GAN may, during the training process, assist the generator in learning the noise distribution of CT scan pictures. Wang et al. (2018) were successful at eliminating metal artifacts from ear CT images by using a conditional GAN (cGAN). Additionally, they were able to cut the average surface error in half compared to what was previously possible. Yang et al. (Yang et al., 2018) developed a GAN with the inclusion of Wasserstein distance and perceptual loss, which, as a consequence, produced a decrease in the CT image noise level. Using a convolutional neural network (cGAN), Yi and Babyn (Yi and Babyn, 2018) were able to solve the blur effect issue that was seen in the results of earlier cGAN denoising approaches. An edge-aware GAN was suggested by Yu et al. (Yu et al., 2019). This GAN takes edge information from multiple modality MR pictures and uses it to generate new images with improved boundaries. The work of Jammes-Floreani et al. (2021), in which they employed enhanced-quality GAN to reduce blur and noise in CT images, is an example of super-resolution. The results of the segmentation done following the super-resolution augmentation were much improved, with a greater number of airways being segmented.

### 7.2.2 Segmentation

The topic of image segmentation is another potential use for the GAN algorithm. One strategy involves using the discriminator network in order to coerce the generator network into producing segmentations that are indistinguishable from those that were manually segmented (Moeskops et al., 2017). Using a simulated and unsupervised GAN combined with a convolutional neural regressor, Nardelli et al. (2020) performed operations such as measuring tiny pulmonary structures. Using their network, the wall thickness of tiny conducting airways may be evaluated. A structure-correcting adversarial network was developed by Dai et al. (2017) in order to efficiently separate lobes from X-ray pictures. The method, on the other hand, is not equipped to segment airways of a smaller diameter. For the purpose of medical picture segmentation, Xue et al. (2018) proposed SegAN, which is based on GAN. The newly developed network fared better than the most advanced U-net-based networks when it was put to the test to segregate brain tumors. Segmentation of retinal vessels (Lahiri et al., 2017; Son et al., 2017) and brain (Kamnitsas et al., 2016; Li et al., 2018) have been the subject of a significant number of investigations in recent years, in contrast to the limited number of studies that have been undertaken on airway segmentation. These studies could lay the groundwork for a GAN-based segmentation of the airways in the future.

## References

Ahookhosh, K., Pourmehran, O., Aminfar, H., Mohammadpourfard, M., Sarafraz, M.M. and Hamishehkar, H. (2020). Development of human respiratory airway models: A review. Eur J Pharm Sci 145: 105233.

Akbar, M., Luettich, J.R., Fleming, J. and Aliabadi, S. (2017). Camel and adcirc storm surge models—a comparative study. Journal of Marine Science and Engineering 5: 35.

Alya. (2022). Alya User Guide. B. S. Center. Barcelona, Spain.

AVL. (2022). Avl fire$^{tm}$ m. AVL. Graz, Austria.

Balashazy, I., Hofmann, W. and Heistracher, T. (2003). Local particle deposition patterns may play a key role in the development of lung cancer. J Appl Physiol 94: 1719–1725.

Calmet, H., Inthavong, K., Eguzkitza, B., Lehmkuhl, O., Houzeaux, G. and Vazquez, M. (2019). Nasal sprayed particle deposition in a human nasal cavity under different inhalation conditions. PLoS One 14: e0221330.

Calmet, H., Kleinstreuer, C., Houzeaux, G., Kolanjiyil, A.V., Lehmkuhl, O., Olivares, E. et al. (2018). Subject-variability effects on micron particle deposition in human nasal cavities. Journal of Aerosol Science 115: 12–28.

CDC. (2022). Leading causes of death. Centers for Disease Control and Prevention. Retrieved April 23 from https://www.cdc.gov/nchs/fastats/leading-causes-of-death.htm.

CFX, A. (2022). Ansys cfx reference guide. I. ANSYS. Canonsburg, USA.

Chen, J., Lu, Y., Yu, Q., Luo, X., Adeli, E., Wang, Y. et al. (2021). Transunet: Transformers make strong encoders for medical image segmentation.

Cheng, K.-H., Cheng, Y.-S., Yeh, H.-C. and Swift, D.L. (1997). Measurements of airway dimensions and calculation of mass transfer characteristics of the human oral passage. Journal of Biomechanical Engineering 119: 476–482.

Clow, H. and Young, I.R. (1978). Britain's brains produce first nmr scans. NewScientist 80: 588.

Comer, J.K., Kleinstreuer, C. and Zhang, Z. (2001). Flow structures and particle deposition patterns in double-bifurcation airway models. Part 1. Air flow fields. Journal of Fluid Mechanics 435: 25–54.

Dai, W., Doyle, J., Liang, X., Zhang, H., Dong, N., Li, Y. et al. (2017). Scan: Structure correcting adversarial network for organ segmentation in chest x-rays.

Damadian, R., Goldsmith, M. and Minkoff, L.A. (1977). Nmr in cancer: Xvi. Fonar image of the live human body. Physiological Chemistry and Physics 9 1: 97–100, 108.

Das, A., Pathak, M.P., Patowary, P. and Das, S. (2021). Introduction to lung physiology from a drug delivery perspective. pp. 1–11. *In*: Pathak, Y. and Islam, N. (1st ed.). Handbook of Lung Targeted Drug Delivery Systems: Recent Trends and Clinical Evidences. Taylor & Francis Group.

Engelman, M.S. (1982). Fidap (a fluid dynamics analysis program). Advances in Engineering Software (1978) 4: 163–166.

Erattakulangara, S. and Lingala, S.G. (2020, 3–7 April 2020). Airway segmentation in speech mri using the u-net architecture (ed.). 2020 IEEE 17th International Symposium on Biomedical Imaging (ISBI). 1887–1890.

Esteban, R.J.R., Mccormick, J., Prieto, D.R.M. and Hernandez, J.D. (2017). Corrosion casting, a known technique for the study and teaching of vascular and duct structure in anatomy. International Journal of Morphology 35: 1147–1153.

Fahad, A.S., Boddu, S.H. and Nesamony, J. (2021). Mechanism and ways of pulmonary drug administration. pp. 29–39. *In*: Pathak, Y. and Islam, N. (1st ed.). Handbook of Lung Targeted Drug Delivery Systems: Recent Trends and Clinical Evidences. Taylor & Francis Group.

Farkas, Á., Balashazy, I. and Szocs, K. (2006). Characterization of regional and local deposition of inhaled aerosol drugs in the respiratory system by computational fluid and particle dynamics methods. Journal of Aerosol Medicine 19: 329–343.

FLUENT. (2021). Fluent user's guide 2021. I. ANSYS. Canonsburg, USA.

GAMBIT. (2004). Gambit 2.2 tutorial guide. I. Fluent. Lebanon, USA.

Garcia-Uceda, A., Selvan, R., Saghir, Z., Tiddens, H. and de Bruijne, M. (2021). Automatic airway segmentation from computed tomography using robust and efficient 3-d convolutional neural networks. Sci Rep 11: 16001.

Garrity, J.M., Segars, W.P., Knisley, S. and Tsui, B.M.W. (2002). Development of a dynamic model for the lung lobes and airway tree in the ncat phantom. pp. 1858–1862. 2002 IEEE Nuclear Science Symposium Conference Record.

Gemci, T., Ponyavin, V., Chen, Y., Chen, H. and Collins, R. (2008). Computational model of airflow in upper 17 generations of human respiratory tract. J Biomech 41: 2047–2054.

Gemci, T., Ponyavin, V., Collins, R., Corcoran, T.E., Saha, S.C. and Islam, M.S. (2022). Cfd study of dry pulmonary surfactant aerosols deposition in upper 17 generations of human respiratory tract. Atmosphere 13.

Geva, T. (2006). Magnetic resonance imaging: Historical perspective. J Cardiovasc Magn Reson 8: 573–580.

Gill, S., Löbenberg, R., Ku, T., Azarmi, S., Roa, W. and Prenner, E.J. (2007). Nanoparticles: Characteristics, mechanisms of action, and toxicity in pulmonary drug delivery—a review. Journal of Biomedical Nanotechnology 3: 107–119.

Guan, S., Khan, A.A., Sikdar, S. and Chitnis, P.V. (2020). Fully dense unet for 2-d sparse photoacoustic tomography artifact removal. IEEE J Biomed Health Inform 24: 568–576.

Hinshaw, W.S., Bottomley, P.A. and Holland, G.N. (1977). Radiographic thin-section image of the human wrist by nuclear magnetic resonance. Nature 270: 722–723.

Hounsfield, G.N. (1980). Computed medical imaging. Science 210: 22–28.

Hu, B., Guo, H., Zhou, P. and Shi, Z.L. (2021). Characteristics of sars-cov-2 and covid-19. Nat Rev Microbiol 19: 141–154.

Huang, H., Lin, L., Tong, R., Hu, H., Zhang, Q., Iwamoto, Y. et al. (2020, 4–8 May 2020). Unet 3+: A full-scale connected unet for medical image segmentation. (ed.). pp. 1055–1059. ICASSP 2020 - 2020 IEEE International Conference on Acoustics, Speech and Signal Processing (ICASSP).

Huang, X., Francis, I., Saha, G., Rahman, M.M. and Saha, S.C. (2024a). Large eddy simulation-based modeling of cold-air inhalation from nasal cavities to the distal lung: Insights for athlete health and performance. Results in Engineering 23: 102475.

Huang, X., Saha, S.C., Saha, G., Francis, I. and Luo, Z. (2024b). Transport and deposition of microplastics and nanoplastics in the human respiratory tract. Environmental Advances 16: 100525.

Huang, X., Yin, Y., Saha, G., Francis, I. and Saha, S.C. (2024c). A comprehensive numerical study on the transport and deposition of nasal sprayed pharmaceutical aerosols in a nasal-to-lung respiratory tract model. Particle & Particle Systems Characterization, p.2400004.

Ibtehaz, N. and Rahman, M.S. (2020). Multiresunet : Rethinking the u-net architecture for multimodal biomedical image segmentation. Neural Netw 121: 74–87.

ICEM. (2012). Ansys icem cfd tutorial manual 14.5. I. ANSYS. Canonsburg, USA.

Inthavong, K., Tian, Z.F., Li, H.F., Tu, J.Y., Yang, W., Xue, C.L. et al. (2006). A numerical study of spray particle deposition in a human nasal cavity. Aerosol Science and Technology 40: 1034–1045.

Inthavong, K., Tian, Z.F., Tu, J.Y., Yang, W. and Xue, C. (2008). Optimising nasal spray parameters for efficient drug delivery using computational fluid dynamics. Comput Biol Med 38: 713–726.

Isensee, F., Jaeger, P.F., Kohl, S.A.A., Petersen, J. and Maier-Hein, K.H. (2021). Nnu-net: A self-configuring method for deep learning-based biomedical image segmentation. Nature Methods 18: 203–211.

Islam, M.S., Saha, S.C., Sauret, E., Gemci, T. and Gu, Y.T. (2017a). Pulmonary aerosol transport and deposition analysis in upper 17 generations of the human respiratory tract. Journal of Aerosol Science 108: 29–43.

Islam, M.S., Saha, S.C., Sauret, E., Gemci, T., Yang, I.A. and Gu, Y.T. (2017b). Ultrafine particle transport and deposition in a large scale 17-generation lung model. Journal of Biomechanics 64: 16–25.

Islam, M.S., Saha, S.C. and Young, P.M. (2018). Aerosol particle transport and deposition in a ct-based lung airway for helium-oxygen mixture (ed.). 21st Australasian Fluid Mechanics Conference, AFMC 2018. Adelaide, Australia.

Islam, M.S., Larpruenrudee, P., Paul, A.R., Paul, G., Gemci, T., Gu, Y. et al. (2021). Sars cov-2 aerosol: How far it can travel to the lower airways? Phys Fluids 33: 061903.

Jammes-Floreani, M., Laine, A.F. and Angelini, E.D. (2021, 13–16 April 2021). Enhanced-quality gan (eq-gan) on lung ct scans: Toward truth and potential hallucinations (ed.). pp. 20–23. 2021 IEEE 18th International Symposium on Biomedical Imaging (ISBI).

Jha, D., Smedsrud, P.H., Riegler, M.A., Johansen, D., Lange, T.D., Halvorsen, P. et al. (2019, 9–11 Dec. 2019). Resunet++: An advanced architecture for medical image segmentation (ed.). pp. 225–2255. 2019 IEEE International Symposium on Multimedia (ISM).

Jha, D., Riegler, M.A., Johansen, D., Halvorsen, P. and Johansen, H.D. (2020, 28–30 July 2020). Doubleu-net: A deep convolutional neural network for medical image segmentation (ed.). pp. 558–564. 2020 IEEE 33rd International Symposium on Computer-Based Medical Systems (CBMS).

Kamnitsas, K., Baumgartner, C., Ledig, C., Newcombe, V., Simpson, J., Kane, A. et al. (2016). Unsupervised domain adaptation in brain lesion segmentation with adversarial networks.

Kannan, R., Chen, Z.J., Singh, N., Przekwas, A., Delvadia, R., Tian, G. et al. (2017). A quasi-3d wire approach to model pulmonary airflow in human airways. Int J Numer Method Biomed Eng 33.

Kazeminia, S., Baur, C., Kuijper, A., van Ginneken, B., Navab, N., Albarqouni, S. et al. (2020). Gans for medical image analysis. Artificial Intelligence in Medicine 109: 101938.

Kimbell, J.S., Segal, R.A., Asgharian, B., Wong, B.A., Schroeter, J.D., Southall, J.P. et al. (2007). Characterization of deposition from nasal spray devices using a computational fluid dynamics model of the human nasal passages. J Aerosol Med 20: 59–74.

Kitaoka, H., Takaki, R. and Suki, B. (1999). A three-dimensional model of the human airway tree. Journal of Applied Physiology 87: 2207–2217.

Lahiri, A., Ayush, K., Biswas, P.K. and Mitra, P. (2017, 21–26 July 2017). Generative adversarial learning for reducing manual annotation in semantic segmentation on large scale miscroscopy images: Automated vessel segmentation in retinal fundus image as test case (ed.). pp. 794–800. 2017 IEEE Conference on Computer Vision and Pattern Recognition Workshops (CVPRW).

Larpruenrudee, P., Islam, M.S., Paul, G., Paul, A.R., Gu, Y.T. and Saha, S.C. (2021). Model for pharmaceutical aerosol transport through stenosis airway. pp. 91–128. *In*: Pathak, Y. and Islam, N. (1st ed.). Handbook of Lung Targeted Drug Delivery Systems: Recent Trends and Clinical Evidences. Taylor & Francis Group.

Li, Z., Wang, Y. and Yu, J. (2018). Brain tumor segmentation using an adversarial network. pp. 123–132. *In*: Brainlesion: Glioma, Multiple Sclerosis, Stroke and Traumatic Brain Injuries: Third International Workshop, BrainLes 2017, Held in Conjunction with MICCAI 2017, Quebec City, QC, Canada, September 14, 2017, Revised Selected Papers 3. Springer International Publishing.

Liu, Y., So, R.M.C. and Zhang, C.H. (2002). Modeling the bifurcating flowin a human lung airway. Journal of Biomechanics 35: 465–473.

Liu, Y., So, R.M.C. and Zhang, C.H. (2003). Modeling the bifurcating flow in an asymmetric human lung airway. Journal of Biomechanics 36: 951–959.

Lou, A., Guan, S. and Loew, M. (2020). Dc-unet: Rethinking the u-net architecture with dual channel efficient cnn for medical images segmentation.

Luo, H.Y. and Liu, Y. (2008). Modeling the bifurcating flow in a ct-scanned human lung airway. J Biomech 41: 2681–2688.

Mead-Hunter, R., King, A.J.C., Larcombe, A.N. and Mullins, B.J. (2013). The influence of moving walls on respiratory aerosol deposition modelling. Journal of Aerosol Science 64: 48–59.

Medical, M.M. (2021). Mimics Medical 24.0 – Instructions for Software Use. Materialise. Leuven, Belgium.

Meng, Q., Roth, H.R., Kitasaka, T., Oda, M., Ueno, J. and Mori, K. (2017). Tracking and segmentation of the airways in chest ct using a fully convolutional network (ed.). pp. 198–207. Medical Image Computing and Computer-Assisted Intervention − MICCAI 2017. Cham.

Michelini, G., Corridore, A., Torlone, S., Bruno, F., Marsecano, C., Capasso, R. et al. (2018). Dynamic mri in the evaluation of the spine: State of the art. Acta Biomed 89: 89–101.

Miyawaki, S., Hoffman, E.A. and Lin, C.-L. (2016). Effect of static vs. Dynamic imaging on particle transport in ct-based numerical models of human central airways. Journal of Aerosol Science 100: 129–139.

Miyawaki, S., Hoffman, E.A. and Lin, C.L. (2017). Numerical simulations of aerosol delivery to the human lung with an idealized laryngeal model, image-based airway model, and automatic meshing algorithm. Comput Fluids 148: 1–9.

Moeskops, P., Veta, M., Lafarge, M.W., Eppenhof, K.A.J. and Pluim, J.P.W. (2017). Adversarial training and dilated convolutions for brain mri segmentation (ed.). pp. 56–64. Deep Learning in Medical Image Analysis and Multimodal Learning for Clinical Decision Support. Cham.

Nardelli, P., Ross, J.C. and San José Estépar, R. (2020). Generative-based airway and vessel morphology quantification on chest ct images. Medical Image Analysis 63: 101691.

Ochs, M., Nyengaard, J.R., Jung, A., Knudsen, L., Voigt, M., Wahlers, T. et al. (2004). The number of alveoli in the human lung. Am J Respir Crit Care Med 169: 120–124.

Octane. (1999). Octane® Workstation Owner's Guide. I. Silicon Graphics. San Francisco, USA.

OpenFOAM. (2022). Openfoam User Guide Version 9. O. Foundation. Leuven, Belgium.

Park, J., Yun, J., Kim, N., Park, B., Cho, Y., Park, H.J. et al. (2020). Fully automated lung lobe segmentation in volumetric chest ct with 3d u-net: Validation with intra- and extra-datasets. Journal of Digital Imaging 33: 221–230.

Pathak, Y. and Islam, N. (2021). Handbook of Lung Targeted Drug Delivery Systems: Recent Trends and Clinical Evidences. Taylor & Francis Group.

Patton, J.S. and Byron, P.R. (2007). Inhaling medicines: Delivering drugs to the body through the lungs. Nat Rev Drug Discov 6: 67–74.

Phalen, R.F., Yeh, H.C., Raabe, O.G. and Velasquez, D.J. (1973). Casting the lungs *in-situ*. Anat Rec 177: 255–263.

Pourmehran, O., Gorji, T.B. and Gorji-Bandpy, M. (2016). Magnetic drug targeting through a realistic model of human tracheobronchial airways using computational fluid and particle dynamics. Biomech Model Mechanobiol 15: 1355–1374.

Qi, S., Zhang, B., Yue, Y., Shen, J., Teng, Y., Qian, W. et al. (2018). Airflow in tracheobronchial tree of subjects with tracheal bronchus simulated using ct image based models and cfd method. J Med Syst 42: 65.

Rahimi-Gorji, M., Gorji, T.B. and Gorji-Bandpy, M. (2016). Details of regional particle deposition and airflow structures in a realistic model of human tracheobronchial airways: Two-phase flow simulation. Comput Biol Med 74: 1–17.

Ronneberger, O., Fischer, P. and Brox, T. (2015). U-net: Convolutional networks for biomedical image segmentation. pp. 234–241. *In*: Medical Image Computing and Computer-assisted Intervention – miccai 2015.

Ruge, C.A., Kirch, J. and Lehr, C.-M. (2013). Pulmonary drug delivery: From generating aerosols to overcoming biological barriers—therapeutic possibilities and technological challenges. The Lancet Respiratory Medicine 1: 402–413.

Russo, J., Robinson, R. and Oldham, M.J. (2008). Effects of cartilage rings on airflow and particle deposition in the trachea and main bronchi. Med Eng Phys 30: 581–589.

Saha, S.C., Francis, I., Huang, X. and Paul, A.R. (2022a). Heat transfer and fluid flow analysis in a realistic 16-generation lung. Physics of Fluids 34.

Saha, S.C., Francis, I., Huang, X. and Rahman, M.M. (2022b). Fluid-particle dynamics of nasal drug delivery in a realistic human airway lung model. Proceedings of the 23rd Australasian Fluid Mechanics Conference (AFMC2023), Sydney, Australia, 4–8 December 2022.

Saha, S.C., Huang, X., Francis, I. and Saha, G. (2024). Airway stability in sleep apnea: Assessing continuous positive airway pressure efficiency. Respiratory Physiology & Neurobiology 325: 104265.

Sanders, M. (2007). Inhalation therapy: An historical review. Prim Care Respir J 16: 71–81.

Schmidt, A., Zidowitz, S., Kriete, A., Denhard, T., Krass, S. and Peitgen, H.O. (2004). A digital reference model of the human bronchial tree. Comput Med Imaging Graph 28: 203–211.

Shang, Y.D., Inthavong, K. and Tu, J.Y. (2015). Detailed micro-particle deposition patterns in the human nasal cavity influenced by the breathing zone. Computers & Fluids 114: 141–150.

Shen, D., Wu, G. and Suk, H.I. (2017). Deep learning in medical image analysis. Annu Rev Biomed Eng 19: 221–248.

Shi, H., Kleinstreuer, C. and Zhang, Z. (2007). Modeling of inertial particle transport and deposition in human nasal cavities with wall roughness. Journal of Aerosol Science 38: 398–419.

Shujaat, S., Jazil, O., Willems, H., Van Gerven, A., Shaheen, E., Politis, C. et al. (2021). Automatic segmentation of the pharyngeal airway space with convolutional neural network. Journal of Dentistry 111: 103705.

Si, X.A., Sami, M. and Xi, J. (2021). Liquid film translocation significantly enhances nasal spray delivery to olfactory region: A numerical simulation study. Pharmaceutics 13.

Siemens. (2017). Nx 12 Release Notes. Siemens. Munich, Germany.

SolidWorks. (2022). Solidworks 2022 sp2 Release Notes. D. Systèmes. Waltham, USA.

Son, J., Park, S.J. and Jung, K.-H. (2017). Retinal vessel segmentation in fundoscopic images with generative adversarial networks.

Soni, B. and Aliabadi, S. (2013). Large-scale cfd simulations of airflow and particle deposition in lung airway. Computers & Fluids 88: 804–812.

Soni, B., Arra, N., Aliabadi, S., Luke, W.H. and Walters, D.K. (2011). Mesh refinement study of flow and particle deposition in human lung airway model. 20th AIAA Computational Fluid Dynamics Conference. Honolulu, Hawaii.

Spencer, R.M., Schroeter, J.D. and Martonen, T.B. (2001). Computer simulations of lung airway structures using data-driven surface modeling techniques. Computers in Biology and Medicine 31: 499–511.

Talaat, M., Si, X.A., Tanbour, H. and Xi, J. (2019). Numerical studies of nanoparticle transport and deposition in terminal alveolar models with varying complexities. Med One.

Tan, S.Y. and Poole, P.S. (2020). Allan macleod cormack (1924–1998): Discoverer of computerised axial tomography. Singapore Med J 61: 4–5.

Tawhai, M.H., Pullan, A.J. and Hunter, P.J. (2000). Generation of an anatomically based three-dimensional model of the conducting airways. Annals of Biomedical Engineering 28: 793–802.

Tawhai, M.H., Hunter, P., Tschirren, J., Reinhardt, J., McLennan, G. and Hoffman, E.A. (2004). Ct-based geometry analysis and finite element models of the human and ovine bronchial tree. J Appl Physiol 97: 2310–2321.

Tian, L. and Ahmadi, G. (2013). Fiber transport and deposition in human upper tracheobronchial airways. Journal of Aerosol Science 60: 1–20.

Tong, X., Dong, J., Shang, Y., Inthavong, K. and Tu, J. (2016). Effects of nasal drug delivery device and its orientation on sprayed particle deposition in a realistic human nasal cavity. Comput Biol Med 77: 40–48.

Vachhani, S. and Kleinstreuer, C. (2021). Comparison of micron- and nano-particle transport in the human nasal cavity with a focus on the olfactory region. Comput Biol Med 128: 104103.

Valanarasu, J.M.J., Sindagi, V.A., Hacihaliloglu, I. and Patel, V.M. (2022). Kiu-net: Overcomplete convolutional architectures for biomedical image and volumetric segmentation. IEEE Trans Med Imaging 41: 965–976.

Van de Moortele, T., Wendt, C.H. and Coletti, F. (2018). Morphological and functional properties of the conducting human airways investigated by *in vivo* computed tomography and *in vitro* mri. J Appl Physiol 124: 400–413.

van Ertbruggen, C., Hirsch, C. and Paiva, M. (2005). Anatomically based three-dimensional model of airways to simulate flow and particle transport using computational fluid dynamics. J Appl Physiol 98: 970–980.

Wang, J., Zhao, Y., Noble, J.H. and Dawant, B.M. (2018). Conditional generative adversarial networks for metal artifact reduction in ct images of the ear. Med Image Comput Comput Assist Interv. 11070: 3–11.

Weibel, E.R. (1963). Geometry and dimensions of airways of conductive and transitory zones. pp. 110–135. *In*: Morphometry of the Human Lung. Springer Berlin Heidelberg.

WHO. (2022a). The top 10 causes of death. World Health Organization. Retrieved April 23 from https://www.who.int/news-room/fact-sheets/detail/the-top-10-causes-of-death.

WHO. (2022b). Who coronavirus (covid-19) dashboard. World Health Organization. Retrieved April 23 from https://covid19.who.int/.

Willis, L., Hayes, Jr., D. and Mansour, H.M. (2012). Therapeutic liposomal dry powder inhalation aerosols for targeted lung delivery. Lung 190: 251–262.

Wolterink, J.M., Leiner, T., Viergever, M.A. and Išgum, I. (2017). Generative adversarial networks for noise reduction in low-dose ct. IEEE Transactions on Medical Imaging 36: 2536–2545.

Wrap, G. (2022). Geomagic Wrap Overview. I. 3D Systems. Canonsburg, USA.

Wu, W., Yu, Y., Wang, Q., Liu, D. and Yuan, X. (2021). Upper airway segmentation based on the attention mechanism of weak feature regions. IEEE Access 9: 95372–95381.

Xi, J. and Si, X.A. (2017). Numerical simulation and experimental testing to improve olfactory drug delivery with electric field guidance of charged particles. Advanced Technology for Delivering Therapeutics.

Xi, J., Talaat, M., Tanbour, H. and Talaat, K. (2018). Airflow and particle deposition in acinar models with interalveolar septal walls and different alveolar numbers. Comput Math Methods Med. 2018: 3649391.

Xiao, X., Lian, S., Luo, Z. and Li, S. (2018). Weighted res-unet for high-quality retina vessel segmentation. pp. 327–331. 2018 9th International Conference on Information Technology in Medicine and Education (ITME).

Xue, Y., Xu, T., Zhang, H., Long, L.R. and Huang, X. (2018). Segan: Adversarial network with multi-scale l(1) loss for medical image segmentation. Neuroinformatics 16: 383–392.

Yang, Q., Yan, P., Zhang, Y., Yu, H., Shi, Y., Mou, X. et al. (2018). Low-dose ct image denoising using a generative adversarial network with wasserstein distance and perceptual loss. IEEE Trans Med Imaging 37: 1348–1357.

Yeh, H.-C. and Schum, G.M. (1980). Models of human lung airways and their application to inhaled particle deposition. Bulletin of Mathematical Biology 42: 461–480.

Yi, X. and Babyn, P. (2018). Sharpness-aware low-dose ct denoising using conditional generative adversarial network. J Digit Imaging 31: 655–669.

Yu, B., Zhou, L., Wang, L., Shi, Y., Fripp, J. and Bourgeat, P. (2019). Ea-gans: Edge-aware generative adversarial networks for cross-modality mr image synthesis. IEEE Transactions on Medical Imaging 38: 1750–1762.

Zhang, B., Qi, S., Yue, Y., Shen, J., Li, C., Qian, W. et al. (2018). Particle disposition in the realistic airway tree models of subjects with tracheal bronchus and copd. Biomed Res Int 2018: 7428609.

Zhang, Z. and Kleinstreuer, C. (2004). Airflow structures and nano-particle deposition in a human upper airway model. Journal of Computational Physics 198: 178–210.

Zhang, Z., Kleinstreuer, C. and Kim, C.S. (2009). Comparison of analytical and cfd models with regard to micron particle deposition in a human 16-generation tracheobronchial airway model. Journal of Aerosol Science 40: 16–28.

Zhou, Z., Siddiquee, M.M.R., Tajbakhsh, N. and Liang, J. (2018). Unet++: A nested u-net architecture for medical image segmentation. Deep Learn Med Image Anal Multimodal Learn Clin Decis Support 11045: 3–11.

# Chapter 2

# Fluid Flow and Particle Transport in Upper Airway Model

## With and Without Stenosis

*Bao Vuong Duong,*[1] *Adnan Munir,*[2] *Shahid Husain,*[3] *Hamidreza Mortazavy Beni,*[4] *Mohammad S. Islam*[1] and *Puchanee Larpruenrudee*[1,*]

## 1. Introduction

There is no doubt that in contemporary society, air pollution has been recognized as global health and contributes to millions of losses in human life in many countries around the world, particularly in developing countries. In other words, in developing countries, in order to adapt with a rapidly growing in economy, they tend to exploit inexpensive energy resources, which potentially results in a host of negative air pollutants, such as Particulate Matter (PM), nitrogen dioxide ($NO_2$), ozone ($O_3$), and Volatile Organic Compounds (VOCs). Apart from this, PM concern is found to be one of the most noticeable aspects of air pollution substances. In addition, PM with different diameters can easily enter and be deposited in the human respiratory

[1] School of Mechanical and Mechatronic Engineering, University of Technology Sydney (UTS), 15 Broadway, Ultimo, NSW-2007, Australia.

[2] School of Mechanical and Manufacturing Engineering, National University of Sciences and Technology, H-12, Islamabad, Pakistan.

[3] Department of Mechanical Engineering, Zakir Husain College of Engineering & Technology, Aligarh Muslim University, India.

[4] Department of Biomedical Engineering, Arsanjan Branch, Islamic Azad University, Arsanjan, Iran.

* Corresponding author: puchanee.larpruenrudee@uts.edu.au

DOI: 10.1201/9781003287636-2

system. In terms of healthy human airways, PM deposition fraction may not have severe impacts. However, PM depositions highly play a vital role in causing many adverse breathing issues or chronic lung diseases Albuquerque-Silva et al. (2014), especially narrowing or stenosis of airways. Therefore, investigating air flow characteristics and particle deposition and transport has become an interesting topic for many researchers.

In relation to airflow study, several papers, including *in silico* (Albuquerque-Silva et al., 2014; Farkas et al., 2008; Ghosh et al., 2020; Inthavong et al., 2011; Islam et al., 2017; Ito et al., 2016; Rahimi-Gorji, 2015) and experimental methods (Heenan et al., 2003; Lizal et al., 2018), have been established to investigate flow structure and particle motion in both realistic and idealized human respiratory. Although a certain understanding from these researchers, they have only focused on healthy human airways or large-scale models of airways (Islam et al., 2018, 2017; Longest et al., 2012; Gemci et al., 2008). One of the most highlighted studies of particle transport and deposition in the healthy human upper airway is conducted by Ghalati et al. (2012). Ghalati et al. (2012) contributed that the particle deposition efficiency numbers were highest in the nasal cavity region compared with the nasopharynx, oropharynx, and trachea by deploying a computational simulation with micro and nano-particles. Regardless of obstructive airways, it still has limitations in particle deposition characterization and flow structure. Furthermore, from the existing literature, in order to tackle these limitations due to the complexity of the human airway, particularly obstructive airways, the computerized method is found to be a reliable solution for simulating airflow patterns and particle movements. In terms of the achievements of Computational Fluid Dynamic studies (CFD), in recent years, thanks to the advanced techniques of the application of CFD, some studies have obtained the influences of stenosis airways on airflow characteristics. Firstly, Johari et al. (2013) established the location and size of stenosis to the probability of level of difficulty breathing by considering the prime causes, which are location and size of stenosis, which result dramatically in breathing adversities. Thus, the highlight finding of Johari et al. (2013) is that it has the potential to assist other studies to have a better understanding of how the significant rise in pressure drops happens in the trachea with distinctive locations and sizes. Moreover, another flow dynamics study in tracheal stenosis was conducted by Brouns et al. (2007). Brouns et al. (2007) stated that the overall pressure drops are influenced in the critical narrowing cases of the airway, which are around 60 and 80% restriction by implementing many CFD simulations with different degrees of stenosis section. On the other hand, it is necessary to improve the knowledge of particle transport and deposition of stenosis airways in depth. There are some available studies which partly obtain knowledge of particle transport and movements. With interest in the analysis of the drug delivery, Inthavong et al. (2010) concluded that particle deposition efficiencies were different between narrowing the airway model and normal airway model with 10 mu m and 1 mu m, which are 13 and 1%, respectively. Moreover, Rajaraman et al. (2020) analyzed the effect of particle growth, which is induced by hygroscopic particles with a diameter from 1 to 8 $\mu m$. Rajaraman et al. (2020) explored that the particle deposition efficiencies were significant for particle sizes from 2 to 4 $\mu m$. The case is true that this finding was necessary to distinguish asthmatic subjects, which helps to adjust the cluster in engineering aerosol size. Last but not least, Singh et al. (2020) found the deposition efficiency tends to rise corresponding to the increase of flow rates and particle sizes.

In this consideration, the objectives of this study are to not merely investigate the airflow characterization but also highly concentrate on particle transport and deposition of stenosis airways of upper human airways by undertaking a Computational Fluid Dynamic (CFD) study. Regarding the human upper airway, Gemci et al. (2008) showed that the human airway consists of up to 25 generations, which include branches of the bronchial tree. The upper airway is supposed to Generation 0 to 3. When it comes to geometry development, the employment of particle transport studies in realistic airways (Isaacs et al., 2006; van Ertbruggen et al., 2005; Fontana et al., 2005) may be valuable for practical results in airflow characterizations. However, it is recognized that due to the instability of the model, this factor results in a wide range of adversities to evaluate the impacts of stenosis sections on particle movement and deposition. Consequently, based on Weibel (1963), a symmetric airway model was developed with up to four generations as a healthy model. Apart from it, there are three other models were constructed with stenosis at the first, second and third generation.

Ultimately, this study concentrates on exploring the impacts of stenosis on the flow field and particle deposition in the human upper airway by utilizing a computerized study with different values of flow rates and particle sizes.

## 2. Geometrical Model Development

As depicted in the previous section, in order to conduct a computational simulation in the upper airway, firstly, it is necessary to construct the airway geometries. Due to the concentration of this study on the symmetrical model, the healthy airway without stenosis is developed by using the dimensions from Weibel (1963), which is revealed in Fig. 1. In addition, there are three stenosis models, which are constructed by

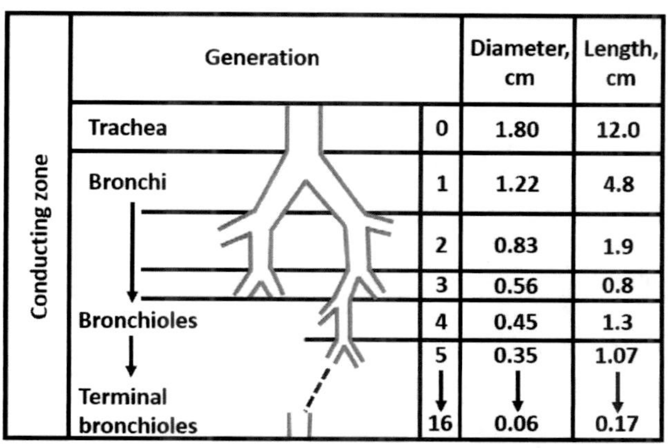


**Figure 1.** The dimensions of the symmetrical 3D model of human respiratory.

**Table 1.** Types, locations, and degrees of stenosis.

| | Generation | Stenosis | Type | Location | Degree |
|---|---|---|---|---|---|
| Geometry 1 | G0 - G4 | Structural | Tapered | G-1 Right Branch | 50% |
| Geometry 2 | G0 - G4 | Structural | Tapered | G-2 Right Branch | 50% |
| Geometry 3 | G0 - G4 | Structural | Tapered | G-3 Left Branch | 50% |
| Geometry 4 | G0 - G4 | Normal airway without stenosis | | | |

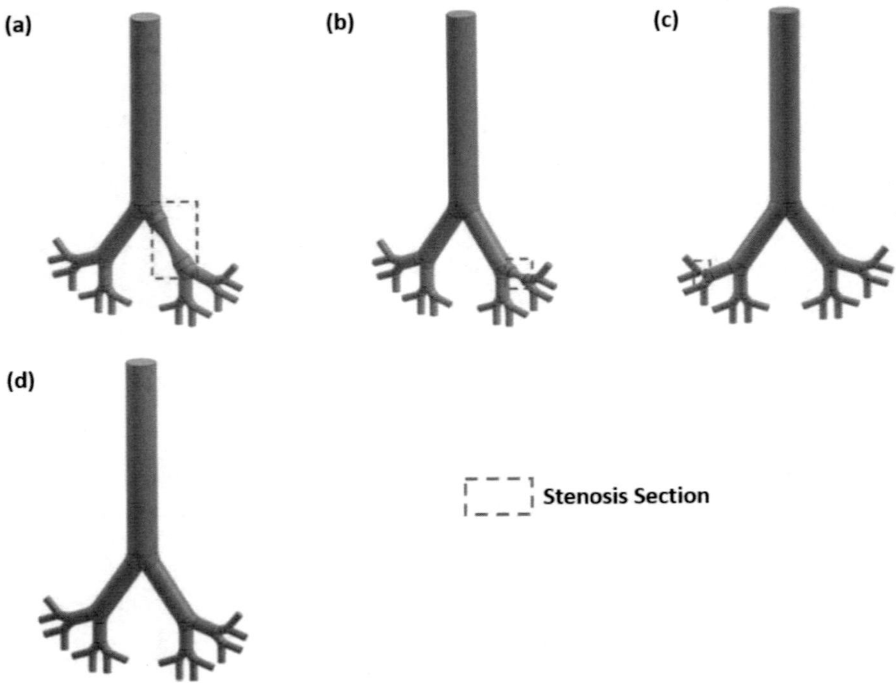


**Figure 2.** The reconstructed geometries of upper airway G0–G4: (a) stenosis section in G1 Right branch, (b) stenosis section in G2 right branch, (c) stenosis section in G3 left branch, and (d) healthy airway.

implanting the stenosis into generations of the healthy airway, which are first, second and third generation (see Fig. 2). Another factor that should be considered is that the dimension of stenosis deriving from Freitag et al. (2007), which is a proposed classification system of airway stenosis. The specifications of stenosis section are provided in Table 1.

## 3. Grid Generation & Refinement Test

It is undeniable that generating a good quality mesh is the facilitator of achieving reliable computation solutions. In other words, good meshing quality enhances the numerical stability and probability of finding a reliable solution. Moreover, regarding CFD (Computational Fluid Dynamics), meshing generation is also known as grid generation, which plays a vital role in influencing the speed, accuracy, and convergence of the solution.

The most striking feature of this problem is that because of the complexity of airway geometry, including multiple generations, especially in this study's four generations, the results of the simulations are highly affected by choosing the meshing software and meshing method. First of all, although there is a lot of software available to create the mesh, ANSYS Meshing is chosen to develop the most appropriate mesh for this particular study. When it comes to ANSYS Meshing, it provides users with a wide range of powerful techniques not merely to generate the mesh but also to control the quality of the mesh. Apart from that, there are two main methods for meshing involving structured mesh and unstructured mesh. There is no doubt that each method has different advantages and disadvantages. For example, in terms of structured mesh, it is simple to get the solution convergence with a small consumption of memory. However, the structured mesh method cannot be applied to complex geometries, such as respiration organs. On the other hand, the unstructured mesh has the capability of dealing with any complex geometries, however, it requires a big computational memory. Particularly, the advanced meshing method, which is a hybrid mesh method, is employed in this study due to the complexity of indicated geometries. It can be explained by the fact that thanks to powerful techniques of ANSYS Meshing, the hybrid mesh is conducted to take advantage of both structured and unstructured mesh methods including appropriate solutions and better convergence. Additionally, in the hybrid grid of this study, geometries are shown in Fig. 3; while the hexahedral cells are generated near solid boundaries, tetrahedral elements are created to fill the remaining volume.

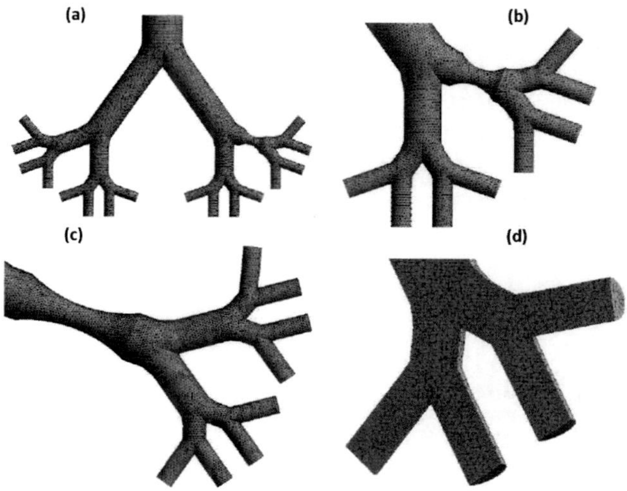


**Figure 3.** The generation of a hybrid mesh of geometries. (a) completed mesh generation, (b) & (c) the mesh of stenosis section, and (d) the performance of inflation layer mesh.

| Element | Node | Middle XZ Plane (Velocity - m/s) | Orthogonal Quality (Min) | Element Size (mm) |
|---|---|---|---|---|
| 526173 | 172547 | 1.33805 | 0.21576 | 1.000 |
| 602157 | 200329 | 1.33912 | 0.26944 | 0.880 |
| 726997 | 246135 | 1.33910 | 0.31068 | 0.750 |
| 805141 | 275076 | 1.33912 | 0.31117 | 0.690 |
| 899691 | 310208 | 1.33987 | 0.35290 | 0.640 |
| 997277 | 346986 | 1.34028 | 0.35397 | 0.600 |
| 1103317 | 386952 | 1.34060 | 0.38283 | 0.560 |
| 1203781 | 424844 | 1.34098 | 0.37601 | 0.530 |
| 1319600 | 479418 | 1.34150 | 0.38934 | 0.515 |
| 1386916 | 505172 | 1.34155 | 0.37666 | 0.500 |
| 1489142 | 545068 | 1.34154 | 0.38122 | 0.480 |

**Figure 4.** The information of mesh generation via using Ansys Meshing for airway model.

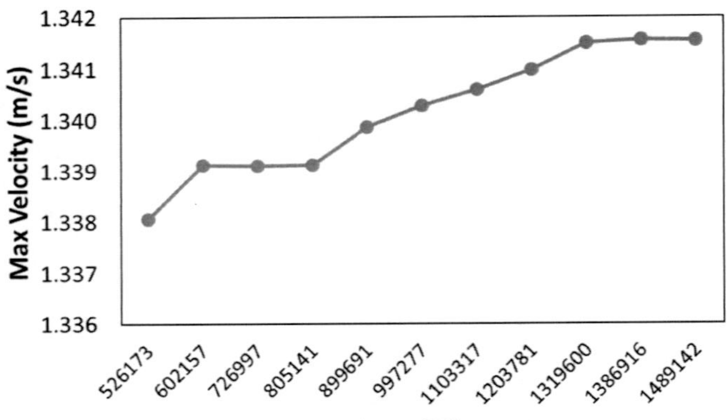


**Figure 5.** Grid refinement – grid-independent (the calculation of velocity at the middle XZ plane of the model) under 30 pm inlet condition.

After the completion of mesh development, it is necessary to implement the grid refinement (Figs. 4 and 5), which aims to enhance the accuracy of simulations. It other words, the objective of grid refinement is to identify the optimal meshes for the simulations. Moreover, the grid refinement is conducted by using Ansys Fluent to run a host of simulation under 30 lpm inlet condition with calculated maximum pressure of the middle XZ plane and the adjustment of element sizes. Parallelly, the values of orthogonal quality are also taken into consideration to ensure the quality of the meshes. Ultimately, the performance of grid refinement test is depicted below.

# 4. Numerical Methodology

## *4.1 Governing Equation*

In this study, based on Weibel's airway model, the upper airway models are developed by implementing of modeling software SolidWorks 2019 (DASSAULT SYSTEMES SolidWorks Corp., Waltham, MA, USA). Fluid flow and particle movements equations are calculated via conservation of mass and momentum supposition:

$$\frac{\partial \rho}{\partial t} + \nabla \cdot (\rho \vec{v}) = S \tag{1}$$

where $S_m$ is the mass source term, and

where $\rho$ is fluid static pressure, $\rho\vec{g}$ is body force because of gravity, and $\vec{F}$ is body force because of external (particle-fluid interactions) force.

$$\frac{\partial}{\partial t}(\rho \vec{v}) + \nabla \cdot (\rho \vec{v}\vec{v}) = -\nabla p + \nabla \cdot \left( \mu \left[ (\nabla \vec{v} + \nabla \vec{v}^T) - \frac{2}{3} \nabla \cdot \vec{v} I \right] \right) + \rho \vec{g} + \vec{F} \tag{2}$$

The $k - \varepsilon$ turbulence model is assumed as a viscous model. The formula for the calculation of turbulence kinetic energy and dissipation rate in an inertial frame.

In terms of turbulence kinetic energy k,

$$\frac{\partial(\rho k)}{\partial t} + \frac{\partial(\rho k u_i)}{\partial x_i} = \frac{\partial}{\partial x_j}\left[ \frac{\mu_t}{\sigma_k} \frac{\partial k}{\partial x_j} \right] + 2\mu_t E_{ij} E_{ij} - \rho \tag{3}$$

And for dissipation,

$$\frac{\partial(\rho \varepsilon)}{\partial t} + \frac{\partial(\rho \varepsilon u_i)}{\partial x_i} = \frac{\partial}{\partial x_j}\left[ \frac{\mu_t}{\sigma_\varepsilon} \frac{\partial \varepsilon}{\partial x_j} \right] + \varepsilon_{1\varepsilon} \frac{\varepsilon}{k} 2\mu_t E_{ij} E_{ij} - \varepsilon_{2\varepsilon} \rho \frac{\varepsilon^2}{k} \tag{4}$$

Where $u_i$ illustrates the velocity element in responding direction, $E_{ij}$ illustrates the element of the rate of deformation, and $\mu_t$ illustrates eddy viscosity. Moreover, in the equation (4), it also includes $\sigma_k$, $\sigma_k$, $C_{1\varepsilon}$ and $C_{2\varepsilon}$ which are based on numerous iterations of common data for turbulent airflows, and supposed to be 1.30, 1.00, 1.44, and 1.92 respectively.

## *4.2 Boundary Conditions*

As mentioned in the previous section, grid generation, three major name selections, which are inlet, outlet, and wall, are defined. Firstly, in the inlet, the parabolic profile is deployed for the inlet conditions. In this study, the values of velocity inlet are 7.5, 15, and 30 lpm (liter per minute). Apart from it, regarding outlet boundary, a zero-pressure is assumed to all outlets. It can be explained by the fact that practically speaking, it should have some certain unknown value of pressure at these outlets, however, in this particular study, the analysis is only first four-generation of the airway. Moreover, the wall surface conditions are defined as 'no-slip', which is derived from Rosin (1933) and Longest and Xi (2007) and the wall of the airway is the stationary wall. Furthermore, particle deposition is attained by implementing a 'trap' condition at the airway wall. In other words, every single time the particle touches the wall, it will be counted as deposited. When it comes to particle information, all particles are developed to inject form inlet surface at once. Each facet includes a single particle with a density value of 1100 kg/m$^3$ and diameters of 1, 2.5, 5, and 10 $\mu m$. Last but not least, the governing equations are calculated by implementing ANSYS Fluent 19.2.

## 5. Result and Discussion

The simulations are carried out in order to investigate flow field and particle deposition and transportation at three different values of flow rates, which are 7.5 lpm, 15 lpm and 30 lpm. Moreover, this study is implemented for monodisperse particle movement in four represented models. Ultimately, there are four different values of particle diameters consisting of 1 $\mu m$, 2.5 $\mu m$, 5 $\mu m$, 10 $\mu m$.

### *5.1 Flow Field Analysis*

In order to analyze the flow field velocity profiles with different values of flow rate, there are four lines, which are sketched cross-sections at four generations in Fig. 6. Moreover, the locations of these lines, 1, 2, 3, and 4, are sketched from trachea to third-generation, respectively. Figure 7 represents the velocity profiles at the indicated lines of the healthy airway. First of all, at the trachea (Fig. 7a), the airflow is fully employed for all cases and is not affected by any factor from the model. However, when airflow goes through lower generations, it becomes locally transitional due to the affections of airflow distribution between the left branch, right branch, and

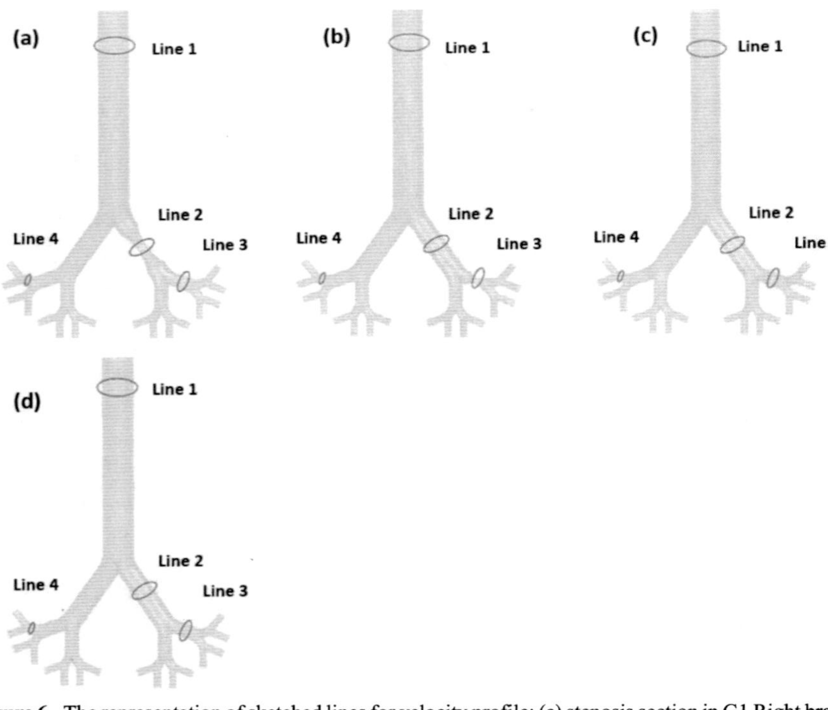


**Figure 6.** The representation of sketched lines for velocity profile: (a) stenosis section in G1 Right branch, (b) stenosis section in G2 right branch, (c) stenosis section in G3 left branch, and (d) healthy airway.

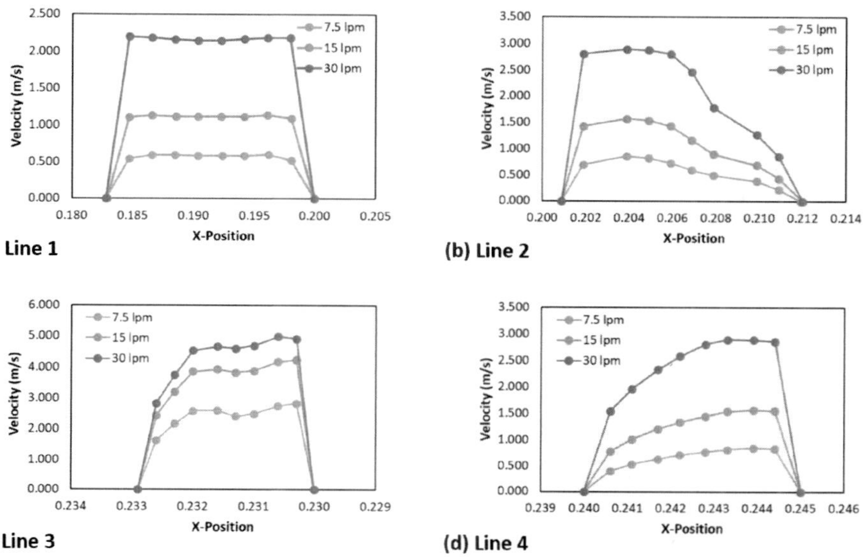


**Figure 7.** Velocity profile of healthy models at three different flow rates at: (a) Line 1 – Generation 0; (b) Line 2 –1st Generation; (c) Line 3 –2nd Generation; (d) Line 4 – 3rd Generation.

narrow sections. At a glance at Fig. 7, the velocity profiles determine the highest distribution of airflow was in the center of the airway, and the value of velocity was reduced near the wall region (Zhang and Kleinstreuer, 2011). According to Singh et al. (2020), they claimed that due to the pressure-driven force and significant changes in the realistic model, the value of the velocity of the throat region is extremely higher than in the third generation. However, in terms of this study, there are small changes in the value of velocity between generations because of the consistency of the symmetrical airway model.

On the other hand, compared the velocity profiles in Fig. 7 to velocity profiles in Fig. 8, which depicts the velocity profile of the stenosis sections in the other three models. The case is true that the major consideration is comparing the impacts of stenosis in each lower generation on the flow field. In other words, there is no investigation between the left and right branches due to the symmetrical characteristic of the airway. Furthermore, the velocity values increase for all models when airflows go through the stenosis sections. It can be explained by the fact that these increases are attributed to the law of mass conservation due to the reduction in the volume of airflow through narrow sections. However, the stenosis section in first-generation significantly rises in velocity due to the great dimensions of this generation, which results in the great size of the stenosis section in first-generation rather than the lower generation, the especially small dimension of the third generation.

In addition, there are nine planes for analyzing velocity contours, which are created in different locations of all models during inhalation to deepen recent findings

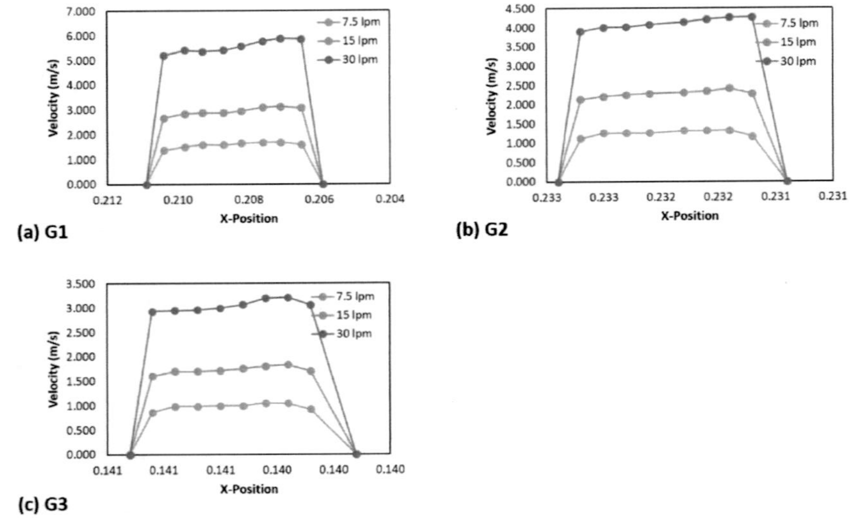


**Figure 8.** Velocity profile of stenosis sections at three different flow rates: (a) stenosis section in G1 Right branch, (b) stenosis section in G2 right branch, and (c) stenosis section in G3 left branch.

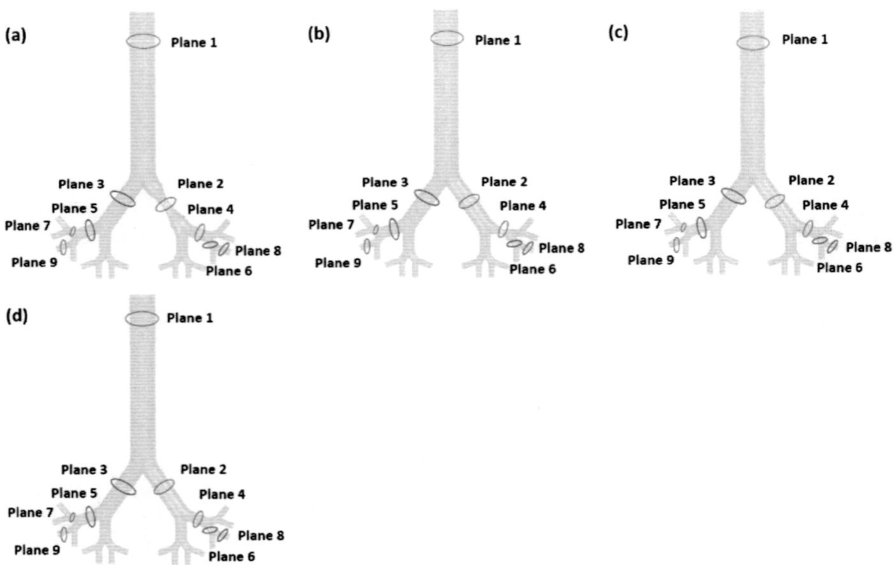


**Figure 9.** The representation of selected planes of models: (a) stenosis section in G1 Right branch, (b) stenosis section in G2 right branch, (c) stenosis section in G3 left branch, and (d) healthy airway.

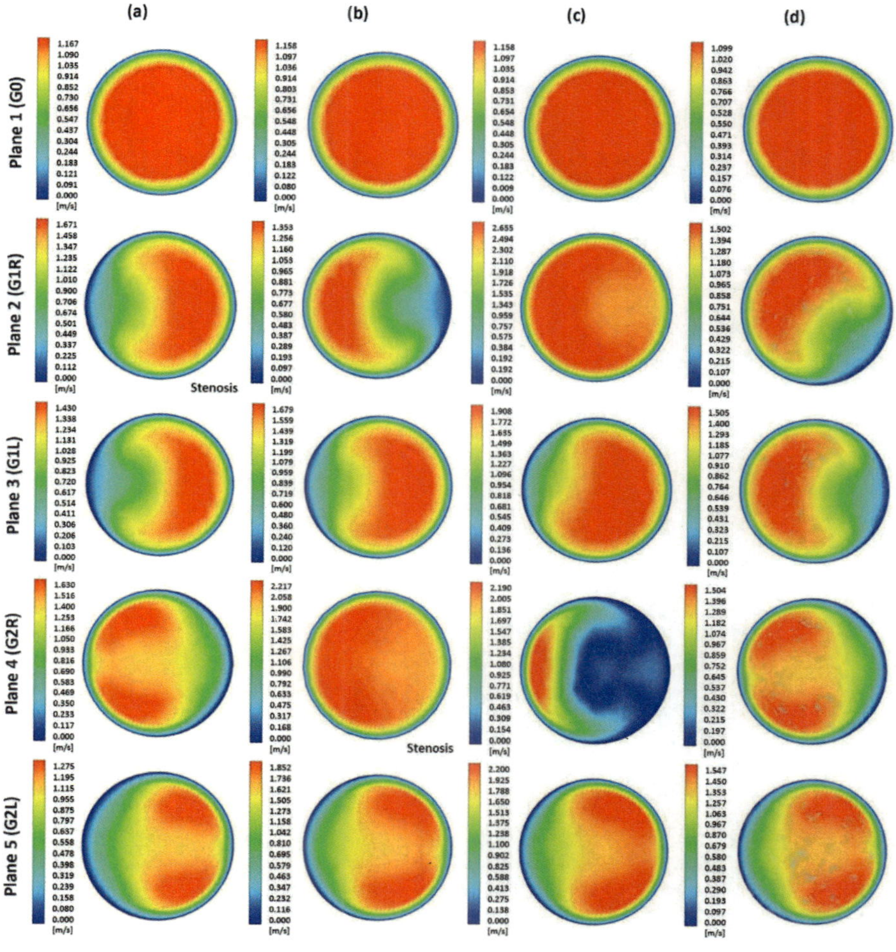


*Figure 10 contd. ...*

*...Figure 10 contd.*

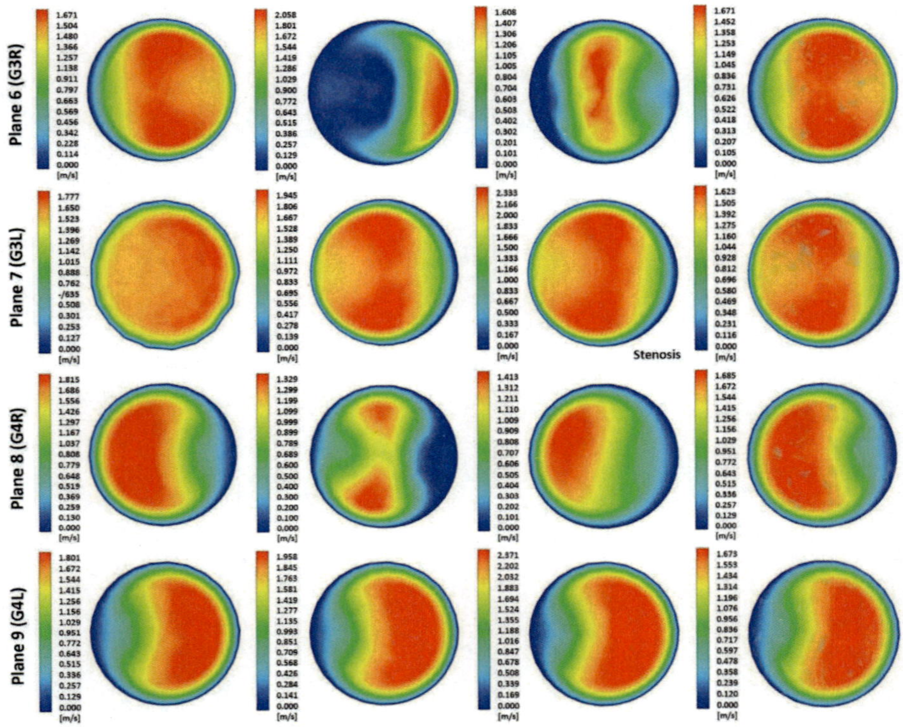


**Figure 10.** Velocity contour at different selected planes of four models at 30 lpm flow rate: (a) stenosis section in G3 left branch, (b) stenosis section in G2 right branch, (c) stenosis section in G1 Right branch, and (d) healthy airway.

in terms of the flow field. Figure 9 provides the representation of selected planes in all models, including the stenosis model. Thus, the airflow velocity contours are calculated in nine different planes for a healthy and narrowing airway. Obviously, airflow contours from the trachea to third-generation bear similar patterns in flow distributions. Another point that should be considered is that, once again, velocity contours in Fig. 10 determine the increase in airflow velocity during inhalation through stenosis sections. Besides, it also indicates there are no differences between the left branch and right branch because nine selected velocity contours are located not only on the right branch but also on the left branch. Lastly, the representation of velocity contours for the whole model in Fig. 11, it specifies that while the velocity tends to go up from the trachea to the third generation in the healthy airway, the velocity in the stenosis airway dramatically rises when airflow goes through the narrowing section.

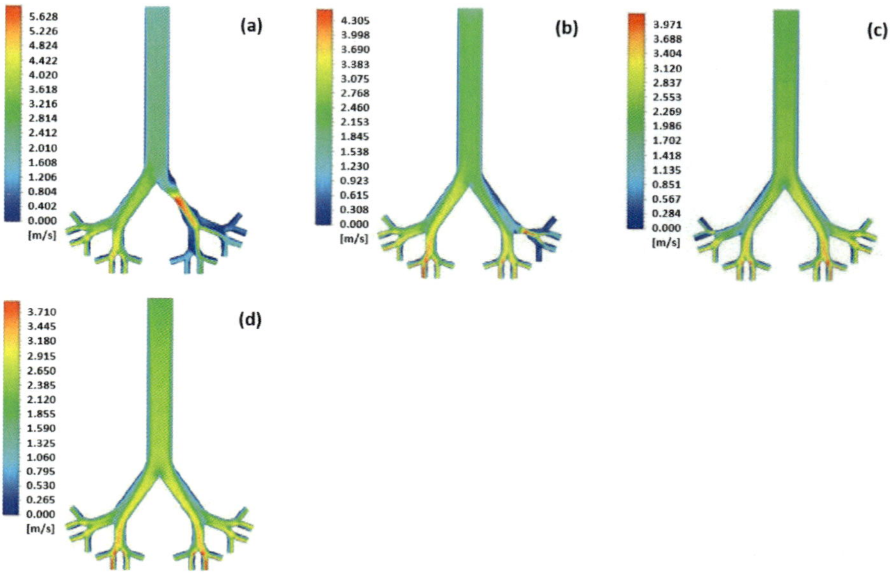


**Figure 11.** Velocity Contour of entire models for four geometries at 30 lpm of flow rate: (a) stenosis section in G1 Right branch, (b) stenosis section in G2 right branch, (c) stenosis section in G3 left branch and (d) healthy airway.

## *5.2 Pressure Drop and Wall Shear*

There is no doubt that the pressure drop is an important factor in the respiratory process. In simple terms, the pressure through the airway has to be smaller than atmospheric pressure. Consequently, airflow is able to be drawn into the lung from the trachea and upper airway. In this consideration, the pressure drop values are calculated from nine different locations in all models (Fig. 12), which are the same locations as nine planes of velocity contours. In relation to the healthy airway model, the values of pressure drop from the trachea to the third generation steadily reduced with three different flow rates due to the consistency of the airway model. In contrast, although the patterns of pressure drop in three stenosis models also reduce from the trachea to the third generation, regarding stenosis airways, the pressure drops are fluctuations, particularly after the stenosis section. Particularly, in the model with stenosis at the first generation, the pressure drop is extremely inconstant, which is caused by the great dimension in the length of stenosis.

Wall Shear Contours are established for whole model contours of healthy airway and stenosis airways in Fig. 13. Overall, Fig. 13d depicts the wall shear contours of a healthy model; it shows that the wall shear values are more likely to be high in the lower generation derived from the reduction of pressure in the airway. On the other hand, concerning stenosis airways, the wall shear values are remarkably high in the middle of the stenosis area.

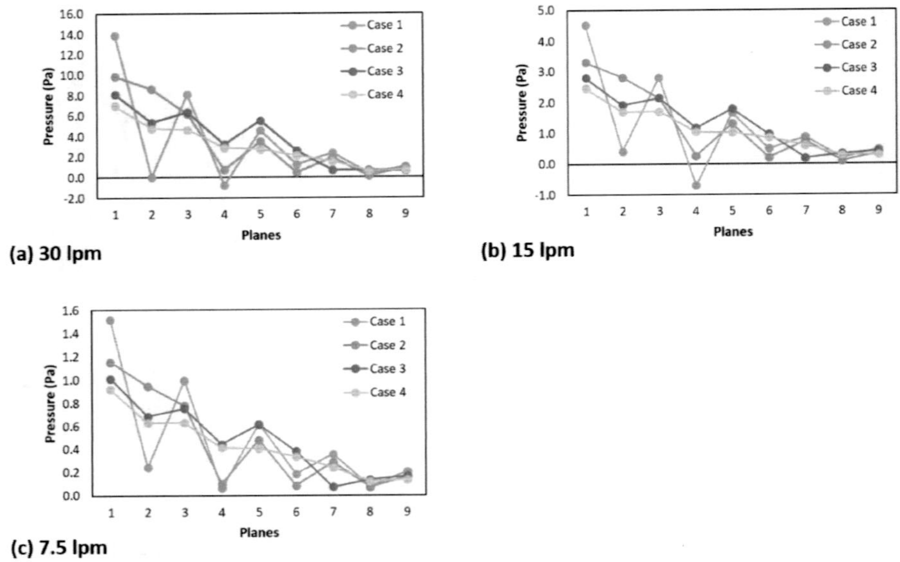


**Figure 12.** Pressure Drop at nine selected planes of four model airways: (a) 30 lpm; (b) 15 lpm; (c) 7.5 lpm.

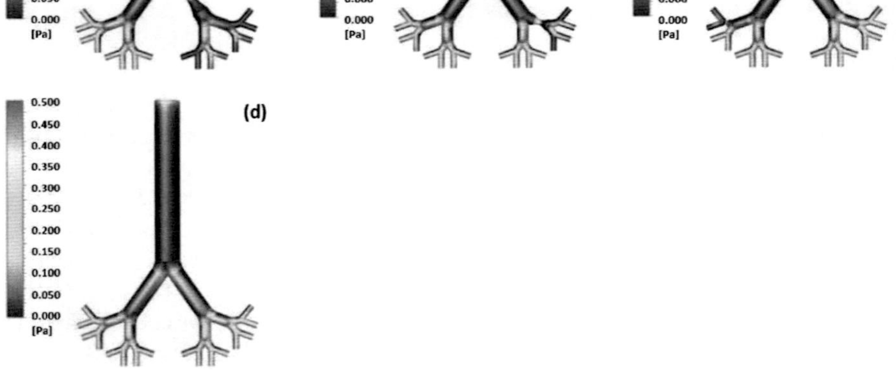


**Figure 13.** Wall Shear Contour for four models at 30 lpm flow rate: (a) stenosis section in G1 Right branch, (b) stenosis section in G2 right branch, (c) stenosis section in G3 left branch, and (d) healthy airway.

## *5.3 Turbulence Intensity*

As described by Zhang and Kleinstreuer (2011), airflow tends to become locally turbulent at 30 lpm of flow rate. In this study, the turbulent intensity contours for volume rendering are employed for all models of the airway at the highest flow rate of 30 lpm in Fig. 14. In terms of the healthy airway, as can be seen in Fig. 14, the high turbulent intensity values are located in the low generation and similar between left branch and right branch, such as partly in the second generation and almost in the third generation due to the symmetrical shape of airway. However, regardless of stenosis airways, the high turbulence intensity potentially occurs after the stenosis section. It can be understood that high-velocity flows are the main influence on driving turbulent flows in airways. Thus, after the stenosis section, there is an expansion in the diameter of the airway, which contributes to the rise of airflow velocity. Apart from that, the values of nine selected planes are revealed for turbulent intensity values of these locations in Fig. 15. The representation examines the differences between four different models in Fig. 15. Obviously, although the degree of stenosis sections in three stenosis airways is the same, 50%, the turbulence intensity through nine locations of stenosis in first-generation is considerable instability. Last but not least, there is consistency in turbulence intensity in not merely in healthy airways but also in stenosis in the third-generation model of the airway. The size of the stenosis section highly influences the turbulence in the airway.

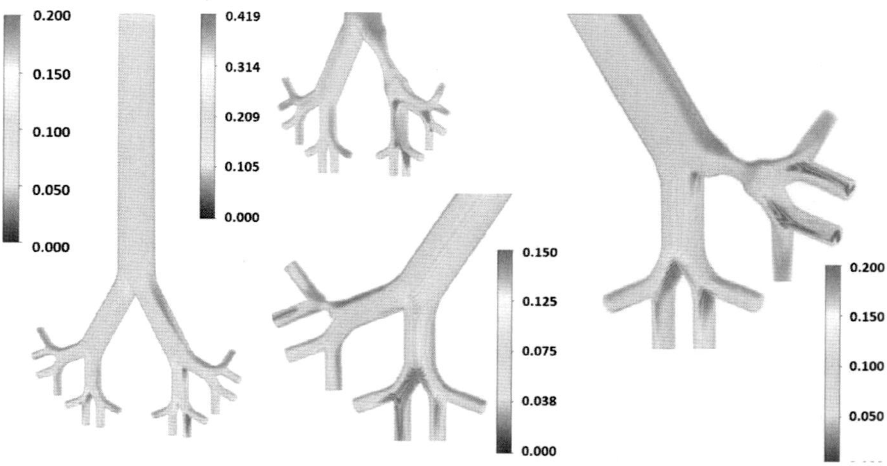


**Figure 14.** Turbulence Intensity at 30 lpm for the entire healthy airway and stenosis sections.

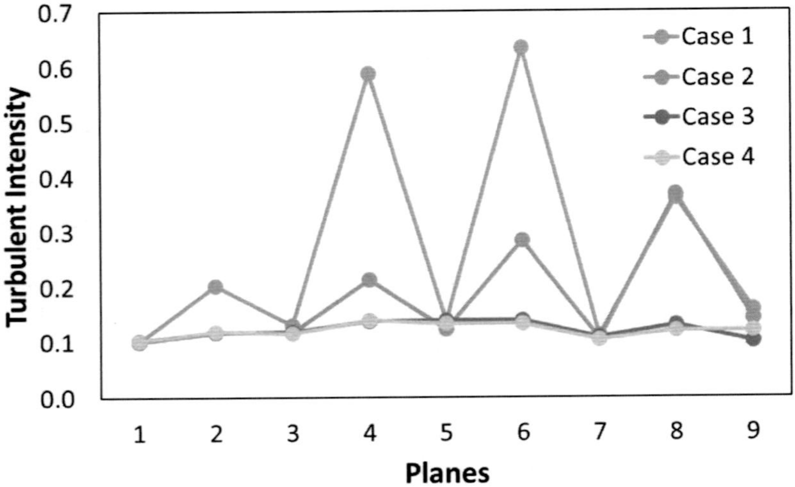


**Figure 15.** Turbulence Intensity at nine selected planes of four models of airway.

### *5.4 Particle Deposition and Movement*

As claimed by Yang et al. (2006), particle deposition mainly relies on the pattern of airflow through the airway. Thus, it is necessary to depict the airflow streamlines from the trachea to the third generation, which are illustrated in Fig. 16. Furthermore, Fig. 17 aims to illustrate the particle fraction of various particle sizes at 30 lpm flow rate for four model airways. Firstly, because of the influences of centrifugal force on the airflow movement, there is more concentration of airflow at the inner walls of branches than the outer walls of branches in Fig. 18(d). This event results in more particles getting deposited at the inner walls than at the outer walls. Moreover, as can be found that the airflow distribution and particle deposition are symmetrical from the trachea to the third generation in the healthy airway. On the other hand, when it comes to the model with stenosis at the first generation, mainly unfolds the impact of the stenosis section on the particle deposition due to the great size of its stenosis area. In Fig. 16(a), the right branch of the first-generation is restricted. Accordingly, the velocity in this branch is higher than in the left branch. Therefore, it is clear, as defined in Fig. 18(a), that there are more particles that go through the left branch, unobstructed area, than the obstructed branch and get deposited in this site. In the narrowing branch, a great number of particles get deposited and happen at the nozzle surface. Fortunately, this finding supports the analysis of drug-aerosol delivery. In Figs. 16(b) and 18(b), the obstructive area is located at G2 right branch. Apparently, the stenosis section at G2 generation does not have a significant impact on upstream areas. The particle deposition at the stenosis section is mostly similar to stenosis at the G1 right branch model. Then, after the stenosis section, there is a small number of particles deposited in the third generation due to the jet effect in the stenosis section. Ultimately, due to the small dimensions of the stenosis section at the third generation and its location at the low area, it does not dramatically influence the particle deposition pattern.

To deepen the understanding of particle movement, various diameters of particles are simulated with three different flow rates: 7.5 lpm, 15 lpm and 30 lpm. The result of particle deposition efficiency is represented in Fig. 17. The deposition efficiency is calculated by following the following formula:

$$Deposition\,Efficiency = \frac{particle\,number\,deposited\,on\,the\,divider}{total\,particle\,number\,entering\,to\,model}$$

Generally speaking, the deposition efficiency increase with the rising of flow rates and particles with a larger diameter is easier to get deposited than a particle with a small diameter. In this study, 10 $\mu m$ particles at 30 lpm flow rate stand for the highest deposition efficiency. Conversely, 1 $\mu m$, and 2.5 $\mu m$ particles tend to account for the lowest of deposition efficiency at 7.5 lpm flow rate. Moreover, as depicted in Fig. 17, the stenosis at the second generation and third generation do not have a strong impact on deposition efficiency when compared with the deposition efficiency of the healthy model. In other words, they bear similar data on particle deposition efficiency, which steadily increases with flow rate and diameter of particles. However, it is interesting

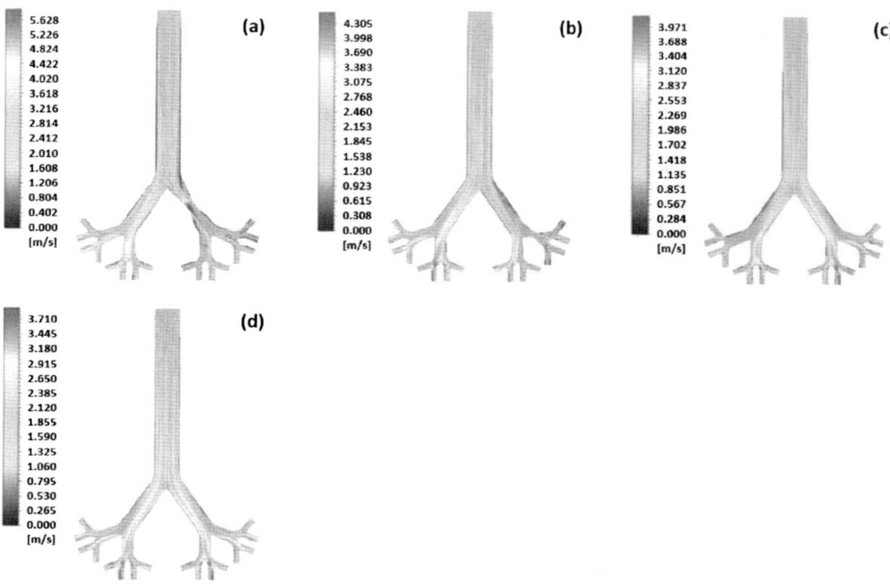


**Figure 16.** Airflow Streamlines of four models airway at 30 lpm flow rate: (a) stenosis section in G1 Right branch, (b) stenosis section in G2 right branch, (c) stenosis section in G3 left branch, and (d) healthy airway.

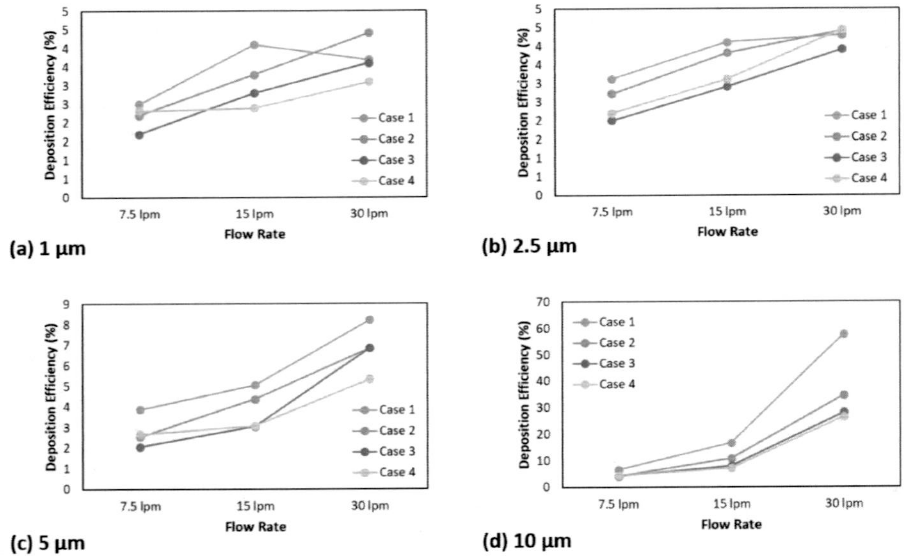


**Figure 17.** Deposition Efficiency of various particle diameters in four models airway: (a) 1 µm; (b) 2.5 µm; (c) 5 µm; (d) 10 µm.

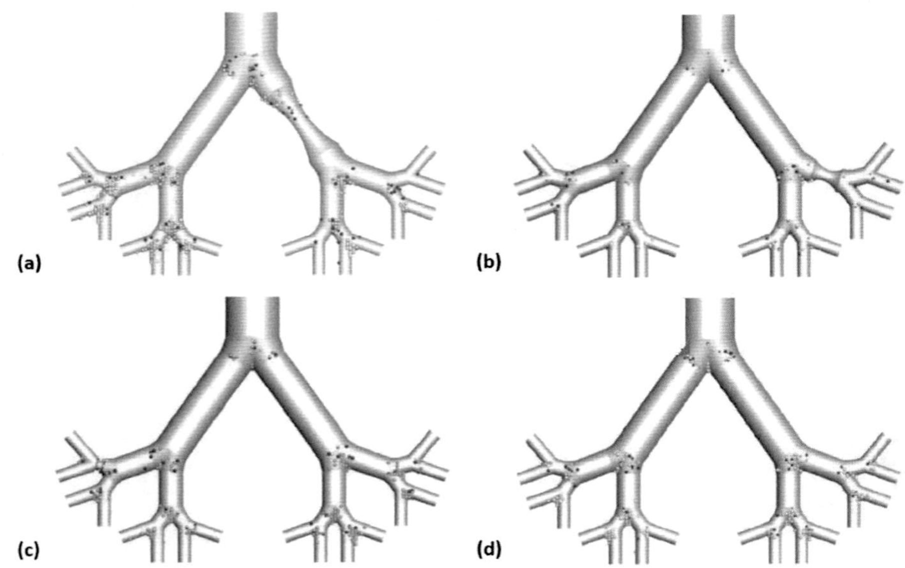


**Figure 18.** Particle Deposition Patterns of four models airway at 30 lpm: (a) stenosis section in G1 Right branch, (b) stenosis section in G2 right branch, (c) stenosis section in G3 left branch, and (d) healthy airway.

to note that stenosis at first generation possesses the ability to cause higher deposition efficiency than other models indicated in Fig. 17.

## 6. Limitations of this Study

- In the boundary condition, the outlet pressures are supposed to be zero. However, in practice, at the end of the fourth generation, there should be a certain value of pressure.
- In boundary conditions, the wall areas are defined as stationary walls. Unfortunately, there is an absence of motion of moving wall boundary in this study's simulations.
- Airway inflammation is not considered.
- The simulations are conducted by considering only one direction of airflow from the trachea to the third-generation.

## 7. Conclusion

After the completion of several simulations, which are implemented with three different flow rates and four different diameters of particles to analyze the impact of the stenosis section on flow field structure and particle characteristics. The key findings of this study are listed below:

- The flow field structures including velocity, pressure drop, turbulence intensity, and wall shear are influenced by the stenosis section. For example, the fluctuations of pressure drop, the high value of turbulence intensity and velocity after stenosis section, and high wall shear at the narrowing area during inhalation.
- The particle transportation and deposition are greatly affected by the stenosis section. Because of the block consequence of obstructive area, this results in the movements of more particles in the normal branch. Likewise, on the stenosis branch, the particles are more likely to be deposited at the inflamed surface due to the creation of nozzle area of particles.
- The dimensions of the stenosis section potentially play an important role in the deposition pattern.
- The obstructive areas at upstream airways have a greater influence on particle deposition than the obstructive area at downstream airways.

Ultimately, some aspects should be taken into consideration for future studies, such as a variety of types, sizes, and locations of the stenosis section in the airway, the motion of moving walls of the airway, and the development of airway models including the increase of the number of generations.

## References

Albuquerque-Silva, I., Vecellio, L., Durand, M., Avet, J., Le Pennec, D., de Monte, M. et al. (2014). Particle deposition in a child respiratory tract model: in vivo regional deposition of fine and ultrafine aerosols in baboons. PloS One 9(4): e95456–e95456. https://doi.org/10.1371/journal.pone.0095456.

Brouns, M., Jayaraju, S.T., Lacor, C., De Mey, J., Noppen, M., Vincken, W. et al. (2007, March 1st). Tracheal stenosis: a flow dynamics study. Journal of Applied Physiology 102(3): 1178–1184. https://doi.org/10.1152/japplphysiol.01063.2006.

Farkas, Á. and Balásházy, I. (2008, April 1st). Quantification of particle deposition in asymmetrical tracheobronchial model geometry. Comput Biol Med 38(4): 508–518. https://doi.org/https://doi.org/10.1016/j.compbiomed.2008.01.014.

Fontana, D., Vanni, M. and Baldi, G. (2005, July). Fluid dynamics modeling of particulate deposition in the lungs. Int J Artif Organs 28(7): 667–677. https://doi.org/10.1177/039139880502800705.

Freitag, L., Ernst, A., Unger, M., Kovitz, K. and Marquette, C. (2007, July 1st). A proposed classification system of central airway stenosis. The European Respiratory Journal : Official Journal of the European Society for Clinical Respiratory Physiology 30: 7–12. https://doi.org/10.1183/09031936.0013280.

Gemci, T., Ponyavin, V., Chen, Y., Chen, H. and Collins, R. (2008, January 1st) Computational model of airflow in upper 17 generations of human respiratory tract. Journal of Biomechanics 41(9): 2047–2054. https://doi.org/https://doi.org/10.1016/j.jbiomech.2007.12.019.

Ghalati, P.F., Keshavarzian, E., Abouali, O., Faramarzi, A., Tu, J. and Shakibafard, A. (2012). Numerical analysis of micro- and nano-particle deposition in a realistic human upper airway. Computers in Biology and Medicine 42(1): 39–49.

Ghosh, A., Islam, M.S. and Saha, S.C. (2020). Targeted drug delivery of magnetic nano-particle in the specific lung region. Computation 8(1): 10. https://doi.org/10.3390/computation8010010.

Heenan, A.F., Matida, E., Pollard, A. and Finlay, W.H. (2003, July 1st). Experimental measurements and computational modeling of the flow field in an idealized human oropharynx. Experiments in Fluids 35(1): 70–84. https://doi.org/10.1007/s00348-003-0636-7.

Inthavong, K., Choi, L.-T., Tu, J., Ding, S. and Thien, F. (2010, December 1st). Micron particle deposition in a tracheobronchial airway model under different breathing conditions. Medical Engineering & Physics 32(10): 1198–1212. https://doi.org/https://doi.org/10.1016/j.medengphy.2010.08.012.

Inthavong, K., Tu, J., Ye, Y., Ding, S., Subic, A. and Thien, F. (2010, June 1st). Effects of airway obstruction induced by asthma attack on particle deposition. Journal of Aerosol Science 41(6): 587–601. https://doi.org/https://doi.org/10.1016/j.jaerosci.2010.03.001.

Inthavong, K., Zhang, K. and Tu, J. (2011, July). Numerical modelling of nanoparticle deposition in the nasal cavity and the tracheobronchial airway. Comput Methods Biomech Biomed Engin 14(7): 633–643. https://doi.org/10.1080/10255842.2010.493510.

Isaacs, K.K., Schlesinger, R.B. and Martonen, T.B. (2006, Fall). Three-dimensional computational fluid dynamics simulations of particle deposition in the tracheobronchial tree. J Aerosol Med 19(3): 344–352. https://doi.org/10.1089/jam.2006.19.344.

Islam, M.S., Saha, S.C., Sauret, E., Gemci, T. and Gu, Y.T. (2017, June 1st). Pulmonary aerosol transport and deposition analysis in upper 17 generations of the human respiratory tract. Journal of Aerosol Science 108: 29–43. https://doi.org/https://doi.org/10.1016/j.jaerosci.2017.03.004.

Islam, M.S., Saha, S.C., Sauret, E., Gemci, T., Yang, I.A. and Gu, Y.T. (2017, November 17th). Ultrafine particle transport and deposition in a large scale 17-generation lung model. Journal of Biomechanics 64: 16–25. https://doi.org/https://doi.org/10.1016/j.jbiomech.2017.08.028.

Islam, S., Saha, S., Gemci, T., Yang, I., Sauret, E. and Gu, Y. (2018, December 1st). Polydisperse microparticle transport and deposition to the terminal bronchioles in a heterogeneous vasculature tree. Scientific Reports 8. https://doi.org/10.1038/s41598-018-34804-x.

Ito, K., Mitsumune, K., Kuga, K., Yoo, S.-J., Kim, J.W., Lu Phuong, N. et al. (2016, May 1st). Prediction of convective heat transfer coefficients for the upper respiratory tracts of rat, dog, monkey, and humans. Indoor and Built Environment 26. https://doi.org/10.1177/1420326X16662111.

Johari, N., Basri, W., Osman, K., Salleh, Z., Helmi, M. and Kadir, M. (2013, October 1st). The effect of tracheal stenosis at different locations and sizes on airflow in the trachea and main bronchi: a numerical modeling analysis. Advanced Science Letters 19: 3041–3046. https://doi.org/10.1166/asl.2013.5100.

Lizal, F., Jedelsky, J., Morgan, K., Bauer, K., Llop, J., Cossio, U. et al. (2018, February 15th). Experimental methods for flow and aerosol measurements in human airways and their replicas. European Journal of Pharmaceutical Sciences 113: 95–131. https://doi.org/https://doi.org/10.1016/j.ejps.2017.08.021.

Longest, P.W. and Xi, J. (2007). Effectiveness of direct lagrangian tracking models for simulating nanoparticle deposition in the upper airways. Aerosol Science and Technology 41(4): 380–397.

Longest, P.W., Tian, G., Walenga, R.L. and Hindle, M. (2012, June). Comparing MDI and DPI aerosol deposition using *in vitro* experiments and a new stochastic individual path (SIP) model of the conducting airways. Pharm Res 29(6): 1670–1688. https://doi.org/10.1007/s11095-012-0691-y.

Rahimi-Gorji, M., Pourmehran, O., Gorji-Bandpy, M. and Gorji, T.B. (2015, September 1st). CFD simulation of airflow behavior and particle transport and deposition in different breathing conditions through the realistic model of human airways. Journal of Molecular Liquids 209: 121–133. https://doi.org/https://doi.org/10.1016/j.molliq.2015.05.031.

Rajaraman, P.K., Choi, J., Hoffman, E.A., O'Shaughnessy, P.T., Choi, S., Delvadia, R. et al. (2020, August 1st). Transport and deposition of hygroscopic particles in asthmatic subjects with and without airway narrowing. Journal of Aerosol Science 146: 105581. https://doi.org/https://doi.org/10.1016/j.jaerosci.2020.105581.

Rosin, P. (1933). The laws governing the fineness of powdered coal. J. Inst. Fuel 7(31): 29–36.

Singh, P., Raghav, V., Padhmashali, V., Paul, G., Islam, M.S. and Saha, S.C. (2020). Airflow and particle transport prediction through stenosis airways. International Journal of Environmental Research and Public Health 17(3): 1119. https://doi.org/10.3390/ijerph17031119.

van Ertbruggen, C., Hirsch, C. and Paiva, M. (2005, March). Anatomically based three-dimensional model of airways to simulate flow and particle transport using computational fluid dynamics. J Appl Physiol (1985) 98(3): 970–980. https://doi.org/10.1152/japplphysiol.00795.2004.

Weibel, E.R. (1963). Morphometry of the Human Lung. Academic Press, Springer, New York.

Yang, X.L., Liu, Y. and Luo, H.Y. (2006, January 1st). Respiratory flow in obstructed airways. Journal of Biomechanics 39(15): 2743–2751. https://doi.org/https://doi.org/10.1016/j.jbiomech.2005.10.009.

Zhang, Z. and Kleinstreuer, C. (2011, December 1st). Laminar-to-turbulent fluid–nanoparticle dynamics simulations: Model comparisons and nanoparticle-deposition applications. International Journal for Numerical Methods in Biomedical Engineering 27(12): 1930–1950. https://doi.org/10.1002/cnm.1447.

# Chapter 3

# Role of CFD in Pulmonary Drug Delivery using Inhalers

*Vivek Kumar Srivastav*,[1] *Akshoy Ranjan Paul*,[2,*] *Anuj Jain*[2] and *Suvash C. Saha*[3]

## 1. Introduction

Rapid industrialization and urbanization in the developing countries causes a surge in respiratory diseases. The Global Burden of Disease (GBD) report published in 2019 estimated that around 300 million people globally are affected with asthma and 392 million in Chronic Obstructive Pulmonary Diseases (COPD) while its prevalence is increased by 50% per decade (Braman, 2006; Dharmage et al., 2019; Paul et al., 2022). World Health Organization's (WHO) Global Health Estimate (GHE) has identified COPD as the third leading cause of death causing the death of 3.23 million of the global population in 2019 (WHO, 2019). The Global Impact of Respiratory Diseases report published by the Forum of International Respiratory Societies (FIRS) in 2017 declared that at least 2 billion global population are exposed to indoor toxic smoke, 1 billion inhale outdoor pollutant air and 1 billion are exposed to tobacco smoke (FIRS, 2017). Therefore, an accurate prediction of local as well as zonal deposition of toxic and drug particles in human airways serves as valuable information for diagnosis, prognosis and treatment of these respiratory diseases.

[1] Applied Science & Humanities, Government Engineering College Siwan, Bihar, India.
[2] Department of Applied Mechanics, Motilal Nehru National Institute of Technology Allahabad, Prayagraj-211004, Uttar Pradesh, India.
[3] School of Mechanical and Mechatronic Engineering, University of Technology Sydney (UTS), 15 Broadway, Ultimo, NSW-2007, Australia.
* Corresponding author: arpaul@mnnit.ac.in

DOI: 10.1201/9781003287636-3

Due to the complexity of geometry and several other aspects, like irregular cross-section, biological aspect, and accurate boundary conditions, the determination of flow characteristics within such a duct remains a very challenging task for engineers. Gradually elevated deposition of allergens and toxic particles in the human airways increases the wall shear stress level with the inner walls of airways leading to acute injury of bronchial tissues and subsequently various pulmonary diseases (Paul et al., 2021). Popular treatment of such pulmonary diseases includes administering drugs into the respiratory system in aerosol form using oral inhalers. Nevertheless, these inhalers have its own limitations as the drugs administered through them often cannot reach to the injured locations in the lungs leading to delay in cure. Therefore, understanding air and drug flows in the Human Respiratory Tract (HRT) is important to design an effective inhaler for effective pulmonary drug delivery.

Although several studies are available on computational simulation of particle deposition in the human respiratory tract but works on the effects of cartilaginous rings (Srivastav et al., 2013), glomus tumour (Shukla et al., 2013; Srivastav et al., 2014; Srivastav et al., 2019a; Singh et al., 2020) and administration of optimum drugs (Tiwari et al., 2021a) are not apparent. The airflow through the oral cavity is targeted in this chapter. Airflow through the oral cavity occurs during coughing, running/exercise, asthma attacks, smoking and drug delivery using an inhaler during an asthma attack.

Normal breathing through the nostrils is cyclic with inhalation, exhalation and little pause. However, most researchers have carried out studies with inhalation (Luo et al., 2004; Zheng et al., 2007; Radhakrishnan and Kassinos, 2009; Koombua and Pidaparti, 2008; Srivastav et al., 2016), a few with exhalation (Green, 2004; Nowak et al., 2003) and limited one with cyclic (Nowak et al., 2003). For the cases of flow through the oral cavity presented in the book, it is enough to consider exhalation for coughing and inhalation for smoking and drug inhaling cases.

As the measurement of flow in human airways is not feasible, Computational Fluid Dynamics (CFD) simulation of air and particle flows offers an understanding of complex airflow fields, shear stress and local deposition of toxic/therapeutic micro-particles in human airways without and with any blockages (lesion/tumor) to assist medical practitioners for better diagnosis, prognosis and treatment of pulmonary diseases (Srivastav et al., 2012; Srivastav et al., 2021). The chapter briefly encompasses the development of oral inhalers, the role of CFD, Multiphysics and multiscale modeling in the development of oral inhalers and drug delivery through pulmonary routes.

## 2. Development of Oral Inhalers

The human population has a long history of inhaling plant-based substances containing medicinal properties to cure respiratory diseases. Historical evidence shows the controlled use of opium for treating asthma symptoms 3500 years ago in China, while ancient Egyptians smoked vapors of the papyrus scrolls to treat several lung diseases (Stein et al., 2017). Later, the English physician John Mudge invented the first inhaler in 1778 to administer drugs for the treatment of coughs and the device was popular for domestic as well as for hospital use. With the development of manufacturing technology, the first nebulizer called 'Pulverisateur' was introduced in the market in the 1880s to treat asthma, tuberculosis and pharyngitis (Bisgaard et al., 2001). In the Pulverisateur, a hand-operated pump enabled the liquid to push through an atomizer to form a vapor. With the development in medicine and pharmaceutics, a pressurized Metered Dose Inhaler (pMDI) was developed in 1956 by the Riker laboratories (USA) that was inspired by a perfume bottle made of glass. Gradually, breath-actuated MDI devices and other variants were introduced to the markets and the same is discussed in the following sub-sections.

## 2.1 Pressurized Metered Dose Inhaler (pMDI)

A pressurized metered dose inhaler is a type of MDI device that contains a pressurized propellant in a chamber. With the rapid increase of respiratory diseases, the global pMDI devices market value is expected to increase to USD 12000 million by the end of 2025. But more than its commercial value, it offers the greater social value of making a healthy society. Pressurized MDIs can be considered as an attractive drug delivery system through the pulmonary route, for effective local therapy, the systemic administration of drugs with irreproducible and poor-to-no bioavailability when administered via the gastrointestinal route (e.g., insulin) and for the drugs for which a rapid onset of action is desired (e.g., morphine). The later applications have challenged pulmonologists to achieve a better-controlled and more reproducible peripheral (deep) lung deposition of the drug.

The pMDI was first used in the late 1950s to deliver asthma medications in a convenient and reliable multi-dose form. Pressurized MDIs consist of a formulation containing dissolved or suspended drugs and hardware needed to contain the formulation and enable efficient and consistent dose delivery to the patient. The device hardware includes a canister that is appropriately sized to contain sufficient formulation for the required number of doses, a metering valve capable of delivering a consistent amount of drug with each dose delivered, an actuator mouthpiece that atomizes the formulation and serves as a conduit to deliver the aerosol to the patient, and often an indicating mechanism that provides information to the patient on the number of doses remaining.

The major components of the pMDI device (viz. propellants, formulation, metering valve and actuator) all have roles to play in the formation of spray, and in determining drug delivery to the lungs. Therefore, the opportunity exists to design a new and better pMDI product by adjusting the formulation, metering-valve size, and actuator nozzle diameter in order to obtain the required spray characteristics and fine-particle dose. Breath-actuated pMDIs, breath-coordinated pMDIs, spray-velocity modifiers, and spacer devices can be useful for patients who cannot use a conventional press-and-breathe pMDI properly.

The pMDI is readily recognized by the majority of patients who have ever received treatment for asthma in developed countries and, increasingly so, in developing countries. Indeed, the relatively low cost of pMDIs has contributed to a significant growth in pMDI use in developing countries and will ensure continued use in developed countries that are facing increased pressure to reduce health care costs. The inherent multi-dose nature of pMDIs makes them more affordable than most competing inhalation delivery systems. Despite many similarities, pMDIs have changed in many ways since the humble starting point in which they were made with glass vials and valves originally designed for perfume bottles.

As has been the case for pMDI formulation development, pMDI hardware technology has been significantly advanced as a result of Chlorofluorocarbon (CFC) to Hydrofluoroalkane (HFA) transition due to atmospheric ozone-layer depletion and also in response to growing competition from Dry Powder Inhalers (DPI). However, use of HFA in pMDIs as a propellant makes its drug deposition performance poor (10–20%) as compared to pMDIs with CFC gases. Therefore, the new-age pMDI devices, containing HFA propellant, has a scope for design modification to make it more efficient in terms of drug delivery via inhalation therapy for a variety of treatment. The principal hardware developments of pMDI that have recently been introduced, or which are currently being brought to the market include spacers, valve-holding chambers, and breath-actuated devices.

Pressured-MDI valves have been redesigned to be compatible with HFA propellants, to have reduced extractable and leachable, to enhance dosing uniformity, and to overcome the loss of prime during storage. Pressurized-MDI canisters have been developed with surface modifications that greatly reduce drug deposition and drug or canister degradation. Pressurized-MDI actuators have been developed to enhance drug delivery efficiency and provide a more aesthetically pleasing user interface for the patient. Breath-actuated technologies and add-on devices have been developed to further enhance drug delivery efficiency and minimize patient coordination challenges. Dose indicators and dose counters are now utilized to provide patients with information so that they know when to replace their pMDI. Furthermore, pMDI device technologies are in development and will continue to enhance pMDI delivery in the future. Many of the future improvements in MDI technology will increase the ability of pMDI systems to meet increased drug delivery in terms of drug deposition and its reach into the respiratory system and regulatory specifications (Tiwari et al., 2021b).

## 2.2 *Dry Powder Inhaler (DPI)*

The huge global population with various respiratory illnesses makes the development and refinement of inhalation treatments a key concern for the healthcare community. The global respiratory inhaler devices market value is expected to increase to USD 43214 million by the end of 2025 and a major share of this will be owned by Dry Powder Inhaler (DPI). Dry powder inhalers can be considered as an attractive drug delivery system through the pulmonary route, for effective local therapy (asthma, chronic obstructive pulmonary disease-COPD or cystic fibrosis), for the systemic administration of drugs with irreproducible and poor-to-no bioavailability when administered via the gastrointestinal route (e.g., peptides and smaller proteins which can pass the alveolar membrane like insulin) and for the drugs (e.g., morphine) for which a rapid onset of action is desired (Tiwari et al., 2022). The later applications has challenged the inhalation specialists, particularly in the past decade, to achieve a better controlled and more reproducible peripheral (deep) lung deposition of the drug.

Patients suffering from cystic fibrosis are treated with inhaled antibiotics. These drugs are given in high doses. These doses require new inhalers that are able to administer doses probably ≥ 100 mg of powder. Such amounts would certainly require special de-agglomeration principles to generate an aerosol with the appropriate particle size. On the other hand, many new drugs are highly potent, highly lipophilic materials that are given at a low dose. The reproducible administration of extremely low doses may also require special dosing and de-agglomeration technology (Larpruenrudee et al., 2022).

A variety of design of DPI exists in the market to meet such diverse needs. In general, DPI consist of a powder formulation, a dose mechanism containing (or measuring) a single drug dose, a powder de-agglomeration principle dispersing the powder into the inhaled air stream, and the inhaler's mouthpiece. Many types of secondary inhaler parts are applied to fulfill a large variety of functions. They are mostly added for safety, ease of handling, signaling to the patient and moisture protection of the drug formulation. In general, multi-dose systems require certain properties of the powder formulation regarding flowability and homogeneity.

In general, the resistance to air flow of DPIs is quite high compared with nebulizers and MDIs. This is a due to its design, which requires flow constriction to increase the kinetic energy of the air flow through the inhaler for adequate dose entrainment and powder de-agglomeration. The de-agglomeration principle is one of the most important parts of the dry powder inhaler as it determines the deposition efficiency of the drug deep into the lung. The de-agglomeration breaks up spherical pellets into primary drug entities or detaches drug particles from the carrier crystals in adhesive mixtures or nucleus agglomerates during inhalation. Its objective is to generate an aerosol that contains more drug particles in the aerodynamic size range of 1–5 μm that can enter the target area for deposition. An ideal DPI would generate a Fine Particle Fraction (FPF) attuned to the inhalation flow of the individual patient. Balancing the adhesive and cohesive forces in the powder formulation with the de-agglomeration forces requires an understanding of the magnitude of these forces and the underlying mechanisms of de-agglomeration.

The de-agglomeration can be increased by increasing the turbulence. The airflow pattern (turbulence) can be altered by changing the DPI structure, leading to enhanced DPI performance. It was found that the turbulence levels in different commercial inhalers were different and hence their performance. A more significant improvement can be obtained by changing the design of the de-agglomeration principle. The recent development of the coupled CFD-DEM model provides an improved method for studying de-agglomeration in DPIs over traditional CFD, which is discussed in detail later in this chapter.

## 3. CFD- A Tool to Revolutionize the Design of Drug Delivery

Any problem which is solved using CFD requires three major steps as shown in Fig. 1.

Due to the complexity of geometry and several other aspects, like irregular cross-section, biological aspect, and accurate boundary conditions, the determination

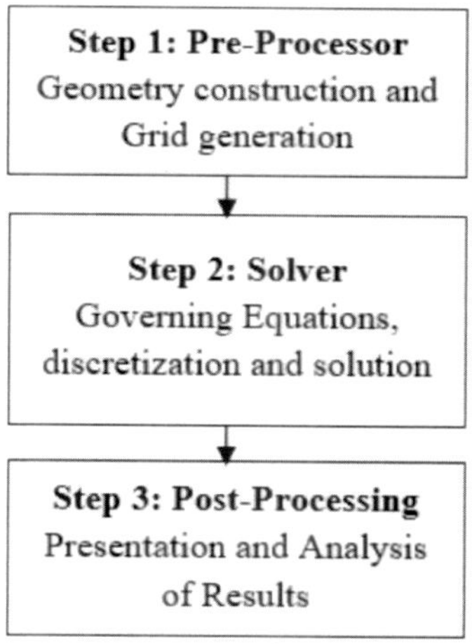


**Figure 1.** Major steps in CFD simulation.

of flow characteristics within such ducts remains a very challenging task for engineers. This section describes in brief the procedures followed for the pre-processing step, the solver and post-processing steps.

### *3.1 Geometry and CAD Modeling*

It was found that most of the respiratory diseases occur in the trachea and bronchi. Hence, geometric models from the trachea to the bronchi are mainly considered.

Two types of airway models, namely, simplified and CT-scan-based realistic models, are constructed for the analysis of airflow and particle deposition in the HRT. The first type of model is constructed considering that the HRT is made of regular pipes of various cross-sections, while the second type of model is constructed based on CT images of the HRT. A first-generation simplified model shown in Fig. 3 is constructed based on the dimensions derived from the Weibel model (1963). Weibel established a geometrical relationship of the bifurcation of the trachea into smaller bronchi based on studies of a number of cadavers. The scientific and engineering community widely accepts the relationships given for the study of the respiratory system. The Weibel model considered the trachea and bronchi made of regular pipes, therefore termed as 'simplified' model. The HRT casting dimensions reported by Schlesinger and Lippmann (1972) are furnished in Table 1.

The three-dimensional reconstruction of HRT from trachea to first-generation bronchi is done using spiral CT-scan images as detailed in Table 2 and downloaded from National Biomedical Imaging Archives (NBIA), NIH, USA. This organization provides images to researchers and doctors for the development of medical research.

The dimension of the realistic model shown in Fig. 1(b) is given in Table 3.

**Table 1.** Dimension of Respiratory Tract Model.

| Region in the HRT | Length (m) | Diameter (m) |
|---|---|---|
| Trachea | $9.2 \times 10^{-2}$ | $2.17 \times 10^{-2}$ |
| Right bronchus | $4.2 \times 10^{-2}$ | $1.7 \times 10^{-2}$ |
| Left bronchus | $5.3 \times 10^{-2}$ | $1.26 \times 10^{-2}$ |

**Table 2.** Details of CT Scan Images used for realistic HRT model.

| Total number of slices | Slice increment | Pixel resolution | HRT generation |
|---|---|---|---|
| 38 | 3 mm | 512 × 512 | 0–1st |

**Table 3.** Dimension of first-generation CT scan model.

| Section | Hydraulic diameter (m) |
|---|---|
| Inlet | $1.88 \times 10^{-2}$ |
| Left outlet | $1.13 \times 10^{-2}$ |
| Right outlet | $1.2 \times 10^{-2}$ |

A tumor in the trachea of the realistic HRT model shown in Fig. 2(b) was created as reported by Watanabe et al. (1998) to study the flow behavior in the presence of a tumor. They studied the lower tracheal glomus tumor of a 43-year-old patient using a bronchoscope, as shown in Fig. 2(a). The size of the polyploidy tumor was 2.01.61.4 cm3 and was located in the membranous portion of the trachea just above the carina, occupying half of the lumen. A similar diploid tumor was constructed at a 2 cm distance above the carina on the posterior wall of the trachea, as shown in Fig. 2(b).

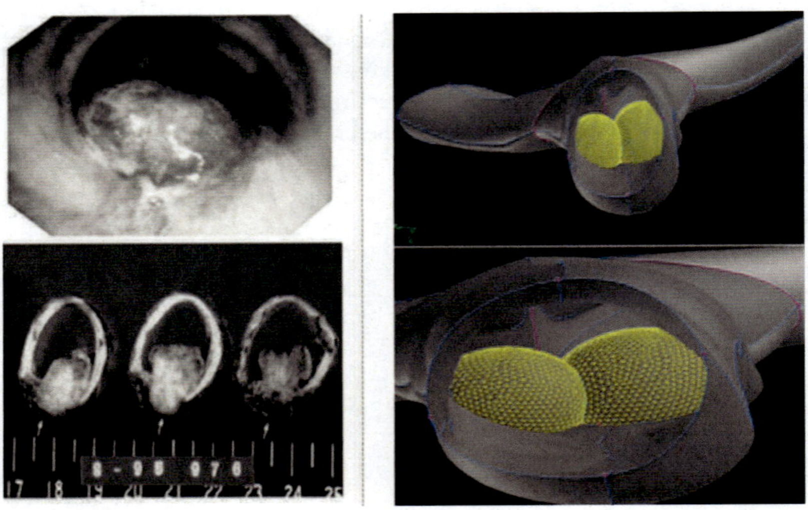

**Figure 2.** (a) Glomus tumor in HRT [Watanabe et al., 1998]. (b) CAD model of glomus tumor in HRT.

Smoking and medication from inhalers are supplied through the oral cavity (mouth cavity). The flow coming through the mouth cavity is not uniform at the inlet of the trachea. Therefore, it is important to construct an oral cavity for the realistic simulation of flow. The oral cavity was constructed by Cheng et al. (1999). The same oral cavity was reconstructed along with the trachea by Kleinstreuer and Zhang (2003). The oral cavity is constructed in Ansys Workbench-14 using the dimensions given by Cheng et al. (1999) as shown in Fig. 5(c). The diameter of the outlet of the oral cavity is taken from the Weibel Model (1963). The diameter of the different cross sections is provided by Cheng et al. (1999) (Table 4). First, circular rings are created using these diameters, and then the surface is generated by connecting these circular rings. Finally flow domain was constructed by connecting the wall surface, inlet and outlet of the oral cavity. These steps are pictorially shown in Fig. 3. Figure 4 shows the inclusion of cartilaginous rings- a ribbed-like structure that affects the flow field in the trachea of the HRT model (Srivastav et al., 2013).

**Table 4.** Dimensions of oral cavity.

| Location | Diameter (m) |
|---|---|
| Inlet | $2.18 \times 10^{-2}$ |
| Outlet | $2.00 \times 10^{-2}$ |

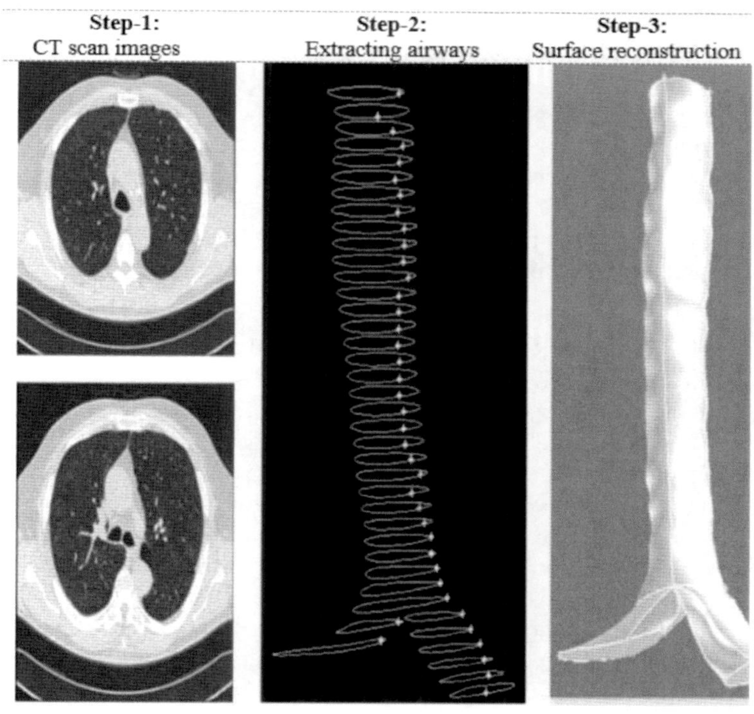


**Figure 3.** Construction steps of HRT model from spiral CT scan.

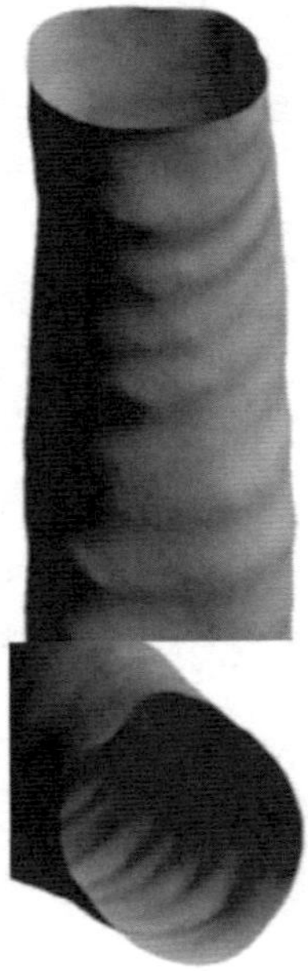

**Figure 4.** Zoomed view of HRT model.

A total of six CAD models of HRT are studied here. These models are

Model-1: Trachea to first generation bronchi weibel (simplified) model
Model-2: Trachea to first generation bronchi CT scan based model
Model-3: Trachea to first generation bronchi CT scan based tumorous model
Model-4: Oral cavity to first generation bronchi CT scan based model
Model-5: Oral Cavity to first generation bronchi simplified model
Model-6: Oral Cavity to sixth generation bronchi CT scan based model

The airflow through the simplified geometry of trachea to first generation bronchi (Model-1) was computationally simulated by Srivastav et al. (2011). Aerosol-based drug delivery was simulated by Srivastav et al. (2014) in a CT-scan tumorous model spanning from trachea to first generation bronchi (Model-3). Besides, the researchers (Shukla et al., 2020; Srivastav et al., 2021) studied the effects of fluid-structure interaction on the airflow in a CT-scan model ranging from the oral cavity to first-generation bronchi (Model-4) by varying the wall elasticity of the human HRT (Paul et al., 2009). Furthermore, Srivastav et al. (2021) investigated aerosolized drug deposition in a CT scan HRT model from the oral cavity to the sixth-generation bronchi (Model-6). Recently, Saha et al. (2022) carried out heat transfer and airflow simulation in a realistic HRT model ranging from the oral cavity to the distal 16-generation.

The three-dimensional reconstruction of the airway model from trachea to sixth-generation bronchi is done using the Spiral CT scan images downloaded from NBIA (USA), as detailed in Table 5.

**Table 5.** Details of CT scan images used for Model-6.

| Total number of slices | Slice increment | Pixel resolution | HRT generation |
|---|---|---|---|
| 519 | 0.624 mm | 300 × 203 | 0–6th |

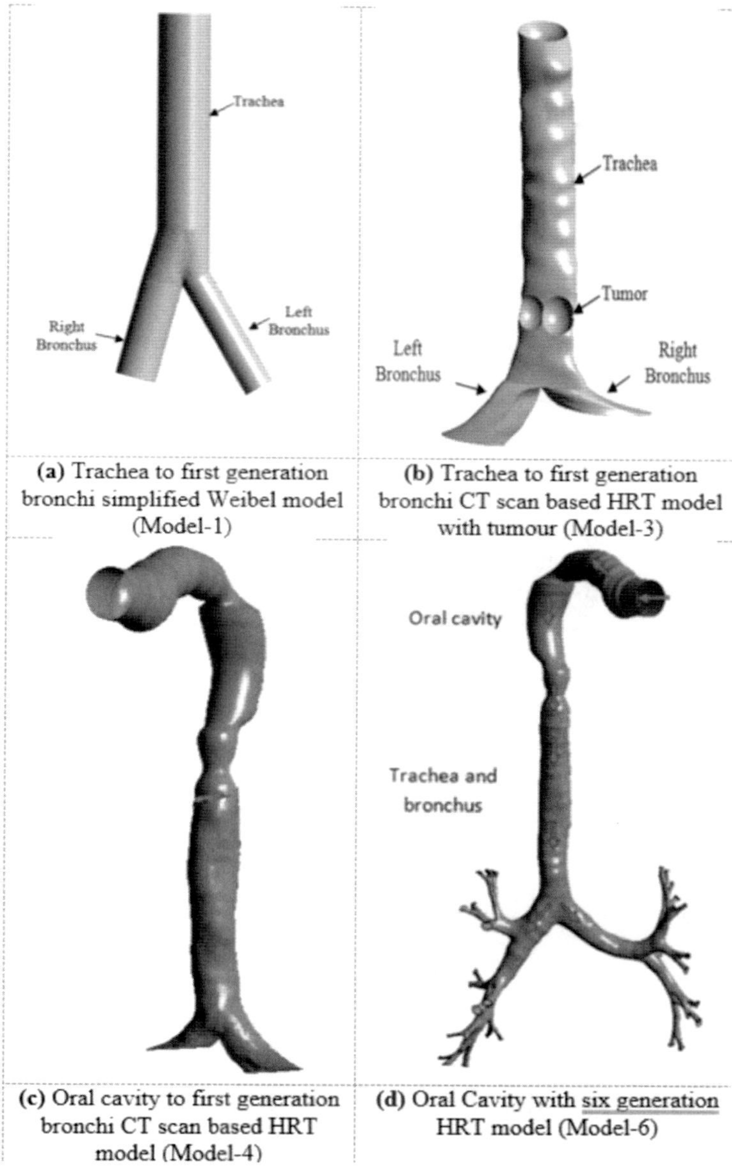


**(a)** Trachea to first generation bronchi simplified Weibel model (Model-1)

**(b)** Trachea to first generation bronchi CT scan based HRT model with tumour (Model-3)

**(c)** Oral cavity to first generation bronchi CT scan based HRT model (Model-4)

**(d)** Oral Cavity with six generation HRT model (Model-6)

**Figure 5.** Various CAD models of HRT.

### *3.2 Grid Generation*

Ansys-Workbench is used for the grid generation. The patch-independent method is adopted for grid generation, which uses a top-down approach (generate mesh from volume to surface). Since the surface of the respiratory tracts is curved therefore, surface meshing is generated using triangular mesh, and volume grid is generated using unstructured tetrahedral elements. Grid is refined near carina junction to capture the flow physics nicely near carina.

CFD simulation is greatly affected by the skewed elements; therefore, grid generation was done in such a way that the highly skewed elements are minimal. Due to the irregular and asymmetrical shape of the human respiratory tract, four different types of elements (tetrahedral, hexahedral, pyramid and wedges) have been used for the grid generation in the model to improve the quality of mesh. For grid generation the size of the elements is taken 0.016 mm. The sizing function is used for the grid generation. As anticipated, the maximum skewness in this model is higher than that of simple; the size of the elements is taken at the model as its geometry is more complex. In order to reduce maximum skewness, the grid is refined. Since highly skewed elements are located near the outlet, these are not likely to affect CFD results significantly.

The same procedure used for the grid generation in model 2 was adopted for model 3. Grid generation has been performed using the tetrahedral elements of size 0.2 mm. The most skewed element, having a skewness of 0.9, has been found at the outlet of the HRT model. Surface meshing is generated by triangular mesh and unstructured tetrahedral elements have been used for the flow volume. Grid generation has been performed using the tetrahedral elements of size range 0.013–2.2 mm. Grid generation of oral cavity was performed in Gambit-Fluent (Fig. 6). For grid generation the size of the tetrahedral elements is taken 1 mm.

Grid generation in trachea to sixth generation bronchi CT scan-based three-dimensional model is done in Gambit-Fluent software using unstructured tetrahedral elements (Fig. 7). For grid generation the size of the elements is taken 0.5 mm.

For better visualization of the generated grid, the file is imported into Ansys-ICEM and is shown in Fig. 8. Further, it is observed that the element with the worst skewness is found at the HRT outlet, which is not likely to affect the CFD simulation.

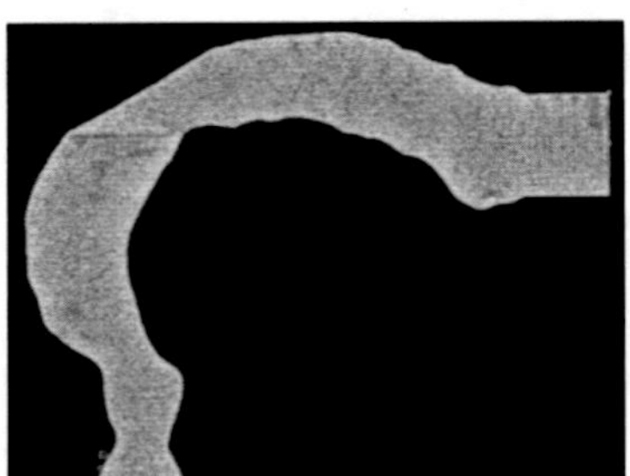

**Figure 6.** Computational grid in the oral cavity.

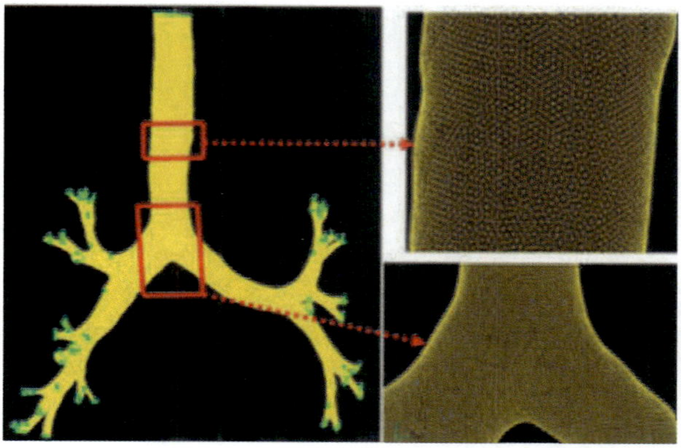

**Figure 7.** Grid generation in sixth generation realistic HRT model.

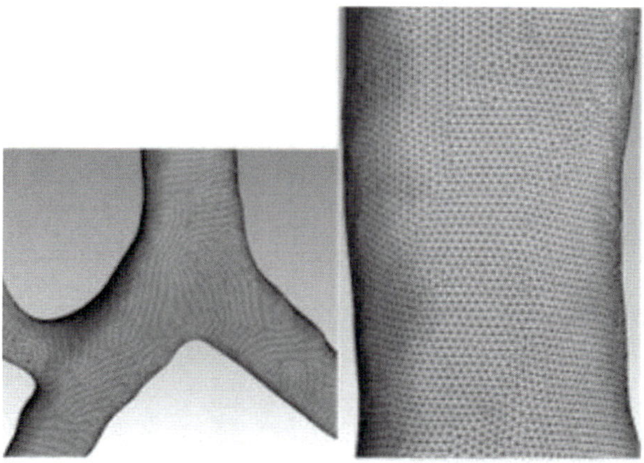

**Figure 8.** Zoomed view of the grids.

### *3.3 Governing Equations*

The governing equations for the airflow are conservation of mass, momentum and energy equations. It is found from literature that most of the researchers applied only conservation of mass and momentum; however, a few have also used energy equations for the study of heat flow in the HRT model.

*Continuity equation*:

<table>
<tr><td rowspan="2">Time-dependence</td><td>Steady</td><td rowspan="2">Steady, as the flow can be treated as steady after the complete development of flow field inside the respiratory tract.</td></tr>
<tr><td>Unsteady</td></tr>
<tr><td rowspan="2">Velocity formulation</td><td>Absolute</td><td rowspan="2">Absolute, since there is no moving meshes or boundary involved</td></tr>
<tr><td>Relative</td></tr>
</table>

Since flow is assumed to be steady and incompressible, therefore,

$$\frac{\partial u_j}{\partial x_j} = 0 \tag{1}$$

*Momentum equation*:

$$\frac{\partial \left(u_i u_j\right)}{\partial x_j} = -\frac{1}{\rho}\frac{\partial p}{\partial x_i} + \frac{\mu}{\rho}\frac{\partial^2 u_i}{\partial x_i \partial x_j} \tag{2}$$

$\mu$ = viscosity coefficient
$u_i$, $u_j$ $(i, j = 1, 2, 3)$ is the velocity component in $x$, $y$ and $z$ direction.
$p$ = pressure
$\rho$ = density of fluid.

## *3.4 Boundary Conditions*

Boundary conditions specify the flow and fluid properties on the computational flow domain for solving the problem. They are therefore, a critical component of the solver and it is important that they must be specified appropriately. Different boundary conditions have been chosen for airflow and two-phase flow, which is briefly discussed in the following.

- **Inspiratory air flow:** Inspiratory air flow refers to the movement of air into the lungs during the process of inhalation. It is the airflow that occurs when a person breathes in, drawing air from the external environment into their respiratory system. The boundary conditions applied during the inspiratory air flow are given below.

  ***Inlet condition***: Two different inlet conditions, namely, velocity inlet and inlet mass flow rate are applied at the inlet of trachea and oral cavity. Velocity inlet is applied at the inlet of trachea for airflow. Inlet mass flow rate is applied for two-phase (aerosol and solid) simulation. If the flow is turbulent then it requires specification of turbulence quantities at the inlet. Ansys-Fluent solver provides turbulence parameters that must be defined for correct prediction of flow physics. The value of turbulence intensity and hydraulic diameter is specified in the boundary condition.

  ***Turbulence intensity***: The turbulence intensity is defined as the ratio of root mean square of fluctuating velocity ($u'$) to the mean flow velocity ($u_{\text{avg.}}$).For internal flow, it is found that turbulence intensity equals 1% or less represents low and turbulence intensities greater than 10% are high turbulence. The turbulence intensity in the fully developed duct can be calculated from the following formula:

$$I = \frac{u'}{u_{avg.}} = 0.16\left(\mathrm{Re}_{D_H}\right)^{-1/8} \tag{3}$$

***Hydraulic diameter:*** Hydraulic diameter is used when the flow occurs in non-circular tubes or channels. Due to the irregular geometry of the CT scan model, hydraulic diameter is used, which is defined as:

$$D_H = \frac{4A}{P} \tag{4}$$

where 'A' and 'P' are the cross-sectional area and perimeter of the cross-section, respectively. For circular duct $D_H = D$, while $D$ being diameter.

***Outlet condition*:** Pressure outlet condition is applied at the outlet of the bronchus. Pressure outlet conditions require defining the static (gauge) pressure at the outlet boundary. Ansys-Fluent uses a zero-gauge pressure (i.e., atmospheric condition) at the outlet, indicating constant pressure conditions at the outlets. Outflow condition is used if the boundary condition is unknown at the outlet. Therefore, both of the conditions (pressure outlet and outflow) are used in the present study but no significant difference is found either using pressure outlet or outflow condition.

***Wall boundary condition*:** No-slip boundary condition is applied at the surfaces of the respiratory tract. According to the no-slip condition at the wall, the relative velocity and the convective flux at any node lying on the wall surface is set to zero. The non-dimensional cell size-based Reynolds number (called y+) plays a critical role in capturing the near-wall flow physics in the HRT model (Srivastav et al., 2019).

| Near-wall treatment | Standard wall functions | *Enhanced Wall treatments* are used for the near wall treatment because it provides better prediction near the wall in comparison of others. |
|---|---|---|
| | Non-equilibrium functions | |
| | Enhanced wall treatments | |

- **Expiratory air flow:** Expiratory airflow refers to the movement of air out of the lungs during the process of exhalation. It is the airflow that occurs when a person breathes out, expelling air from their respiratory system to the external environment. The boundary conditions associated with the expiratory airflow are furnished below.

  ***Inlet*:** In the case of expiratory flow, airflow occurs from the lower bronchus to the trachea and then the oral/nasal cavity; therefore, the outlets of the bronchus are treated as inlet during the expiratory flow.

  ***Outlet*:** The inlet of the trachea is treated as an outlet during the expiratory flow.

  ***Wall*:** The wall condition remains the same in both inspiratory and expiratory flow conditions.

- **Boundary conditions for two-phase flow:** Boundary conditions in CFD analysis play an important role in the accurate prediction of flow physics. Two different boundary conditions for the accurate prediction of the phase model (DPM) were applied at the inlet, outlets and wall of the airway. The 'Escape' option in DPM is defined at the inlet and outlets because particles cross through these planes, while the 'Trap' option is considered at the wall of the airway

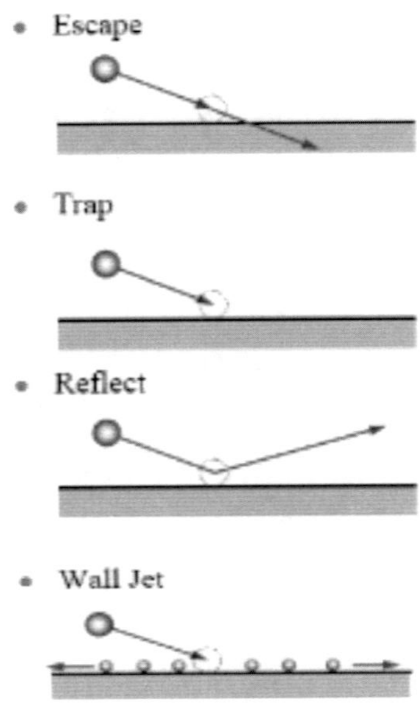


**Figure 9.** DPM wall conditions for particles.

(Fig. 9). Walls of the human airway are layered with mucous and saliva, therefore, as particles come in contact with walls, they are trapped (Steffens, 2010). The walls are highly adhesive; therefore, particles will not detach or reflect from the wall.

### *3.5 Turbulence Modeling*

It is seen that turbulence is formed during the high breathing conditions in the upper few generations of the human respiratory model. Therefore, the selection of a suitable turbulence model is a challenging task to visualize the flow phenomena in the HRT model. In CFD, there are three broad categories of turbulence modeling: namely Direct Numerical Simulation (DNS), Large Eddy Simulation (LES) model, and Reynolds Averaged Navier-Stokes (RANS) model. DNS provides the most accurate results but is also extremely expensive as it solves Navier-Stokes equations for all types of eddies.

LES uses a Sub-Grid Scale (SGS) model over a wide range of turbulence scales and, therefore, is computationally less expensive as compared to DNS. RANS model solves Navier-Stokes equations only for the time-averaged flow and is therefore used by researchers for solving real-life applications. Turbulence models using the RANS equations are classified on the basis of a number of additional transport equations that need to be solved to capture the turbulence.

RANS is used in several models (Standard k-, RNG k-, Realizable k-, SST k-, LRN k-), $k$-$\omega$ but for the modeling of human respiratory, researchers have used few of them depending upon their ability to capture the flow physics. A literature survey comprising of 68 research papers published till 2013 reveals that about 52% of researchers used realizable $k$-$\varepsilon$ turbulence model (Shih et al., 1995) for simulating airflow conditions in HRT model while about 90% researchers used Shear Stress Transport (SST) $k$-$\omega$ and low Reynolds number (LRN) $k$-$\omega$ turbulence models for two-phase flow simulation in HRT.

## 4. Particle Mechanics Modeling

Particle mechanics modeling provides a mathematical framework for understanding and predicting the behavior of particles in different physical systems. By employing concepts from classical mechanics, quantum mechanics, and other disciplines, particle mechanics modeling allows us to study the motion, forces, and interactions of particles in a wide range of scenarios, enabling advancements in various scientific and engineering fields, including classical mechanics, quantum mechanics, fluid dynamics, and astrophysics. In particle mechanics modeling, particles are often considered as point masses with certain properties such as mass, position, and velocity. The behavior of these particles is described by Newton's laws of motion, which state that the net force acting on a particle is equal to the product of its mass and acceleration. By applying these laws and using appropriate mathematical techniques, we can analyze and predict the motion of particles in a variety of situations. One of the key concepts in particle mechanics modeling is the concept of a force. Forces can arise from various sources, such as gravitational attraction, electromagnetic interactions, contact forces, or external fields. These forces can be either conservative or non-conservative, depending on whether they can be expressed as the gradient of a potential energy function. Particle mechanics modeling encompasses various approaches and techniques that are used to describe and analyze the behavior of particles in different physical systems.

Computational fluid dynamics, together with various particle motion modeling tools, such as Discrete Element Modeling (DEM) and Discrete Phase Modeling (DPM), have been increasingly used to optimize DPI design by revealing the details of flow patterns, particle trajectories, de-agglomerations and depositions within a DPI and the delivery paths of aerosols to the lungs (Longest et al., 2019). The following sub-sections discuss both types of particle mechanics modeling commonly used for particle deposition and drug delivery in HRT systems.

### *4.1 Discrete Element Model (DEM)*

A discrete element model (DEM) is a computational method used to simulate and analyze the behavior of granular materials and particulate systems. It is a type of particle mechanics modeling that focuses on individual particles and their interactions rather than treating the material as a continuous medium. DEM is particularly useful for studying systems consisting of large numbers of discrete particles, such as powders, grains, rocks, and bulk solids. In DEM, each particle within the system is represented as an individual entity with its own set of properties, such as mass, size, shape, and material properties. The particles interact with each other through contact forces, which can include normal forces, tangential forces, friction, and cohesion. These contact forces are typically based on contact laws that govern the interaction between particles. The behavior of the particles is simulated over time by numerically integrating the equations of motion for each particle. This involves calculating the forces acting on each particle based on its interactions with neighboring particles and applying Newton's laws of motion. The particle positions, velocities, and accelerations are updated at discrete time intervals to simulate the dynamic behaviour of the system. DEM simulations can provide valuable insights into the mechanical behavior and properties of granular materials. They can be used to study phenomena such as particle packing, compaction, flow, segregation, mixing, and fragmentation. DEM models are also used to analyze the behavior of particulate systems in various industrial processes, such as bulk material handling, mining, pharmaceutical manufacturing, and food processing.

One of the advantages of using a DEM is its ability to capture the individual particle-level interactions and dynamics, which is crucial in granular systems where particle-scale phenomenon play a significant role. However, it is computationally demanding due to the large number of particles and the complexity of the contact forces involved. As a result, efficient algorithms and parallel computing techniques are often employed to simulate DEM models. In recent years, DEM has been coupled with other computational methods, such as CFD or Finite Element Analysis (FEA), to study the interaction between particles and fluid or the interaction between particles and solid structures, respectively. These coupled simulations provide a more comprehensive understanding of complex systems involving both particles and continuous media. DEM is increasingly used in design of effective inhalers for pulmonary drug delivery. The following paragraphs describes the various applications of DEM in the pharmaceutical industry in design of efficient drugs and drug delivery devices.

- ***DEM in Inhaler Design*:** DEM has been utilized in the design and optimization of inhaler devices for pulmonary drug delivery. It enables researchers to investigate the behavior of particles during aerosolization, dispersion, and deposition within the respiratory system. DEM-based simulations help in understanding the interplay between inhaler design parameters (such as nozzle geometry, particle size, and formulation properties) and drug delivery efficiency. In their study, Ponzini et al. (2021) examined the aerosolization process of a dry powder inhaler (DPI) using various simulation techniques. They compared the experimental data found in literature with the outcomes of simulations conducted using CFD, Lagrangian tracing, and CFD-DEM methods. The researchers conducted a comprehensive investigation by varying the DEM and coupling parameters to evaluate their influence on the aerosolization process.
- ***Aerosolization and Dispersion*:** DEM has been employed to study the aerosolization and dispersion of inhalable particles in different types of inhalers, including pressurized Metered-Dose Inhalers (pMDIs), dry powder inhalers (DPIs), and nebulizers. Simulations provide insights into the fluid-particle interactions, particle-particle collisions, and deposition patterns within the respiratory tract. Due to the complex nature of dry powder inhaler (DPI) particle transport, which involves particle-particle interactions, varying particle sizes and shapes, and different forces such as electrostatic and Van der Waals forces, it poses a significant challenge for CFD modelers to accurately

simulate regional lung deposition from DPIs. To address this challenge, Badhan (2019) introduced a novel and high-fidelity approach called CFD Discrete Element Modeling (CFD-DEM), which combines CFD with discrete element modeling. Additionally, a sensitivity analysis was performed to assess the influence of different factors on the transport and deposition of particles from DPIs. The results of the analysis indicate that drug particles within the size range of 1–3.66 μm exhibit better dispersion and aerosolization behavior when combined with carrier particles in the range of 51–63 μm, as determined by their residence time.

- ***Inhalation Technique Optimization*:** DEM simulations have been used to evaluate the impact of inhalation techniques (e.g., flow rate, inhalation profile, breath-hold duration) on drug delivery efficacy. By studying the particle trajectories and deposition patterns, researchers can assess the effect of inhalation parameters on drug targeting to specific regions of the respiratory system.
- ***Particle-Particle Interactions*:** During the inhalation process using a dry powder inhaler (DPI), the dispersion of particles can be influenced by several incidents. These incidents include: (i) the separation or detachment of an agglomerate from a larger agglomerate due to forces related to the flow; (ii) the collision of a particle or agglomerate with other particles or agglomerates, resulting in either fragmentation or a significant change in movement; (iii) the impact of a particle or agglomerate on the device wall, causing it to break into smaller pieces that are either carried back into the airflow or captured by the wall; and (iv) the joining of a particle or agglomerate with another particle or agglomerate to form a new agglomerate. The first three incidents contribute to particle de-agglomeration, while the fourth incident, known as particle re-agglomeration, has received less attention in DPI design discussions (Zheng et al., 2021).

The discrete element method (DEM) allows for the modeling of cohesive forces, agglomeration, and deagglomeration phenomena that occur between particles. This is particularly relevant when studying the behavior of adhesive or cohesive particles in inhalers. DEM simulations provide valuable insights into particle interactions and how cohesive forces influence aerosolization, dispersion, and deposition processes. The combination of CFD and DEM, known as CFD-DEM, has gained popularity for simulating micro-scale deagglomeration and agglomeration behaviors in both carrier-free and carrier-based formulations (Zheng et al., 2021). It has shown great potential in the design of high-performance dry powder inhalers (DPIs). In DPIs, deagglomeration is driven by both aerodynamic forces and impaction. Powder dispersion is mainly governed by particle-wall impacts, while the contribution of internal shearing induced by airflow within the system is minor (Tong et al., 2010).

- ***Drug Formulation Optimization*:** A mechanistic model is a mathematical tool for modeling that has the ability to enhance the understanding of processes, reduce the cost of experimentation, and improve the quality of products (Yeom et al., 2019). Popular mechanistic model, such as DEM-based studies have focused on optimizing drug formulations for enhanced pulmonary delivery. By considering particle size distribution, shape, density, and material properties, simulations can predict the behavior of different formulations within inhalers and provide guidance for formulation design to achieve desirable aerosol characteristics and deposition patterns (Capecelatro et al., 2022).

- ***Inhaler Performance Evaluation*:** The effectiveness of aerosolization in a dry powder inhaler (DPI) is influenced by both the design of the inhaler and the specific dry powder formulation used. CFD-DEM modeling plays a crucial role in predicting the aerosolization process in DPIs. The objective is to develop an accurate model of powder aerosolization that can be utilized to design DPIs with optimal aerosolization efficiency, considering the amount of input energy provided (Capecelatro et al., 2022). DEM simulations also offer a means to assess the performance of inhaler devices by predicting drug deposition patterns and quantifying drug delivery efficiency. Comparisons between different inhaler designs or formulations can be made to identify optimal configurations that maximize drug delivery to the desired target sites while minimizing deposition in non-target areas.
- ***Validation and Experimental Correlation*:** Many studies (Tong et al., 2017; Nguyen et al., 2018; Ponzini et al., 2021) have aimed to validate DEM simulations by comparing the predicted results with experimental data from *in-vitro* or *in-vivo* experiments. Correlations between simulation outputs and experimental measurements help to establish the accuracy and reliability of DEM-based predictions for inhaler performance evaluation.

On the whole, the use of DEM in pulmonary drug delivery using inhalers has provided valuable insights into the behavior of inhalable particles, the impact of inhaler design and technique, and the optimization of drug formulations. DEM-based CFD simulations have the potential to aid in the development of more efficient and targeted pulmonary drug delivery systems, contributing to improved therapeutic outcomes in respiratory diseases. In CFD, it is however, important to mention that the accuracy of the simulated results depends on the appropriate selection of numerical schemes available in a CFD solver (Kumar et al., 2019).

## *4.2 Discrete Phase Model (DPM)*

The developments of CFD have provided the basis for further insight into the dynamics of multi-phase flows. There are two approaches for the numerical simulation of multi-phase flows, namely the Euler-Lagrange approach and the Euler-Euler approach. For a low volume fraction of solid/liquid (less than 10–12% of the gas phase), an Eulerian-Lagrangian approach is used. A fundamental assumption made in this model is that the dispersed phase occupies a low volume fraction as in the present case. In this model, particle or droplet trajectories are computed separately at the specified intervals during the fluid phase estimation. This makes the model appropriate for the modeling of spray dryers, coal and liquid fuel combustion and some particle-laden flows.

The soot particles travel in the respiratory tract during smoking. The two famous drugs, Methacholine and Albuterol, are used through inhalers for the treatment of respiratory diseases (Baoshun and Kenneth, 2009). These drugs are in the form of micron-sized droplets and aerosol particles. These micron-sized droplets are spherical in shape (Jayaraju, 2009; Steffens, 2010). Since the size of the aerosol is of micron, Brownian motion has been neglected.

In the human respiratory study, both types of particles, namely drugs (aerosol) and solid (toxic), are used. The developments of CFD have provided the basis for further insight into the dynamics of multi-phase flows. There are two approaches for the numerical simulation of multi-phase flows, namely the Euler-Lagrange approach and the Euler-Euler approach. For a low volume fraction (less than 10–12% of the gas phase) an Eulerian-Lagrangian approach is used. A fundamental assumption made in the Eulerian-Lagrangian approach is that the dispersed phase occupies a low volume fraction. In this model, particle or droplet trajectories are computed separately at the specified intervals during the fluid phase estimation. This makes the model appropriate for the modeling of spray dryers, coal and liquid fuel combustion and some particle-laden flows. The discrete phase model (DPM) is used to simulate the particle motion in human respiratory tract models where air phase equations will remain the same as described earlier, and particle phase equations will be as described below.

- ***Particle Force Balance***: The force balance equates the particle inertia with forces acting on the particles and is written as

$$\frac{\partial u_p}{\partial t} = F_D(u - u_p) + \frac{g_x(\rho_p - \rho)}{\rho_p} \tag{5}$$

The first term on the right-hand side represents drag force, and the second term is the gravitational force. $F_D(u-u_p)$ is the drag force per unit particle mass where $F_D$ is defined as:

$$F_D = \frac{18}{{}_p d_p^2} \frac{C_D \mathrm{Re}}{24} \tag{6}$$

$u$ is the air velocity, $u_p$ is the particle velocity, $\mu$ is the molecular viscosity of air, $\rho$ is the air density, $\rho_p$ is the density of the particle, and $d_p$ is the particle diameter. *Re* is the relative Reynolds number, which is defined as:

$$\mathrm{Re} = \frac{\rho d_p \left| u_p - u \right|}{\mu} \tag{7}$$

The drag coefficient ($C_D$) can be calculated from

$$C_D = a_1 + \frac{a_1}{\mathrm{Re}} + \frac{a_3}{\mathrm{Re}^2} \tag{8}$$

where $a_1$, $a_2$ and $a_3$ are constants that apply to smooth spherical particles over several ranges of Reynolds number.

## 5. Roads Ahead

The vast population in the world suffers from chronic ailments ranging from asthma, COPD, and tuberculosis to diabetes, etc., requiring long treatment. The challenge is to build an alternative, non-invasive, efficient drug delivery system for the drugs that offer irreproducible and poor-to-no bioavailability or have side effects when administered via the gastrointestinal tract. The goal is to use the pulmonary route for the treatment of not only local diseases but for the systemic administration of drugs (e.g., peptides and smaller proteins that can pass the alveolar membrane like insulin), for the drugs for which a rapid onset of action is desired (e.g., morphine) and even for vaccination and gene therapy. This will require targeted administration of low to high doses of a variety of drug formulations, which must be patient-independent. In this context, the researchers need to focus on the design and development of smart and efficient inhalers for pulmonary drug delivery. This will encompass recent techniques of drug delivery strategies that will be further strengthened by swirl flow-based novel methods to improve the targeted drug delivery. The development of such efficient devices will have a greater impact on the way medicines are administered today. This will encourage the development of new formulations that can be administered through the pulmonary route, resulting in improved efficiency of treatment of various diseases and reduced side effects and leading to a healthier society.

In developing pMDI systems, there are two major areas that need to be considered: the device hardware and the drug formulation. The hardware consists of the vial (i.e., aluminium canister or plasticized glass vial), metering valve, actuator, and for newer pMDIs usually a dose counter. The formulation comprises primarily of the propellant, drug, and often other excipients. Aeosolized drug particle size, nozzle design, injection velocity, aerosol characteristics, aerosol size distribution, spray temperature, and spray impact force, as well as the mass of drug available per shot play vital roles in contributing better drug deposition efficiency and targeted drug delivery of pMDIs. The DEM-coupled CFD, spray dynamics, and aerosol kinetics studies, along with *in-vitro* experimentation, would be carried out in the proposed project to optimize the novel pMDI offering a comprehensive solution for a smart and effective inhaler for pulmonary drug delivery.

It is worthwhile to note that innovations that improve the functionality of pMDIs continue to be balanced by the fact that a key advantage of pMDI system is their low cost per dose. The expansion of the health care market in developing countries and the increased focus on health care costs in many developed countries will ensure that pMDIs remain a cost-effective crucial delivery system for treating pulmonary conditions for many years to come.

On the other hand, the existing DPIs are suitable for specific drug formulation for which they have been designed and tested. Moreover, their drug deposition efficiency is found low, variable and patient dependent. Its higher efficiency in delivering API requires fewer drugs, creating the potential not only for cost savings but also for preventing patients from unnecessary exposure to drugs to reduce the side effects. As the chapter reveals the flow inside the DPI one of the major factors that governs deagglomeration and hence the deposition efficiency. The research needs to be directed to fulfill this need by producing an effective jet of flow inside the DPI, aiding deagglomeration. The DEM-coupled CFD can be used to model this phenomenon, which can be optimized and implemented to improve the efficacy of the existing DPI, leading to a comprehensive solution for a smart and effective inhaler for pulmonary drug delivery.

## References

Badhan, A. (2019). CFD Dem Analysis of A Dry Powder Inhaler. Open Access Theses & Dissertations. 2928. University of Texas at El Paso (USA). https://scholarworks.utep.edu/open_etd/2928.

Baoshun, M. and Kenneth, R.L. (2009). CFD simulation of aerosol deposition in an anatomically based human large–medium airway model. Annals of Biomedical Engineering 37: 271–285.

Bisgaard, H., O'Callaghan, C. and Smaldone, G.C. (ed.). Drug Delivery to the Lung, First Edition, 2001. CRC Press, Boca raton, Florida (USA). ISBN 9780367396879.

Braman, S.S. (2006). The global burden of asthma. Chest. 2006 Jul; 130(1 Suppl): 4S–12S. doi: 10.1378/chest.130.1_suppl.4S.

Capecelatro, J., Longest, W., Boerman, C., Sulaiman, M. and Sundaresan, S. (2022). Recent developments in the computational simulation of dry powder inhalers. Advanced Drug Delivery Reviews 188: 114461. https://doi.org/10.1016/j.addr.2022.114461.

Cheng, Y.S., Zhou, Y. and Chen, B.T. (1999). Particle deposition in a cast of human oral airways. Aerosol Science and Technology 12: 286–300.

Dharmage, S.C., Perret, J.L. and Custovic, A. (2019). Epidemiology of asthma in children and adults. Frontiers in Pediatrics 7: 246. https://doi.org/10.3389/fped.2019.00246.

Forum of International Respiratory Societies (FIRS). The Global Impact of Respiratory Disease – Second Edition. Sheffi eld, European Respiratory Society, 2017.

Green, A.S. (2004). Modelling of peak-flow wall shear stress in major airways of the lung. Journal of Biomechanics 37: 661–667.

Jayaraju, S.T. (2009). Study of the Air Flow and Aerosol Transport in The Human Upper Airways using LES and DES Methodologies. Ph.D thesis. Vrije Universiteit Brussel.

Kleinstreuer, C. and Zhang, Z. (2003). Targeted drug aerosol deposition analysis for a four-generation lung airway model with hemispherical tumors. Journal of Biomechanical Engineering 125: 197–206.

Koombua, K. and Pidaparti, R. M. (2008). Inhalation induced stresses and flow characteristics in human airways through fluid-structure interaction analysis. Modelling and Simulation in Engineering. Volume 2008, Article ID 358748, 8 pages.

Kumar, B., Srivastav, V.K., Jain, A. and Paul, A.R. (2019). Study of numerical schemes for the CFD simulation of human airways. International Journal of Integrated Engineering 11(8): 32–40.

Larpruenrudee, P., Islam, M.S., Paul, G., Paul, A.R. and Saha, S.C. (2022). Model for pharmaceutical aerosol transport through stenosis airway. Chapter-8. *In*: Yashwant Pathak and Nazrul Islam (eds.). Handbook of Lung Targeted Drug Delivery Systems: Recent Trends and Clinical Evidences. CRC Press, Florida (USA), ISBN: 978-03-6749-067-6.

Longest, P.W., Bass, K., Dutta, R., Rani, V., Thomas, M.L., El-Achwah, A. et al. (2019). Use of computational fluid dynamics deposition modeling in respiratory drug delivery. Expert Opin. Drug Deliv. 16: 7–26. doi: 10.1080/17425247.2019.1551875.

Luo, X.Y., Hinton, J.S., Liew, T.T. and Tan, K.K. (2004). LES modelling of flow in a simple airway model. Medical Engineering & Physics 26: 403–413.

Nguyen, D., Remmelgas, J., Björn, I.N., van Wachem, B. and Thalberg, K. (2018). Towards quantitative prediction of the performance of dry powder inhalers by multi-scale simulations and experiments. Int. J. Pharm. 547: 31–43. https://doi.org/10.1016/j.ijpharm.2018.05.047.

Nowak, N., Kakade, P.P. and Annapragada, A.V. (2003). Computational fluid dynamics simulation of airflow and aerosol deposition in human lungs. Annals of Biomedical Engineering 31: 374–390.

Paul, A.R., Jain, A. and Saha, S.C. (2022). Exposure assessment of air pollution in lungs. Atmosphere 13: 1–22. https://doi.org/10.3390/atmos13111767.

Paul, A.R., Khan, F., Jain, A. and Saha, S.C. (2021). Deposition of smoke particles in human airways with realistic waveform. Special issue on 'Heat Wave, Bush Fire and Air-Quality: Impacts on Respiratory Health' in Atmosphere 12(7): 1–22. https://doi.org/10.3390/atmos12070912.

Paul, A.R., Roy, P. and Mukherjee, S. (2009). Mechanical Sciences: Engineering Mechanics & Strength of Materials. Prentice-Hall of India Pvt. Ltd., New Delhi (India), ISBN: 978-81-203-2611-8.

Ponzini, R., Da Vià, R., Bnà, S., Cottini, C. and Benassi, A. (2021). Coupled CFD-DEM model for dry powder inhalers simulation: Validation and sensitivity analysis for the main model parameters. Powder Technology 385L: 199–226. https://doi.org/10.1016/j.powtec.2021.02.044.

Radhakrishnan, H. and Kassinos, S. (2009). CFD Modeling of Turbulent Flow and Particle Deposition in Human Lungs. 31st Annual International Conference of the IEEE EMBS. Minneapolis, Minnesota, USA, September 2–6.

Saha, S.C., Francis, I., Huang, X. and Paul, A.R. (2022). Heat transfer and fluid flow analysis of realistic 16-generation lung. Physics of Fluids 34: 1–12. https://doi.org/10.1063/5.0093912.

Schlesinger, R.B. and Lippmann, M. (1972). Particle deposition in casts of the human upper trachea bronchial tree. American Industrial Hygiene Association Journal 33: 237–251.

Shih, T.H., Liou, W.W., Shabbir, A., Yang, Z. and Zhu, J. (1995). A new k-$\epsilon$ eddy viscosity model for high reynolds number turbulent flows. 24(3): 227–238.

Shukla, R.K., Srivastav, V.K., Paul, A.R. and Jain, A. (2020). Fluid structure interaction studies of human airways. Sādhanā 35(229): 1–6. https://doi.org/10.1007/s12046-020-01460-9.

Shukla, S.K., Srivastav, V.K., Paul, A.R. and Jain, A. (2013). Computational study of aerosol drug deposition on tumorous human airway model. Proceedings of 40th National Conference on Fluid Mechanics & Fluid Power (FMFP-2013), 12–14 December 2013, NIT Hamirpur (India). Paper code: 148, ISBN: 978-93-8288-090-5.

Singh, D., Jain, A. and Paul, A.R. (2020). Numerical study on particle deposition in healthy human airways and airways with glomus tumor. Chapter 31, *In*: Rizvanov, Albert A., Singh, Bikesh Kumar and Ganasala, Padma (eds.). Lecture Notes in Bioengineering, Albert A. Rizvanov et al. (eds.). Advances in Biomedical Engineering and Technology. Springer, Singapore. https://doi.org/10.1007/978-981-15-6329-4_31.

Srivastav, V.K. and Anuj Jain. (2011). Wall shear stress study of airflow in realistic human airways model: A CFD analysis. International Conference AMOC-2011. Dec. 5–7, IIT Roorkee, India. 362–367. ISBN: 9788186224748.

Srivastav, V.K., Jain, A., Paul, A.R. and Joshi, S. (2011). CFD modeling of airflow in human respiratory system. International Conference on Mechanical Engineering (ICME-2011), BUET, Dhaka (Bangladesh).

Srivastav, V.K., Paul, A.R. and Jain, A. (2012). Computational fluid dynamics study of airflow and particle transport in third to sixth generation human respiratory tract. International Journal of Emerging Multidisciplinary Fluid Sciences 3(4): 227–234.

Srivastav, V.K., Paul, A.R. and Jain, A. (2013). Effects of cartilaginous rings on airflow and particle transport through simplified and realistic human upper respiratory tracts. Acta Mechanica Sinica 29(6): 883–892. https://doi.org/10.1007/s10409-013-0086-2.

Srivastav, V.K., Amit Kumar, Shashank K. Shukla, Paul, A.R., Amba D. Bhatt and Jain, A. (2014). Airflow and aerosol-drug delivery in a CT scan based human respiratory tract with tumor using CFD. Journal of Applied Fluid Mechanics 7(2): 345–356, April 2014. https://doi.org/10.36884/jafm.7.02.20282.

Srivastav, V.K., Bholu Kumar, Jain, A. and Paul, A.R. (2016). A computational study on airflow in human respiratory tract at normal and heavy breathing conditions. Proceedings of 6th International and 43rd National Conference on Fluid Mechanics and Fluid Power (FMFP-2016) 15–17 Dec. 2016, MNNIT Allahabad (India). Paper ID: 589, ISBN: 978-93- 5267-407-7.

Srivastav, V.K., Jain, A. and Paul, A.R. (2019a). Computational study of drug delivery in tumorous human airways. International Journal of Computing Science and Mathematics 10(5): 459–475. https://doi.org/10.1504/IJCSM.2019.103676.

Srivastav, V.K., Paul, A.R. and Anuj Jain. (2019). Capturing the wall turbulence in CFD simulation of human respiratory tract. Mathematics and Computers in Simulation 160: 23–28. https://doi.org/10.1016/j.matcom.2018.11.019.

Srivastav, V.K., Shukla, R.K., Paul, A.R. and Anuj Jain. (2021). Computational study on the effect of human respiratory wall elasticity. International Journal of Biomedical Engineering and Technology 37(1): 46–64. https://dx.doi.org/10.1504/IJBET.2021.117514.

Srivastav, V.K., Jain, A. and Paul, A.R. (2021). Computational studies of aerosolized drug deposition in human respiratory tract. *In*: Venkatakrishnan, L., Majumdar, S., Subramanian, G., Bhat, G.S.,

Dasgupta, R. and Arakeri, J. (eds.). Proceedings of 16th Asian Congress of Fluid Mechanics. Lecture Notes in Mechanical Engineering. Springer, Singapore. https://doi.org/10.1007/978-981-15-5183-3_2.

Steffens, J. (2010). Comparison of particle deposition for realistic adult and adolescent upper Airway geometries using unsteady Computational fluid dynamics. Masters Thesis. Department of Mechanical Engineering, Rochester Institute of Technology (USA).

Stein, S.W. and Thiel, C.G. (2017). The history of therapeutic aerosols: a chronological review. Journal of Aerosol Medicine and Pulmonary Drug Delivery 30(1): 20–41. https://doi.org/10.1089/jamp.2016.1297.

Tiwari, A., Jain, A., Paul, A.R. and Saha, S.C. (2021a). Computational evaluation of drug delivery in human respiratory tract under realistic inhalation. Physics of Fluids 33(7): 083311. https://doi.org/10.1063/5.0053980.

Tiwari, A., Sharma, S., Srivastav, V.K., Jain, A. and Paul, A.R. (2021b). Computational study of atomization models and optimal design of a pressurized inhaler. Journal of Biomimetics, Biomaterials and Biomedical Engineering 50: 123–134. https://doi.org/10.4028/www.scientific.net/JBBBE.50.123.

Tiwari, A., Paul, A.R., Jain, A. and Saha, S.C. (2022). Evaluation of the design of efficient dry powder inhalers. Chapter-9, *In*: Yashwant Pathak and Nazrul Islam (eds.). Handbook of Lung Targeted Drug Delivery Systems: Recent Trends and Clinical Evidences. CRC Press, Florida (USA). https://doi.org/10.1201/9781003046547-9.

Tong, Z.B., Yang, R.Y., Chu, K.W., Yu, A.B., Adi, S. and Chan, H.K. (2010). Numerical study of the effects of particle size and poly-dispersity on the agglomerate dispersion in a cyclonic flow. Chem. Eng. J. 164: 432–441. doi: 10.1016/j.cej.2009.11.027.

Tong, Z.B., Yang, R.Y. and Yu, A.B. (2017). CFD-DEM study of the aerosolisation mechanism of carrier-based formulations with high drug loadings. Powder Technol. 314: 620–626. https://doi.org/10.1016/j.powtec.2016.10.004.

Watanabe, M., Keigo, T., Katsuaki, O., Teruhiro, A., Susumu, T., Hideyuki, S. et al. (1998). Successful resection of a glomus tumour arising from the lower trachea: report of a case. Japanese Journal of Surgery 28: 332–334.

Weibel, E.R. (1963). Morphometry of the Human Lung. New York, US: Academic Press.

World Health Organization (WHO). Global Health Estimates (GHE) 2000–2019. Retrieved from https://www.who.int/data/global-health-estimates.

Yeom, S.B., Ha, E.S., Kim, M.S., Jeong, S.H., Hwang, S.J. and Choi, D.H. (2019). Application of the discrete element method for manufacturing process simulation in the pharmaceutical industry. Pharmaceutics 11(8): 414. doi: 10.3390/pharmaceutics11080414.

Zheng Li, Kleinstreuer, C. and Zhang Z. (2007). Simulation of airflows fields and microparticle deposition in realistic human lung airway models. Part I: Airflow patterns. European Journal of Mechanics B/Fluids 26: 632–649.

Zheng, Z., Leung, S.S.Y. and Gupta, R. (2021). Flow and particle modeling of dry powder inhalers: methodologies. Recent Development and Emerging Applications. Pharmaceutics 13(2): 189. doi: 10.3390/pharmaceutics13020189.

# Chapter 4

# Drug Delivery Through Naso-pharyngeal Routes

*Nazrul Islam*[1,2,*] and *Shafiqur Rahman*[3]

## 1. Introduction

The nasal route of drug delivery, where medications are delivered into the nasal mucosa for local and systemic effects has received significant attention for the treatment of various diseases. The nasal mucosa consists of the nasal epithelium which is readily accessible for the absorption of drugs. Compared to that of the oral route, faster and higher levels of drug absorption from nasal mucosa have been obtained due to the low enzymatic environment of the nasal cavity, high permeability, and avoidance of the hepatic first-pass metabolism of the administered drugs. Thus, the nasal delivery of a low dose of drugs produces better systemic bioavailability compared to that of the oral route. The intranasal route is convenient, and safe and enhances compliance for patients (Table 1). Additionally, the drug formulations for nasal delivery also encompass components to degrade enzymatic activity and increase the required therapeutic effects of many drugs. The intranasal route is considered to be an attractive opportunity to efficiently improve the management of various diseases such as Alzheimer's disease, Parkinson's disease, epilepsy, schizophrenia, psychosis, central and peripheral pains, brain cancer, and numerous Central Nervous System (CNS) disorders (Rodrigues et al., 2022) with various types of drug formulations including polymeric nanoparticles (Laffleur and Bauer, 2021; Rabiee et al., 2021).

[1] Pharmacy Discipline, Faculty of Health, Queensland University of Technology, Brisbane, Queensland 4000. Australia.

[2] Centre for Immunology and Infection Control (CIIC), Queensland University of Technology, Brisbane, Queensland, Australia.

[3] Department of Pharmaceutical Sciences, South Dakota State University, 1055 Campanile Avenue, Avera Health and Science Centre, SAV 265, Brookings, SD 57007, United States.

* Corresponding author: nazrul.islam@qut.edu.au

DOI: 10.1201/9781003287636-4

**Table 1.** Advantages and disadvantages of drug delivery via nasopharyngeal route.

| | **Advantages** | **Disadvantages** |
|---|---|---|
| 1 | Non-invasive | Limited dose administrable |
| 2 | Self-administrable | Low bioavailability |
| 3 | Rapid absorption | Pathophysiological conditions affect the drug absorption |
| 4 | Avoid fast-pass metabolism | Drug dose loss due to improper use |
| 5 | Direct brain delivery of drugs via nasal route | Mechanism of drug absorption is not clear |
| 6 | High rate of drug absorption | Irritation may occur |
| 7 | Patience compliance (not absolute) | Small volume of drugs can be administered |

## 2. Anatomy, Physiology of the Nasal Route and Areas of Drug Deposition

The nose is an organ that allows air to the respiratory tract for respiration; however, it has been identified as a potential drug delivery route. The nose has various parts such as nasal cavity, atrium/septum, respiratory area, nasopharynx, and olfactory regions. The nasal cavity, covered by epithelial mucosa is divided into two symmetrical halves by the septum and extends to the nasopharynx (Anderson et al., 2008), which is a part of the upper airway system behind the nose and is located at the end of the nasal septum and apertures. The pharynx of the nasal route consists of three main regions, i.e., the nasopharynx, oropharynx, and hypopharynx (Emura et al., 2013), as presented in Fig. 1. The interior part of the nasal cavity expands from the nostrils to the back of the nasopharynx and is divided into two similar halves by the nasal septum (Akel et al., 2020). This important organ represents the uppermost region of the pharynx associated with the nasal passages to the larynx and trachea (David A. Morton et al., 2019). The nasopharynx is constructed by muscle and fascia with mucosal surface and the structure also serves as the posterior extension of the nasal cavity. It is the continuation of the nasal cavity through the nasal apertures, and, at the bottom, it communicates with the oropharynx (David A. Morton et al., 2019; Norton, 2016). Functionally the nasopharynx acts as a passage and connects the nasal cavity to the larynx and trachea the oropharynx (David A. Morton et al., 2019; Norton, 2016). The nasally administered drugs and pathogens are caught by the vibrissae, which are enclosed by the nasal mucus and transported by the ciliary movement to the nasopharynx, where they are harmless by swallowing into the stomach (Jain et al., 2019). Nasopharynx contains the lymphoid tissue, which is composed of adenoid, lingual and palatal tonsils. This region is an appealing site of drug target to stimulate mucosal immune response following vaccination (Lycke, 2012).

The nasally administered drugs are deposited in different areas for absorption and thus it is important to understand the exact location where the drugs will be retained for absorption. The drug formulation is initially deposited in the atrium and respiratory region, where it remains for long time to be absorbed. The drug formulations deposited in the nasopharynx and olfactory areas, get short residence

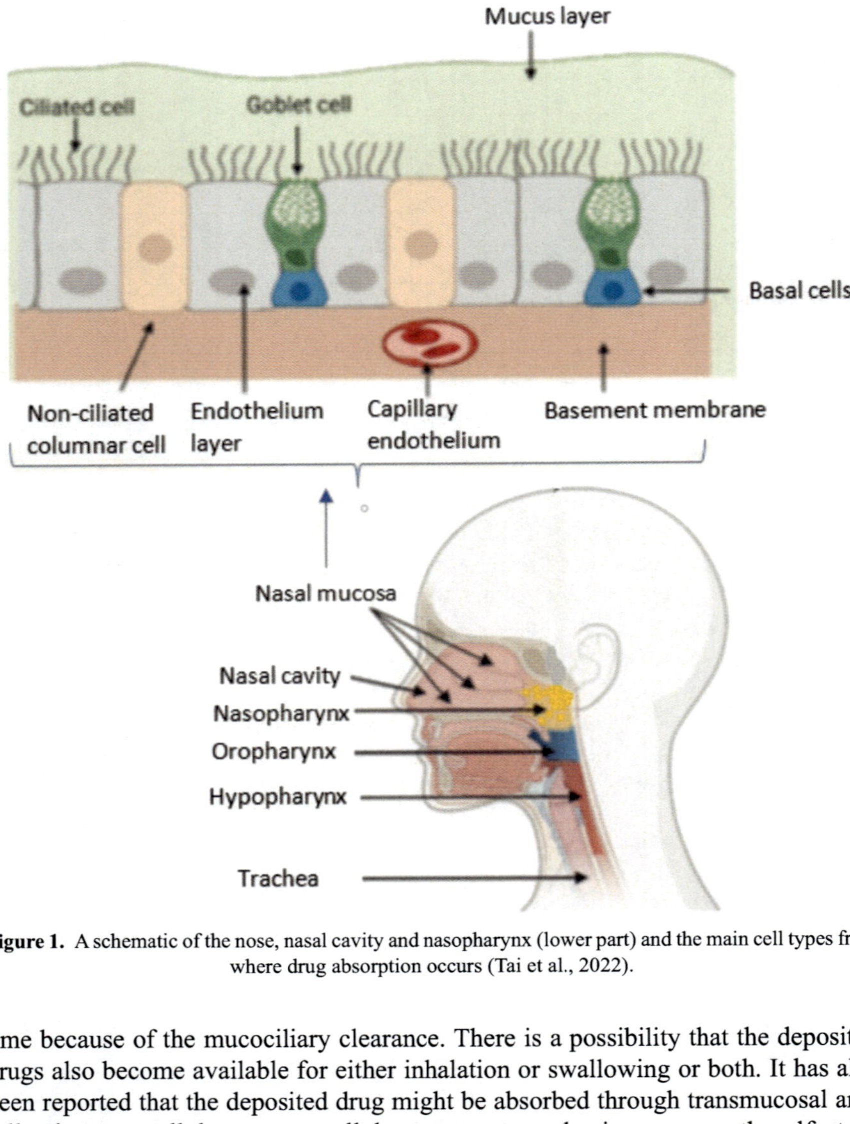


**Figure 1.** A schematic of the nose, nasal cavity and nasopharynx (lower part) and the main cell types from where drug absorption occurs (Tai et al., 2022).

time because of the mucociliary clearance. There is a possibility that the deposited drugs also become available for either inhalation or swallowing or both. It has also been reported that the deposited drug might be absorbed through transmucosal area following paracellular or transcellular transport mechanisms across the olfactory region (Bahadur and Pathak, 2012). Although the actual mechanism of drug transport to the lungs following the nasal delivery is not clear, the fraction of the deposited drug in the postnasal area is probably captured into the lung during inhalation. Interestingly, the drug targeting cerebral and lymphoid release, the olfactory bulb region and the nasopharynx, respectively are favoured. Therefore, the drug's physicochemical properties, formulations, and technique of administration have direct impacts on the deposition as well as absorption (Scherliess, 2020).

## 3. Mechanism of drug absorption following nasal administration

The nasally administered drug may get absorbed mostly into the brain, partly the systemic circulation and other areas of the body as demonstrated in the Fig. 2. The nasally administered drugs enter the brain via three routes, i.e., the olfactory pathway, the trigeminal pathway, and the peripheral area. Drugs enter the systemic circulation via vascular absorption and finally enter the brain cells through the BBB. The delivery of drugs through the nasal route may also reach other areas of the body such as the gastrointestinal tract (GIT), lymphatic system and lungs. Furthermore, the delivered drugs through the nasal route are possibly swallowed via the oral cavity and are absorbed by the GIT. Moreover, the fractions of nasally administered drugs enter the lungs following inhalation and reach systemic circulation. Drugs may also be absorbed by the lymphatic system (Quintana et al., 2015). Drug absorption following inhalation during nasal delivery is not common. The absorption of drugs from the intranasal route primarily occurs from the nasal cavity, which contains

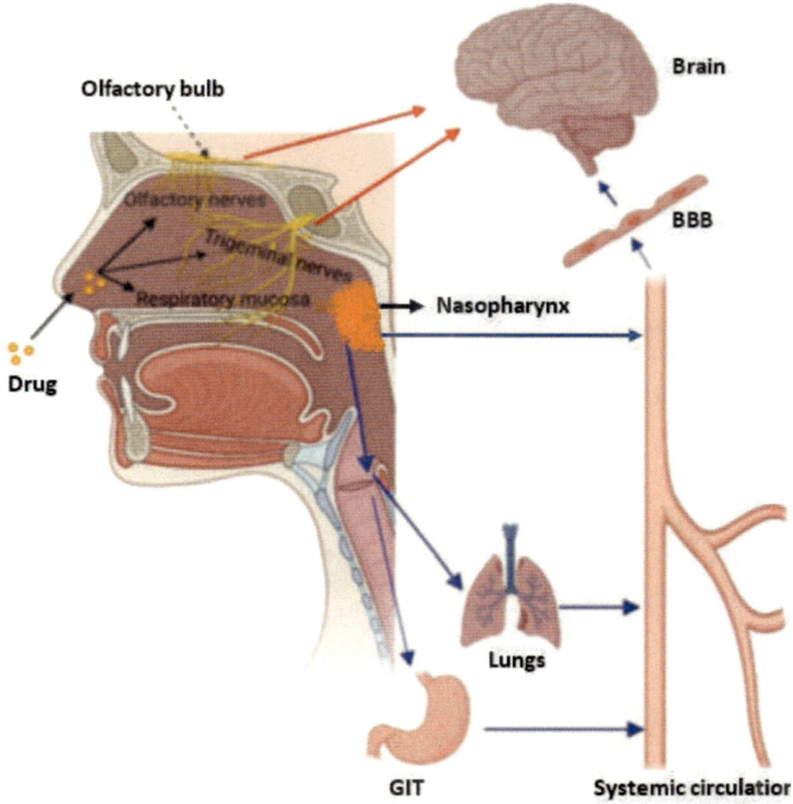


**Figure 2.** Drug transport from the nasal route to the brain, systemic circulation, lungs, and GIT. BBB: Blood Brain Barrier; GIT: Gastrointestinal Tract; Red arrow: direct route of drug absorption; Blue arrow: indirect route of drug absorption following nasal administration. This picture has been adapted with minor modifications from the ref (Duong et al., 2023).

mucin (a protein) that binds with drug molecules and diffuses into the cell. The drug diffusion can be simple diffusion, transcellular or paracellular. Depending on the physicochemical properties, i.e., the solubility and molecular weight of the drug are a major factor in drug absorption from nasal epithelia (Faraj et al., 1990; Huang et al., 1985). Thus, absorption or permeability enhancers, that induce reversible modification of the epithelial structure for enhanced drug absorption, are used to increase the bioavailability of some nasally administered drugs.

## 4. Barriers of Inhaled Drug Delivery

The natural defence mechanism and physical barrier of the lungs affect the inhaled drugs to reach the target site as the deposited drugs are either inactivated or cleared off by mucociliary escalation. Drug delivery into the lungs needs some barriers such as (i) anatomical barrier, (ii) mucous barrier, and (iii) immunological and metabolic barriers. As the lungs are composed of a very complex branched bronchial tree, where particle impactions occurred and drug deposition is affected depending on the particle size. The mucociliary clearance mechanism is responsible to clear off the deposited particle from the lung cells. Another important barrier, i.e., mucous produced by some diseased conditions, i.e., asthma, Chronic Obstructive Pulmonary Disease (COPD) and Cystic Fibrosis (CF), caused to prohibit penetration of the inhaled drugs to the affected area. The inhaled drug delivery is also severely affected by immunological and metabolic barriers. The metabolic barriers in the lungs consisted of proteolytic enzymes (endopeptidase and cathepsin), that caused the inactivation of drugs. The alveolar macrophages and phagocytes can engulf and remove particles that reach the alveoli. Therefore, to obtain effective therapeutic efficacy, the delivered drug needs to overcome these barriers.

## 5. Overcoming Absorption Barrier

The drug absorption from the nasal route is dependent on many factors like the molecular weight of the drug, physiology of the nasal cavity, physicochemical characteristics of the administered drug and the formulation of the drugs, which can be manipulated based on the condition of the absorption site (Ruigrok and de Lange, 2015). The drug's molecular weight is another factor for appropriate absorption. For example, absorption of drugs with molecular weight more than 1000 Dalton was found to be low through endothelial basement membrane (Jintapattanakit et al., 2010). Recently, comprehensive reviews on the critical parameters that influence drug absorption from the formulation in the nose emphasizing the importance of various excipients required for the development of the formulations have been reported (Laffleur and Bauer, 2021; Protopapa et al., 2022).

To make efficient absorption of drugs, the following strategies are taken into account:

a. *Enhance drug residence time*: For the efficient absorption of some drugs from the nasal route, enhancing the residence time is crucial (Lohakare et al., 2020). As the mucociliary clearance mechanism removes the foreign particles from nasal mucosa immediately after the delivery, the drugs need to be formulated with some polymers such as methylcellulose, hydroxy propyl methyl cellulose or polyacrylic acid, which increase the viscosity of the products and enhance the contact time of the drug to the absorption site. Alternatively, biodegradable polymer encapsulated drug microspheres, which take time to swell in the nasal fluid, increase the viscosity that leads to enhance the residence time.

b. *Enhance nasal absorption*: The tight junctions of the nasal epithelium especially the olfactory and their protective mucus layer decreases the permeability and diffusion of some drug (Ruigrok and de Lange, 2015). The lipophilic drugs can easily be absorbed via passive diffusion; however, for active transportation of the hydrophilic drug, the residence time needs to be increased. Drugs with high water solubility may have poor permeability across the nasal epithelia and lead to poor bioavailability, which can be improved by adding the permeation enhancers in the formulations. Some polar drugs and macromolecules also showed poor absorption and bioavailability due to rapid clearance and enzymatic degradation in the nasal cavity (Xu et al., 2020). The most common permeation enhancers like surfactants, cationic polymers, ulator and cyclodextrin are used in nasal formulations (Rabiee et al., 2021). Mucoadhesive agents like chitosan, carbopol and starch enhance drug absorption by opening the intercellular tight junctions and increasing the nasal residence time (Erdo et al., 2018). For example, the addition of trimethyl chitosan in the formulations significantly enhanced the absorption of insulin in rat models (Jintapattanakit et al., 2010).

c. *P-glycoprotein efflux transport and enzyme inhibitor*: The P-glycosylated membrane proteins, commonly known as P-glycoproteins, which act as multidrug resistance pumps throughout the nasal mucosa and BBB, remove the administered drugs by efflux pump. Nanocarriers are a promising strategy to bypass this barrier (Trevino et al., 2020). Drug-loaded polymer nanocarriers can protect the drug form this efflux mechanism and efficiently enhance the administered drug absorption (Mittal et al., 2014). Drug metabolism after nasal administration is not common; however, proteins and peptides are susceptible to some metabolic enzymes such as cytochromeP450 enzymes, endopeptidase, and exopeptidases available in the respiratory and olfactory mucosa that cause enzymatic metabolism of the delivered drugs and significantly reduce their absorption. To overcome this problem, enzyme inhibitors such as phospholipids, bestatin, boroleucine, amastatin, and fusidic acids, in the formulation to protect the administered drug from degradation and increase the absorption (Hinchcliffe and Illum, 1999; Wong and Zuo, 2010).

Overall, the efficient drug absorption after nasal administration is related to the characteristics of the drugs and their formulations, position of the nasal route during administration, nasal surface and the location of the drug deposition during administration (Costa et al., 2019).

## 6. Nasal Delivery Formulations

*Liquid formulations*: Most of the nasally administered drugs are formulated as liquid dosage forms such as aqueous solutions, suspensions, or emulsions. A humidifying agent is used in the formulation to keep the mucosa hydrated otherwise allergy or nasal irritation may occur due to dryness of the membrane. Additional care should be taken to maintain the stability formulations by adjusting the pH of the liquid formulation containing pH sensitive drugs and maintaining the pH ranging between 4 to 5.5. As the formulations are mainly water based, antimicrobial preservatives are used to prevent the degradation of the formulation by microbial growth. The liquid formulations are generally administered at a dose of 25–200 μL (Ghori et al., 2015). There is a possibility that the formulation will get limited resident time because of sneezing by the patient due to irritation.

*Semisolid formulations*: The semisolid nasal formulations are gels and ointments, which are viscous and enhance the drug contact time for improving maximum drug absorption. The polymers used in preparing the semisolid nasal preparations are poloxamers and chitosan, which are mostly thermosensitive polymers. There are some pH-sensitive gels consisting of polymers such as Carbopol derivatives, which increase the viscosity of semi-solid nasal formulations. The semisolid preparations are reported to cause nasal obstruction and reduce the spread ability of the formulation in the nasal cavity (Salade et al., 2019).

*Dry powder formulations*: Dry powder formulations for nasal delivery are not very common; however, it has attracted significant consideration due to the dry form of particles that has high stability and sustained residence time in the nasal cavity compared to those of liquid formulations. The most important parameters of the dry powder formulations are the addition of some excipients such as water fillers, i.e., mannitol, trehalose, lactose, and sorbitol to enhance the solubility of the drug in the site of absorption. Additionally, powder flow promoters like talc, magnesium stearate, etc., are used in the powder formulations to spray the powder in the nasal cavity; however, the particle size is important. Additionally, the muco adhesive agents (cellulose derivatives, cationic chitosan and its derivatives and gelatin), solubilizers (surfactants), permeation and absorption enhancers (fatty acids, cyclodextrins bile salts and surfactants) are added to the formulations for obtaining maximum therapeutic efficacy of the delivered drugs (Alabsi et al., 2022; Illum et al., 2001; Tiozzo Fasiolo et al., 2018). Nasal delivery of spray-dried dexamethasone sodium phosphate microspheres blended with lactose or mannitol showed promising results in brain deposition (Nodilo et al., 2021). Microspheres blended with mannitol showed better flow compared to that of the mixtures with lactose with enhanced mucoadhesive and drug permeability across the model epithelial barrier. Approximately 17% drugs were found to be deposited in the olfactory region of the brain suggesting the potential targeted olfactory delivery of the nasally administered drug as powder form. Nasal delivery of spray-dried donepezil loaded in mucoadhesive chitosan and excipient mannitol microparticles showed 65.5% drug deposition in the olfactory region in model nasal cavity produced by a 3D printer (Perkusic et al., 2022). The authors emphasized that the developed formulation with promising swelling, muco adhesion, drug release and permeation-enhancing properties, could be useful for the efficient management of Alzheimer's disease; however, further studies with animal model and clinical trials are warranted. The readers are encouraged to go through the published article that demonstrated different methods and formulation for the nasally administered drugs and vaccines (Tai et al., 2022).

## 7. Direct Nose-to-Brain Delivery

The intranasal route of drug delivery offers a promising direct brain delivery of drugs as it delivers drugs via the neuronal pathways, i.e., olfactory and trigeminal nerves and reduces the exposure of the drug to the systemic circulation and reduces the systemic side effects of the CNS acting drugs (Alexander et al., 2020; Rabiee et al., 2021). Because of the direct connection between the nasal cavity and the brain, the nasal route is a reliable system to bypass the Blood Brain Barrier (BBB) and drug molecules reach the brain for CNS effects (Trevino et al., 2020). For example, the delivery of antidepressant drugs via nose showed efficient absorption, systemic bioavailability and therapeutic benefits (Xu et al., 2020). The nasal cavity has a highly vascularized epithelial layer and has connectivity with the cerebrospinal fluid (CSF) and CNS and thus nasal delivery of drugs can directly reach the brain following bypassing the BBB. The disorders in the brain such as neurodegeneration, psychiatric disorders, stroke, brain cancers and spinal cord injury can be treated by nose to brain delivery of proteins and peptides (Alabsi et al., 2022) because it can bypass the systemic circulation. Using the nasal-brain-lymphatic system, intranasal delivery of liposomal anticancer drugs to the brain tissues has been demonstrated (Semyachkina-Glushkovskaya et al., 2022). This technology overcomes the BBB and maintained the therapeutic drug concentration in the mouse brain cancer cells. This finding reflected the application of the nasal route for the treatment of life-threatening brain cancers. Nasal delivery of polymeric donepezil film for the treatment of Alzheimer's disease has been investigated (Papakyriakopoulou et al., 2022). The authors used hydroxypropyl methylcellulose, polyethylene glycol 400 and methyl-β-Cyclodextrin to prepare the film, from where the drug release was conducted. Using an animal model, the brain pharmacokinetic study of the optimized film is ongoing. Nasal delivery of insulin for the treatment of diabetes has been demonstrated (Von Zuben et al., 2022). The authors used liposomal hydroxyethyl cellulose based hydrogels of insulin for nasal delivery. Recently, Meirinho et al. demonstrated the lipidic nanosystems known as Self-Emulsifying Drug Delivery Systems (SEDDS) for enhancing the bioavailability of nasally administered poorly water soluble drugs (Meirinho et al., 2022). Trevino and colleagues have published a comprehensive review article on the nose to brain drug delivery (Trevino et al., 2020).

## 8. Intranasal Vaccines: COVID 19 and Mucosal Immunity

As the nose is exposed to the environment, bacteria, viruses, fungi, and other pathogenic organisms get direct entry to the respiratory mucosa via the nose and infect the host cells. The nasally administered vaccines could be a useful strategy to combat these infections (Nian et al., 2022). Recently, because of the COVID pandemic, efficacious vaccines with the adequate safety profile for nasal delivery has progressed. Mucosal immunity, a complex network of the nasal mucosa, consisted of tissues, non-lymphoid cells, lymphocytes, and effector molecules, which are activated by nasally administered vaccines especially the COVID 19 vaccines (Alu et al., 2022). Alcantara and others found enhanced nasal deposition of favipiravir-loaded chitosan-alginate nanoparticles, which showed promising effects against COVID 19 (Alcantara et al., 2022). The optimized nanoparticulate formulation of favipiravir showed higher deposition nasal deposition, superior muco adhesion, higher permeation and over 35-fold inhibition of viral replication compared to that of free drug. Nasal vaccines can produce strong mucosal immunity against the pathogens and help avoid further entry of mucosal pathogens. The first line of defence governed by the immunity components in the upper airways fights against the invaded pathogens or induces mucosal immunity to destroy pathogens. The immune cells including macrophages, neutrophils, killer cells, and dendritic cells form a unified mechanism to destroy mucosal pathogens. In the case of viruses, which reach deeper airways of the lungs, are recognized by the alveolar macrophages and halt the viral growth (Villena and Kitazawa, 2020). The immunity from the nasal mucosa and viral protection occurs via the later mechanism. Thus, the nasal deliveries of vaccines, which do not reach the deep lung cells, enhance the immunity from the nasal mucosa and viral protection occurs via the later mechanism.

Vaccines delivered via the intranasal route are prone to induce local mucosal and systemic immune responses and thus, suitable for respiratory diseases caused by viral infections. It has been reported that intramuscular (IM) vaccines are less efficient in controlling viral replication in the mucosal area. The viral entry through the nasal route and shedding in the upper respiratory tract are the causative factors in leading to asymptomatic viral infection that can transmit the pathogens to other areas. Whereas, nasally administered vaccines have the potential to induce immunity against mucosal pathogens everywhere in the nasal area as well as the respiratory tract and thus ensure efficient protection against viral attack by inducing potent immune responses at nasal and distant mucosa (Hassan et al., 2020). Therefore, intranasal vaccines are better in terms of lower doses with the safety of vaccines compared to that of intramuscular vaccines, which is less patient compliance. Intranasal delivery of liposomal drugs, gene and vaccine delivery has been reported (Duong et al., 2023). Although currently several approved intramuscular vaccines are used worldwide, global mass vaccination, especially in low socioeconomic countries, has not been achieved yet. The reasons behind this were due to some limitations such as a lack of appropriate cold chain storage facilities at the site of vaccination and the limited number of expert people to administer the injection-based vaccine. As the COVID-19 pathogen invades the body via oral and nasal routes, an oral or nasal vaccine can efficiently stimulate the mucosal immune system to combat this deadly virus. Although various approaches for developing mucosal vaccines are being studied, more vigorous consideration is required to make efficient vaccines available (Kar et al., 2022). Alu et al., reviewed various types of intranasal vaccines investigated and their possible applications against COVID-19 (Alu et al., 2022). Very recently, nasally administered adenovirus vaccines showed a safer immune response than that of intramuscular vaccines (Bruessow, 2023).

## 9. Computational Fluid Dynamic (CFD) Technology and Nasal Drug Delivery

Traditional drug delivery via the nasal route is primarily achieving enhanced nasal distribution of the delivered drugs and limiting deposition of drugs into the lower airways of the lungs. To address this issue, Djupesland et al. (Djupesland et al., 2004), introduced a novel nasal delivery concept for efficient delivery of drugs into the nasal area. The authors emphasized that the nasal delivery of drugs minimizes the risks and problems associated with the lung deposition that occurs during conventional inhalation from a nebulizer. This finding reveals a new opportunity for intranasal delivery of medicines; however, no further improvements of this technology are investigated. Later, Kleven et al. (Kleven et al., 2005), developed a device called OptiNose by using CFD. They used the simulations to visualize and demonstrate the basic features of the developed technique and discussed the application of the optimized design for drug delivery. Although the authors claimed promising results; however, no further development is available. It is evident that the CFD method is very useful for developing various devices for drug delivery; however, there is a big gap between the real-life situation and *in-vitro* models. More detailed studies are required to understand the mechanism for developing a nasal or lung drug delivery system following the CFD method. By combining the CFD and numerical models, there might be a solution; however, extensive studies are warranted to get a successful outcome. Using CFD, delivering the nose to lung delivery of ciprofloxacin aerosol formulation with via Dry Powder Inhaler (DPI) was studied (Longest et al., 2015). The *in vitro* model produced a FPF of 95.9%, which was not ideal for the patient model. The authors demonstrated that the developed DPI formulation with high aerosolization could be used to enable the nose to lung delivery of drugs; however, no further studies were carried out to validate this outcome.

## 10. Conclusions and Future Directions

The advantages of nasally administered drugs including good patient compliance, rapid and high drug permeability, bypassing of hepatic first-pass metabolism, and potential direct drug delivery to the brain along the olfactory nerves are obvious. Drug delivery through the nasal route is promising by overcoming the BBB and enhancing targeting ability. Carrier-based nasal drug delivery is encouraging. Mucosal vaccination following nasal routes is a breakthrough for many diseases especially the COVID 19 infection. As the COVID-19 pathogen invades the body via oral and nasal mucosal surfaces, an oral or nasal vaccine could be a promising approach to stimulate the immune system and combat this deadly virus at the mucosal level and thus prevent further transmitting it via a strong mucosal and systematic immune response. Although some developments have been achieved; however, it is challenging, and more studies are required to fully explore all types of CNS drugs with carriers. Some limitations including rapid drug elimination prompted by mucociliary clearance, limited administration volume, and enzymatic degradation is a major factor. The recent progress, challenges, and future perspectives of nasally administered drugs are on the way to bringing new hope for the management of various diseases. Numerical and CFD studies should be incorporated in future research to understand the drug residence time, release, and absorption of the drug from the formulation in the nasal route. The nasopharyngeal route of drug delivery could be the unique leading research direction in the future and could be used for local and systemic treatment of brain and lung diseases and systemic treatment of brain and lung diseases.

## References

Akel, H., Ismail, R. and Csoka, I. (2020). Progress and perspectives of brain-targeting lipid-based nanosystems via the nasal route in Alzheimer's disease. European Journal of Pharmaceutics and Biopharmaceutics 148: 38–53. DOI: 10.1016/j.ejpb.2019.12.014.

Alabsi, W., Eedara, B.B., Encinas-Basurto, D., Polt, R. and Mansour, H.M. (2022). Nose-to-brain delivery of therapeutic peptides as nasal aerosols. Pharmaceutics 14(9): 1870. DOI: 10.3390/pharmaceutics14091870.

Alcantara, K.P., Nalinratana, N., Chutiwitoonchai, N., Castillo, A.L., Banlunara, W., Vajragupta, O. et al. (2022). Enhanced nasal deposition and anti-coronavirus effect of favipiravir-loaded mucoadhesive chitosan-alginate nanoparticles. Pharmaceutics 14(12): 2680. DOI: 10.3390/pharmaceutics14122680.

Alexander, A., Agrawal, M., Chougule, M.B., Saraf, S. and Saraf, S. (2020). Nose-to-brain drug delivery: an alternative approach for effective brain drug targeting. Nanopharmaceuticals 175–200. DOI: 10.1016/b978-0-12-817778-5.00009-9.

Alu, A., Chen, L., Lei, H., Wei, Y., Tian, X. and Wei, X. (2022). Intranasal COVID-19 vaccines: From bench to bed. EBioMedicine 76: 103841. DOI: 10.1016/j.ebiom.2022.103841.

Anderson, K.J., Henneberg, M. and Norris, R.M. (2008). Anatomy of the nasal profile. Journal of Anatomy 213(2): 210–216. DOI: 10.1111/j.1469-7580.2008.00924.x.

Bahadur, S. and Pathak, K. (2012). Physicochemical and physiological considerations for efficient nose-to-brain targeting. Expert Opin. Drug Delivery 9(1): 19–31. DOI: 10.1517/17425247.2012.636801.

Bruessow, H. (2023). Do we need nasal vaccines against COVID 19 to suppress the transmission of infections. Microb. Biotechnol. 16(1): 3–14. DOI: 10.1111/1751-7915.14181.

Costa, C., Moreira, J.N., Amaral, M.H., Sousa Lobo, J.M. and Silva, A.C. (2019). Nose-to-brain delivery of lipid-based nanosystems for epileptic seizures and anxiety crisis. Journal of Controlled Release 295: 187–200. DOI: 10.1016/j.jconrel.2018.12.049.

David, A., Morton, K., Bo Foreman and Albertine, K.H. (2019). The Big Picture. Pharynx, The McGraw-Hill Companies, Inc.

Djupesland, P.G., Skretting, A., Winderen, M. and Holand, T. (2004). Bi-directional nasal delivery of aerosols can prevent lung deposition. Journal of Aerosol Medicine 17(3): 249–259. DOI: 10.1089/jam.2004.17.249.

Duong, V.-A., Nguyen, T.-T.-L. and Maeng, H.-J. (2023). Recent advances in intranasal liposomes for drug, gene, and vaccine delivery. Pharmaceutics 15(1). DOI: 10.3390/pharmaceutics15010207.

Erdo, F., Bors, L.A., Farkas, D., Bajza, A. and Gizurarson, S. (2018). Evaluation of intranasal delivery route of drug administration for brain targeting. Brain Research Bulletin 143: 155–170. DOI: 10.1016/j.brainresbull.2018.10.009.

Emura, F., Baron, T.H. and Gralnek, I.M. (2013). The pharynx: examination of an area too often ignored during upper endoscopy. Gastrointest. Endosc 78: 143–149, doi:10.1016/j.gie.2013.02.021.

Faraj, J.A., Hussain, A.A., Aramaki, Y., Iseki, K., Kagoshima, M. and Dittert, L.W. (1990). Mechanism of nasal absorption of drugs. III: Nasal absorption of leucine enkephalin. Journal of Pharmaceutical Sciences 79(8): 698–702. DOI: 10.1002/jps.2600790810.

Ghori, M.U., Mahdi, M.H., Smith, A.M. and Conway, B.R. (2015). Nasal drug delivery systems: an overview. Am. J. Pharmacol. Sci. 3(5): 110–119. DOI: 10.12691/ajps-3-5-2.

Hassan, A.O., Kafai, N.M., Dmitriev, I.P., Fox, J.M., Smith, B.K., Harvey, I.B. et al. (2020). A single-dose intranasal ChAd vaccine protects upper and lower respiratory tracts against SARS-CoV-2. Cell (Cambridge, MA, U. S.) 183(1): 169–184 e113. DOI: 10.1016/j.cell.2020.08.026.

Hinchcliffe, M. and Illum, L. (1999). Intranasal insulin delivery and therapy. Adv. Drug Delivery Rev. 35(2,3): 199–234.

Huang, C.H., Kimura, R., Nassar, R.B. and Hussain, A. (1985). Mechanism of nasal absorption of drugs. I: Physicochemical parameters influencing the rate of *in situ* nasal absorption of drugs in rats. Journal of Pharmaceutical Sciences 74(6): 608–611. DOI: 10.1002/jps.2600740605.

Illum, L., Jabbal-Gill, I., Hinchcliffe, M., Fisher, A.N. and Davis, S.S. (2001). Chitosan as a novel nasal delivery system for vaccines. Adv. Drug Delivery Rev. 51(1-3): 81–96. DOI: 10.1016/s0169-409x(01)00171-5.

Jain, A., Hurkat, P., Jain, A., Jain, A., Jain, A. and Jain, S.K. (2019). Thiolated polymers: pharmaceutical tool in nasal drug delivery of proteins and peptides. International Journal of Peptide Research and Therapeutics 25(1): 15–26. DOI: 10.1007/s10989-018-9704-y.

Jintapattanakit, A., Peungvicha, P., Sailasuta, A., Kissel, T. and Junyaprasert, V.B. (2010). Nasal absorption and local tissue reaction of insulin nanocomplexes of trimethyl chitosan derivatives in rats. Journal of Pharmacy and Pharmacology 62(5): 583–591. DOI: 10.1211/jpp.62.05.0004.

Kar, S., Devnath, P., Emran, T.B., Tallei, T.E., Mitra, S. and Dhama, K. (2022). Oral and intranasal vaccines against SARS-CoV-2: Current progress, prospects, advantages, and challenges. Immun., Inflammation Dis. 10(4): e604. DOI: 10.1002/iid3.604.

Kleven, M., Melaaen, M.C., Reimers, M., Roetnes, J.S., Aurdal, L. and Djupesland, P.G. (2005). Using computational fluid dynamics (CFD) to improve the bi-directional nasal drug delivery concept. Institution of Chemical Engineers.

Laffleur, F. and Bauer, B. (2021). Progress in nasal drug delivery systems. Int. J. Pharm. (Amsterdam, Neth.) 607: 120994. DOI: 10.1016/j.ijpharm.2021.120994.

Lohakare, T.S., Kajale, A.D., Channawar, M.A. and Gawande, S.R. (2020). Intra nasal drug delivery system - a review. World J. Pharm. Pharm. Sci. 9(7): 1225–1240. DOI: 10.20959/wjpps20207-16584.

Longest, P.W., Golshahi, L., Behara, S.R.B., Tian, G., Farkas, D.R. and Hindle, M. (2015). Efficient nose-to-lung (N2L) aerosol delivery with a dry powder inhaler. J. Aerosol Med. Pulm. Drug Delivery 28(3): 189–201. DOI: 10.1089/jamp.2014.1158.

Lycke, N. (2012). Recent progress in mucosal vaccine development: potential and limitations. Nature Reviews: Immunology 12(8): 592–605. DOI: 10.1038/nri3251.

Meirinho, S., Rodrigues, M., Santos, A.O., Falcao, A. and Alves, G. (2022). Self-emulsifying drug delivery systems: an alternative approach to improve brain bioavailability of poorly water-soluble drugs through intranasal administration. Pharmaceutics 14(7): 1487. DOI: 10.3390/pharmaceutics14071487.

Mittal, D., Ali, A., Md, S., Baboota, S., Sahni, J.K. and Ali, J. (2014). Insights into direct nose to brain delivery: current status and future perspective. Drug Delivery 21(2): 75–86. DOI: 10.3109/10717544.2013.838713.

Nian, X., Zhang, J., Huang, S., Duan, K., Li, X. and Yang, X. (2022). Development of nasal vaccines and the associated challenges. Pharmaceutics 14(10): 1983. DOI: 10.3390/pharmaceutics14101983.

Nodilo, L.N., Ugrina, I., Spoljaric, D., Klaric, D.A., Brala, C.J., Perkusic, M. et al. (2021). A dry powder platform for nose-to-brain delivery of dexamethasone: formulation development and nasal deposition studies. Pharmaceutics 13(6): 795. DOI: 10.3390/pharmaceutics13060795.

Norton, N. (2016). Netter's Head and Neck Anatomy for Dentistry, Elsevier.

Papakyriakopoulou, P., Rekkas, D.M., Colombo, G. and Valsami, G. (2022). Development and *in vitro-ex vivo* evaluation of novel polymeric nasal donepezil films for potential use in alzheimer's disease using experimental design. Pharmaceutics 14(8): 1742. DOI: 10.3390/pharmaceutics14081742.

Perkusic, M., Nizic Nodilo, L., Ugrina, I., Spoljaric, D., Jakobusic Brala, C., Pepic, I. et al. (2022). Tailoring functional spray-dried powder platform for efficient donepezil nose-to-brain delivery. Int. J. Pharm. (Amsterdam, Neth.) 624: 122038. DOI: 10.1016/j.ijpharm.2022.122038.

Protopapa, C., Siamidi, A., Pavlou, P. and Vlachou, M. (2022). Excipients used for modified nasal drug delivery: a mini-review of the recent advances. Materials 15(19): 6547. DOI: 10.3390/ma15196547.

Quintana, D.S., Alvares, G.A., Hickie, I.B. and Guastella, A.J. (2015). Do delivery routes of intranasally administered oxytocin account for observed effects on social cognition and behavior? A two-level model. Neuroscience and Biobehavioral Reviews 49: 182–192. DOI: 10.1016/j.neubiorev.2014.12.011.

Rabiee, N., Ahmadi, S., Afshari, R., Khalaji, S., Rabiee, M., Bagherzadeh, M. et al. (2021). Polymeric nanoparticles for nasal drug delivery to the brain: relevance to Alzheimer's disease. Adv. Ther. (Weinheim, Ger.) 4(3): 2000076. DOI: 10.1002/adtp.202000076.

Rodrigues, M., Ali, J. and Alves, G. (2022). Editorial: Intranasal delivery of central nervous system active drugs: opportunities and challenges. Frontiers in Pharmacology 13: 927812. DOI: 10.3389/fphar.2022.927812.

Ruigrok, M.J.R. and de Lange, E.C.M. (2015). Emerging insights for translational pharmacokinetic and pharmacokinetic-pharmacodynamic studies: towards prediction of nose-to-brain transport in humans. AAPS J. 17(3): 493–505. DOI: 10.1208/s12248-015-9724-x.

Salade, L., Wauthoz, N., Goole, J. and Amighi, K. (2019). How to characterize a nasal product. The state of the art of *in vitro* and *ex vivo* specific methods. Int. J. Pharm. (Amsterdam, Neth.) 561: 47–65. DOI: 10.1016/j.ijpharm.2019.02.026.

Scherliess, R. (2020). Nasal formulations for drug administration and characterization of nasal preparations in drug delivery. Ther. Delivery 11(3): 183–191. DOI: 10.4155/tde-2019-0086.

Semyachkina-Glushkovskaya, O., Shirokov, A., Blokhina, I., Telnova, V., Vodovozova, E., Alekseeva, A. et al. (2022). Intranasal delivery of liposomes to glioblastoma by photostimulation of the lymphatic system. Pharmaceutics 15(1). DOI: 10.3390/pharmaceutics15010036.

Tai, J., Han, M., Lee, D., Park, I.-H., Lee, S.H. and Kim, T.H. (2022). Different methods and formulations of drugs and vaccines for nasal administration. Pharmaceutics 14(5): 1073. DOI: 10.3390/pharmaceutics14051073.

Tiozzo Fasiolo, L., Manniello, M.D., Tratta, E., Buttini, F., Rossi, A., Sonvico, F. et al. (2018). Opportunity and challenges of nasal powders: Drug formulation and delivery. European Journal of Pharmaceutical Sciences 113: 2–17. DOI: 10.1016/j.ejps.2017.09.027.

Trevino, J.T., Quispe, R.C., Khan, F. and Novak, V. (2020). Non-invasive strategies for nose-to-brain drug delivery. J Clin Trials 10(7).

Villena, J. and Kitazawa, H. (2020). The modulation of mucosal antiviral immunity by immunobiotics: could they offer any benefit in the SARS-CoV-2 pandemic. Frontiers in Physiology 11: 699. DOI: 10.3389/fphys.2020.00699.

Von Zuben, E.d.S., Eloy, J.O., Inacio, M.D., Araujo, V.H.S., Baviera, A.M., Gremiao, M.P.D. et al. (2022). Hydroxyethylcellulose-based hydrogels containing liposomes functionalized with cell-penetrating peptides for nasal delivery of insulin in the treatment of diabetes. Pharmaceutics 14(11): 2492. DOI: 10.3390/pharmaceutics14112492.

Wong, Y.C. and Zuo, Z. (2010). Intranasal delivery-modification of drug metabolism and brain disposition. Pharmaceutical Research 27(7): 1208–1223. DOI: 10.1007/s11095-010-0127-5.

Xu, J., Tao, J. and Wang, J. (2020). Design and application in delivery system of intranasal antidepressants. Front Bioeng Biotechnol 8: 626882. DOI: 10.3389/fbioe.2020.626882.

# Chapter 5

# Analysis of Newtonian Physiological Blood Flow through an Abdominal Aortic Aneurysm

*Md. Anwar Hossain,*[1] *Mohammad Matiur Rahman,*[1] *Most. Nasrin Akhter,*[1,*] *Md. Mamun Molla*[2] and *Suvash C. Saha*[3]

## 1. Introduction

Blood flow over normal physiological conditions is a vital field of study, as blood flow under diseased circumstances. A greater number of deaths are caused by cardiovascular diseases in developed countries, and most of them are connected to some form of irregular blood flow in arteries. Hemodynamics is the study of physical forces involved in blood circulation. The blood circulatory system refers to physiological factors as governing. Aneurysm is a balloon-like dilation found on the walls of a blood vessel or a sac formed by the localized dilatation of the wall of an artery or in a vein, or in the heart. A ruptured aneurysm can cause massive internal bleeding, which is usually fatal. Around 8 out of 10 people with a rupture either die before they reach the hospital or do not survive the surgery. The most common symptom of a ruptured aortic aneurysm is sudden and severe pain in the abdomen

[1] Department of Mathematics, Dhaka University of Engineering and Technology, Gazipur-1700, Bangladesh.
[2] Department of Mathematics & Physics, North South University, Dhaka-1229, Bangladesh.
[3] School of Mechanical and Mechatronic Engineering, University of Technology Sydney, Sydney, New South Wales 2007, Australia.
Emails: anwarmath23@gmail.com; matiurrahman1975@gmail.com; mamun.molla@northsouth.edu; suvash.saha@uts.edu.au
* Corresponding author: nasrin@duet.ac.bd

DOI: 10.1201/9781003287636-5

(Ernst 1993). An unprocessed aneurysm may rupture under insistent internal pressure, causing fatality or severe disability. Even an unruptured aneurysm can lead to damage by interrupting the flow of blood on the wall of the vessel, in some cases eroding nearby blood vessels, organs, or bones. The abdominal aorta is the final section of the aorta, the largest artery in the body. Hemodynamic parameters are calculated to be responsible for starting growth and rupture of aneurysms. In arterial walls, if decreased blood flow occurs, it leads to enlargement of vessel diameter and reduction of shear forces. To understand aneurysm behavior, flow dynamics have been studied in a multitude of experimental models. Multiple aneurysms can grow from the same location of an artery, and the interaction between these aneurysms raises the risk of rupture.

Several authors have studied patients with various aneurysm geometries, including Fukushima et al. (1989), Taylor and Yamaguchi (1994), and Elger et al. (1995). Blood flow has been numerically analyzed by Wille (1981) in moderately dilated rigid blood vessels to trace streamlines of the flow. Similarly, the path lines of the flow particles were analyzed by Perktold et al. (1984). Egelhoff et al. (1997, 1999) have also analyzed the single aneurysm for different dilations. Knowledge of flow partition, pressure and shear stress may provide a better understanding of the relationship between the fluid dynamics in pulsatile blood flow and arterial diseases.

As the flow in aneurysms is complex with the presence of vortices, secondary flows, and strong amplification of instability so the aim of this study is to investigate the flow dynamics for single and multiple abdominal aneurysms using Computational Fluid Dynamics (CFD). The numerical simulation of Newtonian and non-Newtonian blood flow patterns in AAA would be simulated. It is performed in the 3D model of AAA with an aneurysm along with its peripheral branches for the systolic and diastolic cardiac phase. Limitation on the amount of dilation is ignored. Since the arterial wall is gently elastic, we neglect the wall dispensability.

## 2. Model Description

### *2.1 Geometry*

Three-dimensional Double Abdominal Aortic Aneurysms (DAAA) and Single Abdominal Aortic Aneurysms (SAAA) are used as geometry for this study shown in Fig. 1(a). The diameter of the blood vessel AAA (1.7 cm), the diameter of the first aneurysm (3.5 cm), the diameter of the second aneurysm (4.6 cm), the length of the artery from the inlet (i) 0.0 cm are at position (j) 4.6875 cm, (k) 9.35 cm, (l) 19.975 cm. Figure 1(b) shows the mesh in the cross-sectional plane of an abdominal aortic aneurysm.

### *2.2 Blood Properties*

The density of the blood is 1050 kg/m$^3$. In a Newtonian model for the blood viscosity, the value is treated as a constant, usually set to $\mu = 3.45 \times 10^{-3}$ *pas*.

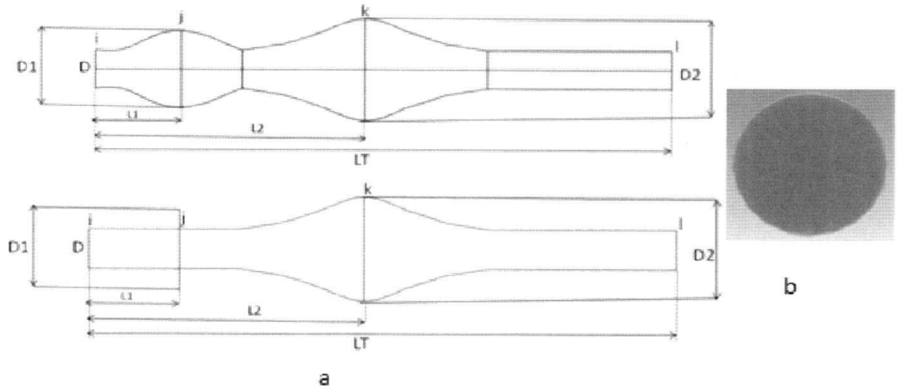


**Figure 1.** (a) model of SAAA and DAAA; (b) mesh in the cross-sectional plane of a AAA.

# 3. Governing Equation and Numerical Method

## *3.1 Governing Equation*

Due to enlargement, blood passes through the aneurysm region with low velocity, creating high pressure. Flow velocity at the aneurysm region increases, but the pressure of that region decreases. Neither laminar flow modeling nor standard two-equation models are suitable for this kind of blood flow. So, the Wilcox low-turbulence model is more acceptable for flow analysis found by Varghese and Frankel (2003). Therefore, low Re $k-\omega$ turbulent model is taken for the calculation.

Now, the Navier-Stokes equation can be given by

$$\frac{\partial u_i}{\partial x_i} = 0 \tag{1}$$

$$\frac{\partial u_i}{\partial t} + u_j \frac{\partial u_i}{\partial x_j} = -\frac{1}{\rho}\frac{\partial p}{\partial x_i} + \frac{\partial^2 u_i}{\partial x_j \partial x_i} \tag{2}$$

Since each term of this equation is time-averaged, the equation is referred to as a Reynolds-Averaged Navier-Stokes (RANS) equation. During this procedure, several additional unknown parameters appear, which require additional equations to be introduced as turbulence models. The set of RANS equations is-

$$\frac{\partial \rho}{\partial t} + \frac{\partial(\rho u_i)}{\partial x} = 0 \tag{3}$$

$$\frac{\partial(\rho u_i)}{\partial t} + \frac{\partial(\rho u_i u_j)}{\partial x_j} = -\frac{\partial p}{\partial x_i} + \frac{\partial}{\partial x_j}\left[\mu\left(\frac{\partial u_i}{\partial x_j} + \frac{\partial u_j}{\partial x_i} - \frac{2}{3}\delta_j\left(\frac{\partial u_1}{\partial x_1}\right)\right)\right] + \frac{\partial}{\partial t}\left(-\rho\overline{u_i^{\prime} u_j^{\prime}}\right) \tag{4}$$

In this equation $-\rho\overline{u_i^{/} u_j^{/}}$ is an additional term known as the Reynolds's stress tensor, which can be approximated by using Boussinesq hypothesis-

$$-\rho\overline{u_i^{/} u_j^{/}} = \mu_t\left(\frac{\partial u_i}{\partial x_j}+\frac{\partial u_j}{\partial x_i}\right)-\frac{2}{3}\left(\rho k+\mu_t\frac{\partial u_k}{\partial x_k}\right) \tag{5}$$

Eddy viscosity can be modeled as a function of the turbulence kinetic energy ($k$) and specific dissipation rate ($\omega$); therefore, it is referred to as the two-equation turbulent model. The turbulence kinetic energy $k$ and specific dissipation rate $\omega$ of standard $k-\omega$ model are determined by following two equations:

The $k$ equation: $$\frac{\partial}{\partial t}(\rho k)+\frac{\partial}{\partial x_i}(\rho k u_i)=\frac{\partial}{\partial x_j}\left(\Gamma_k\frac{\partial k}{\partial x_i}\right)+G_k-Y_k+S_k \tag{6}$$

The $\omega$ equation: $$\frac{\partial}{\partial t}(\rho\omega)+\frac{\partial}{\partial x_i}(\rho\omega u_i)=\frac{\partial}{\partial x_j}\left(\Gamma_\omega\frac{\partial \omega}{\partial x_i}\right)+G_\omega-Y_\omega+S_\omega \tag{7}$$

In these equations, $G_k$ represents the generation of turbulence kinetic energy due to mean velocity gradients. $G_\omega$ represents the generation of $\omega$. $\Gamma_k$ and $\Gamma_\omega$ represent the effective diffusivity of $k$ and $\omega$, respectively. $Y_k$ and $Y_\omega$ represent the dissipation of $k$ and $\omega$ due to turbulence. $S_k$ and $S_\omega$ are user-defined source terms.

A low Reynolds number correction factor controls the influence on the overall structure of the flow field depending upon local conditions, and it is given as-

$$\alpha^* = \alpha_\infty^*\left(\frac{\alpha_0^* + \mathrm{Re}_t/R_k}{1+\mathrm{Re}_t/R_k}\right) \tag{8}$$

where, $\mathrm{Re}_t = \dfrac{\rho k}{\mu\omega}, R_k = 8, \alpha_0^* = \dfrac{\beta_i}{4}, \beta_i = 0.072, \alpha_\infty^* = 1.$

The closure Coefficient for the Transitional $k-\omega$ Model is-

$\alpha_\infty^* = 1, \alpha_\infty = 0.52, \alpha_0 0.1111, \beta_\infty^* = 0.09, \beta_i = 0.072, R_k = 8 \, and \, R_\beta = 8$

Detailed descriptions of sixteen harmonic coefficients are shown in Table 1. Here, Reynolds's number varies from 77 to 894. Since the cardiac pulse cycle is 0.82 sec $\omega$ is pulsation, found from the calculation $\omega = \dfrac{2\pi}{0.82} = 7.66\, rad/\sec$.

**Table 1.** Harmonic coefficients for pulsatile waveform shown in Fig. 2(a).

| n | $A_n$ | $B_n$ | n | $A_n$ | $B_n$ | n | $A_n$ | $B_n$ |
|---|---|---|---|---|---|---|---|---|
| 0 | 0.166667 | 0 | 6 | –0.01735 | 0.01915 | 12 | –0.00341 | 0.005463 |
| 1 | –0.03773 | 0.0985 | 7 | –0.00648 | 0.002095 | 13 | –0.00194 | 0.000341 |
| 2 | –0.10305 | 0.012057 | 8 | –0.01023 | –0.0078 | 14 | –0.00312 | –0.00017 |
| 3 | 0.007745 | –0.06763 | 9 | 0.008628 | –0.00663 | 15 | 0.000157 | –0.00299 |
| 4 | 0.025917 | –0.02732 | 10 | 0.002267 | 0.001817 | 16 | 0.001531 | 0.000226 |
| 5 | 0.037317 | 0.024517 | 11 | 0.005723 | 0.003352 | | | |

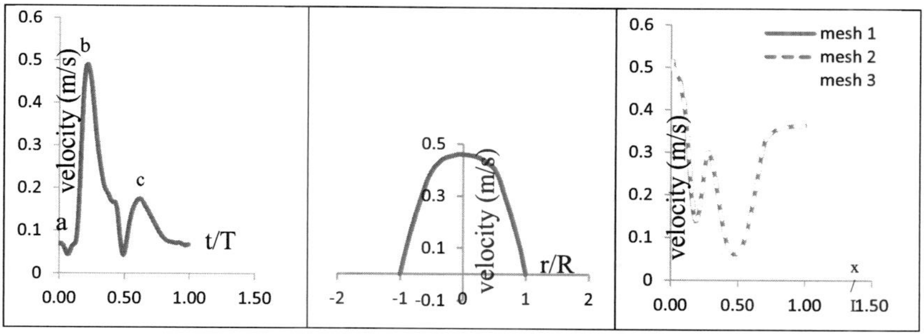


**Figure 2.** (a) oscillatory physiological waveform, (b) parabolic inlet velocity profile, and (c) velocity distribution in AAA from different mesh sizes.

Figure 2(a) and 2(b) show oscillatory physiological waveform and parabolic inlet velocity profile respectively. In Figure 2a, b, and c represent the positions of early systole (0.0615 sec), peak systole (0.1804 sec), and diastole (0.533 sec), respectively.

### *3.2 Grid Independence Check*

An extensive test is carried out with different sizes of the mesh such as mesh1 (200259 elements), mesh2 (325536 elements) and mesh3 (450556 elements) respectively. Figure 2(c) shows the velocity distributions for the Abdominal Aortic Aneurysm with the mentioned mesh sizes. In all cases, the velocity distributions are the same. It implies that the solution is grid independence.

### *3.3 Validation*

Before starting the present investigation, the numerical simulation needs to be validated. Validation of the present numerical computation is done by plotting the steady velocity profile at downstream from the stenosis throat and comparing it with the velocity profile of Varghese and Frankel (2003). For this case, a parabolic velocity profile is assumed as an inlet boundary condition. The mean inlet velocity corresponds to Reynolds number 500, and the flow is assumed to be steady. The results are shown in Fig. 3, where a good agreement can be found with Varghese and Frankel (2003).

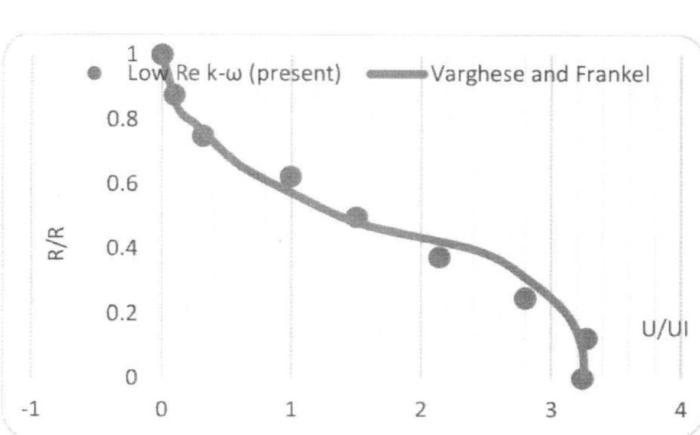


**Figure 3.** Validation with the velocity profile of Varghese and Frankel (2003).

### *3.4 Numerical Scheme*

The numerical simulations are performed by the well-known software ANSYS Fluent 14.5. A pressure-based algorithm is chosen as the solver type. This solver is generally selected for an incompressible fluid. As there is no heat transfer in the blood flow process energy equation is not solved. In solution methods, the SIMPLE algorithm is selected for pressure-velocity coupling. The Order upward scheme is employed as a numerical scheme for the discretization of the momentum equation. The time step is set to 0.00041 with 2000 number of total time steps. A maximum of 10 iterations are performed per time step.

## 4. Results and Discussion of Physiological Blood Flow

The computational results are conducted to study the influence of aneurysm on the flow behavior. Physiological blood flow is more realistic than sinusoidal blood flow. For this reason, the physiological condition is used as inlet velocity profile. The flow parameters like velocity, pressure, WSS and streamline are observed from longitudinal contours at specific instants of pulse cycle for comparing the flow variation. The discussion is categorized with the observations of flow variation starting from early systole, peak systole, and diastole respectively.

### *4.1 Wall Shear Stresses*

The wall shear stress is proportional to the shear rate at the wall and the fluid dynamic viscosity $\mu$. The measurements of the velocity gradient near the wall are technically difficult. The gradient depends highly on the shape of the velocity profile and on the measurement accuracy of distance from the wall. The endothelial cells respond to shear stress. At the luminal surface, the shear stress can be sensed directly as a force on an endothelial cell. High WSS elongates the endothelial cells and forces them to align in the direction of the flow. Low WSS had a negligible effect on the cell but increased intercellular permeability and consequently increased plaque formation in the region.

Figure 4(b) shows the distribution of WSS in single abdominal aortic aneurysms (SAAA) and double abdominal aortic aneurysms (DAAA). In all cases, the shear rate decreases rapidly in the aneurysm area, reaching a peak just at the throat of the aneurysm. In the normal region the shear rate is low. Both high and low WSS have been associated with aneurysm groth and rupture.

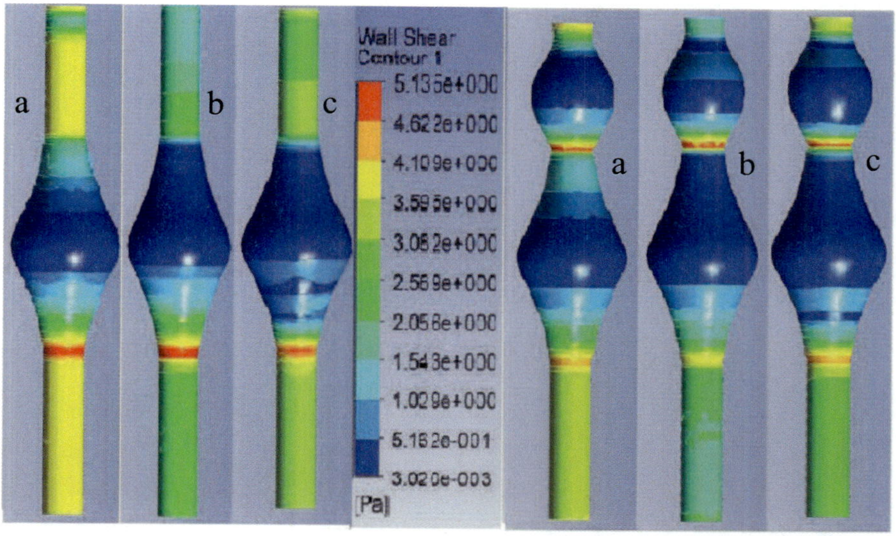


**Figure 4(a).** Comparison between SAAA and DAAA WSS contour at (a) early systole (b) peak systole and (c) diastole and WSS.

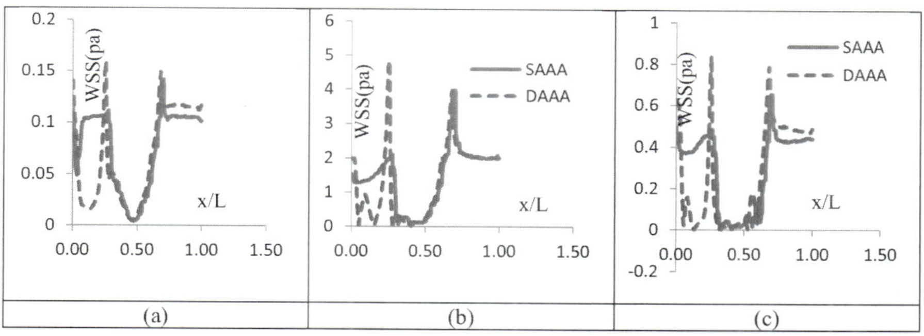


**Figure 4(b).** Comparison between SAAA and DAAAWSS at (a) early systole (b) peak systole and (c) diastole.

## *4.2 Velocity Distribution*

At every time step, centerline velocity decreases at the aneurysm region and increases at the pre and post-aneurysm. The blue color denotes zero or less than zero velocity, and in these sites is the possible growth of atherosclerosis if the patient has already begun stadium of the disease. Figure 5 displays the velocity contour of SAAA, DAAA, and at cross-section form inlet (i) 0.0 cm, (j) 4.6 cm, (k) 9.35 cm, (l) 19.975 cm. Figures 6 and 7 show the comparison of velocity distribution along the centerline and diameter at different cross-section respectively. At every pulse cycle velocity is lower in case of AAA at the cross section (j), (k) and (l). Due to aneurysm velocity is decreased in the inner side of SAAA and DAAA at peak systole and diastole. Figure 8; display the velocity streamline. It shows during peak systole and diastole strong recirculation flow is dominant at the aneurysm area in the AA.

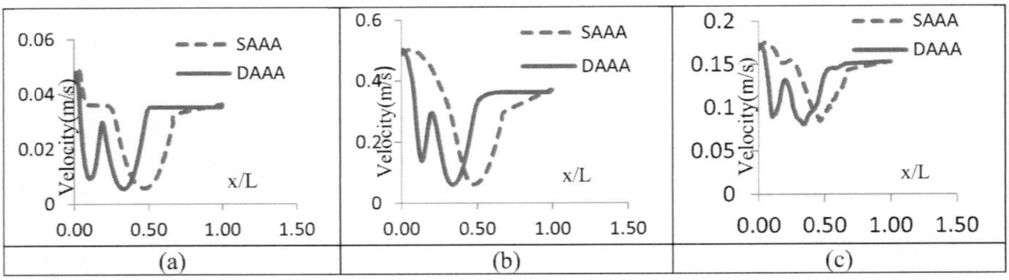


**Figure 5.** Comparison centerline velocity distribution of SAAA and DAAA at (a) early systole (b) peak systole and (c) diastole.

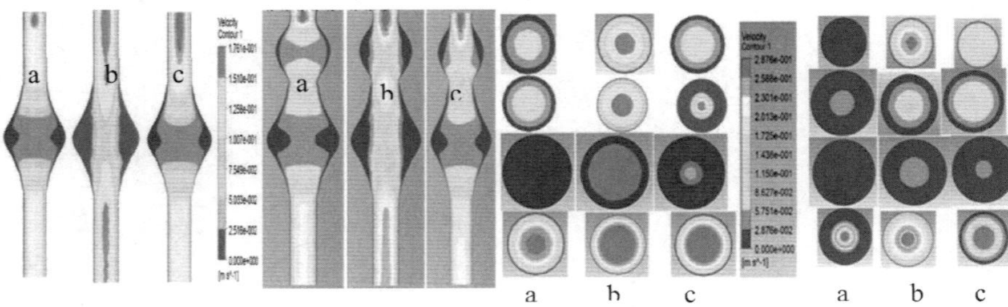


**Figure 6.** Comparison between SAAA and DAAA velocity contour at (a) early systole (b) peak systole and (c) diastole along different cross section.

### *4.3 Pressure Distribution*

Figure 9; shows the pressure contour, Figs. 10 and 11 compared the centerline pressure distribution along the centerline and diameter at cross-section at (a) early systole (b) peak systole and (c) diastole.

## 5. Conclusion

Numerical simulations are performed on pulsatile blood flow through SAAA and DAAA. Physiological velocity profiles are imposed at the inlet for the simulations because physiological blood flow is more realistic. In case of single abdominal aortic aneurysms (SAAA) wall shear stress distribution shows that as the severity increases shear stress of flow increases for all cases and minimum stress is exerted in aneurysm region at early systole, peak systole, and diastole. Pressure distribution demonstrate that at every case aneurysm creates more force than normal aorta can cause fatal death by blasting the aneurysm region of the patient of high blood pressure. Due to aneurysm enlargement, the flow behavior changes abruptly in the downstream of the aneurysm. During peak systole the blood flow decreases at the aneurysm region forming blood recirculation due to pressure drop in downstream side. Analysis of

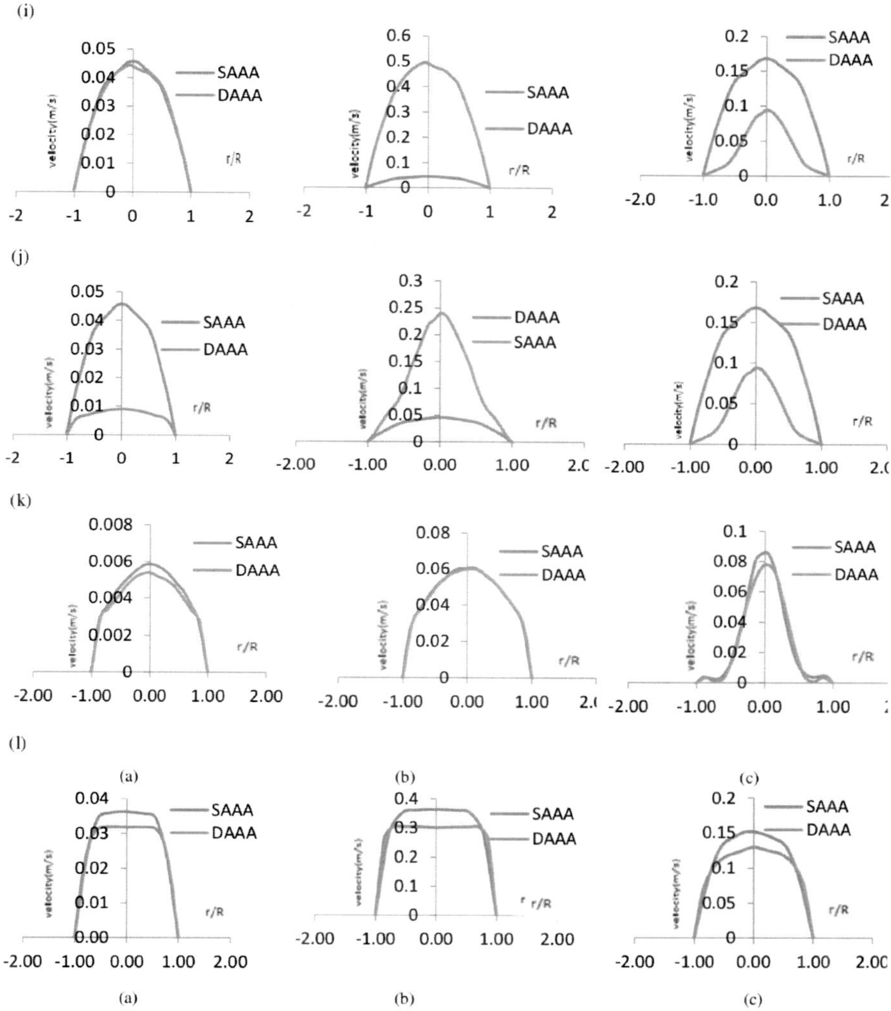


**Figure 7.** Comparison between SAAA and DAAA velocity distribution at cross-section along diameter (a) early systole (b) peak systole and (c) diastole at distance from velocity-inlet (i) 00 cm, (j) 4.6 cm (1st aneurysm), (k) 9.35 cm (2nd aneurysm) (l) 19.975 cm.

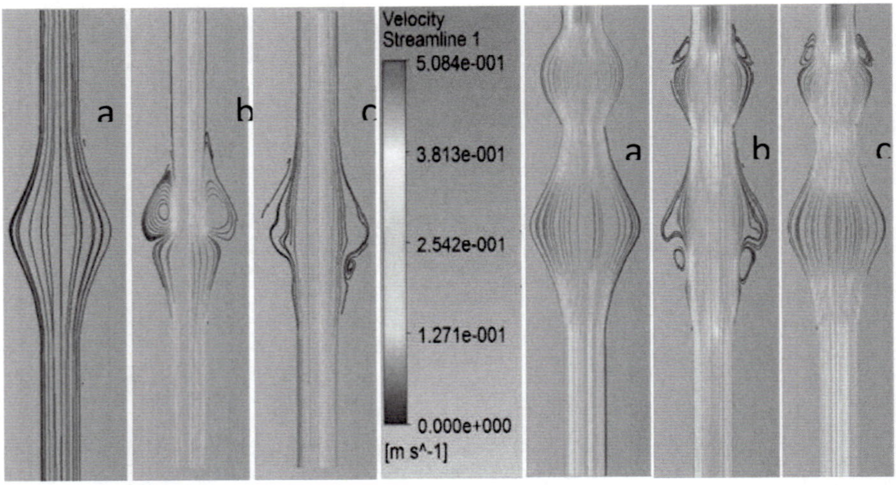


**Figure 8.** Comparison between SAAA and DAAA velocity streamline at (a) early systole (b) peak systole and (c) diastole.

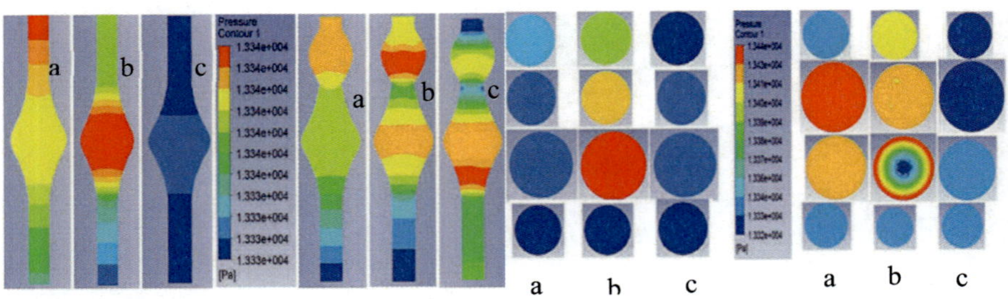


**Figure 9.** Comparison between SAAA and DAAA pressure contour at (a) early systole (b) peak systole and (c) diastole.

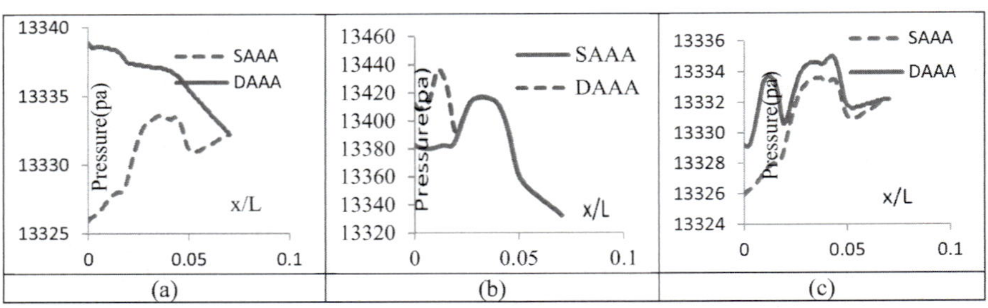


**Figure 10.** Comparison between the centerline pressure distribution of SAAA and DAAA at (a) early systole (b) peak systole and (c) diastole.

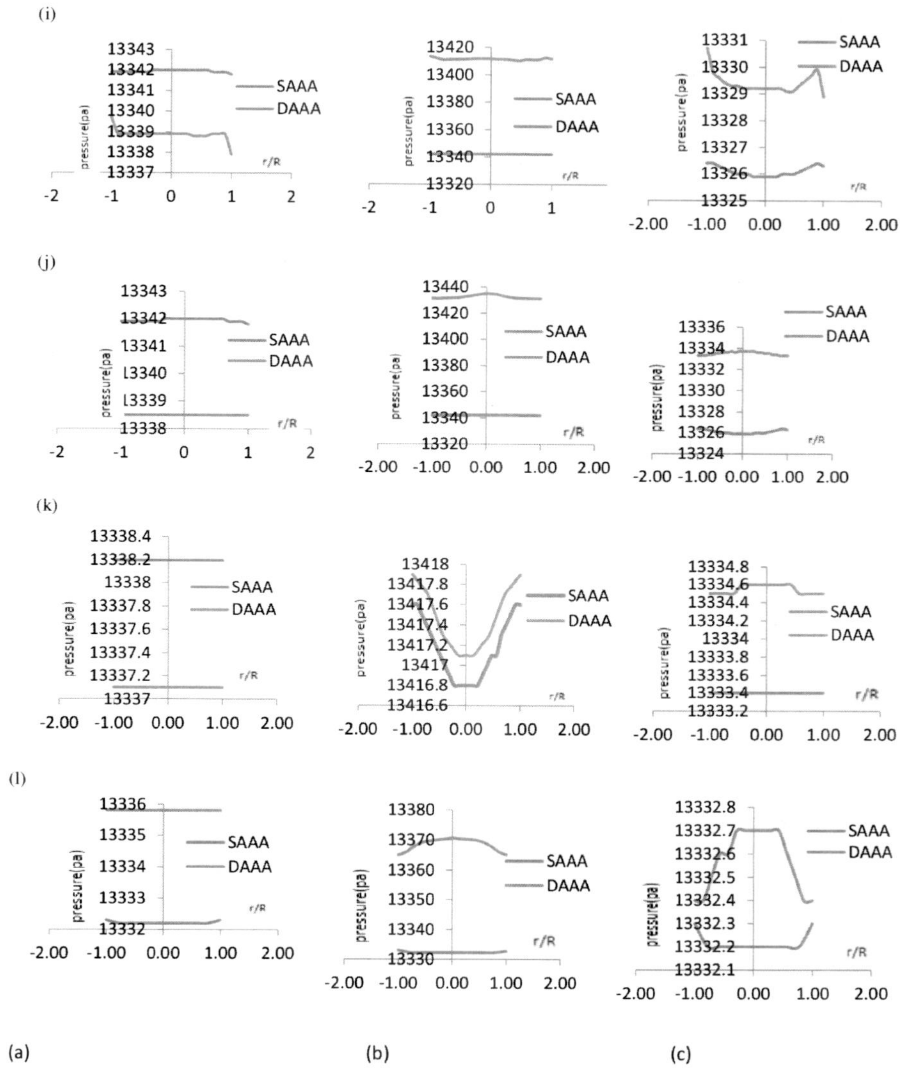


**Figure 11.** Comparison between SAAA and DAAA pressure distribution at cross-section along diameter (a) early systole (b) peak systole and (c) diastole at distance from velocity-inlet (i) 00 cm, (j) 4.6 cm (1st aneurysm), (k) 9.35 cm (2nd aneurysm) (l) 19.975 cm.

streamline shows that a recirculation region is visible at the aneurysm region for all cases. Finally, significant flow changes are noticed in abdominal aortic aneurysm.

## Nomenclature

| | | | |
|---|---|---|---|
| AA | Abdominal aorta | AAA | Abdominal aortic aneurysm |
| SAAA | Single abdominal aortic aneurysm | DAAA | Double Abdominal aortic aneurysm |
| D | Diameter of the healthy artery | x | Axial location of the flow field |
| L | Length of the artery | g | Acceleration due to gravity |
| r | Radial location of the flow field | t | Time period of the inlet flow cycle |
| R | Radius of the healthy artery | u | Instantaneous velocity |
| U | Average velocity | WSS | Wall shear stress |
| $\mu$ | Constant viscosity of blood | $\rho$ | Density of blood |
| LDL | Lower density Lipoprotein | 3D | Three Dimension |
| PDE | Partial Differential Equation | Re | Reynolds Number |
| $\tau$ | Shearing stress | $v$ | Kinematic viscosity of the fluid |
| $\gamma$ | Shear rate | $k$ | Turbulent kinetic energy |
| $\omega$ | Dissipation of turbulent kinetic energy | UDF | User Defined Function |

## References


Amon, C., Guzma´n, A. and Morel, B. (1996). Lagrangian Chaos, Eulerian Chaos, and mixing enhancement in converging–diverging channel flows. Phys. Fluids 8(5): 1192–1206.

Bluestein, D., Niu, L., Schoephoerster, R. and Dewanjee, M. (1996). Steady flow in an aneurysm model: correlation between fluid dynamics and blood platelet deposition. ASME J. Biomech. Eng. 118: 280–286.

DePaola, N., Gimbrone, M., Davies, P. and Dewey, C. (1992). 'Vascular endothelium responds to fluid shear stress gradients. Arterioscler. Thromb. 12(11): 1254–1257.

Egelhoff, C., Budwig, R., Elger, D. and Khraishi, T. (1997). A model study of pulsatile flow regimes in abdominal aortic aneurysms. Proc. 1997 ASME Fluids Engineering Division Summer Meeting, FED 21: 1–8.

Egelhoff, C., Budwig, R., Elger, D., Khraishi, T. and Johansen, K. (1999). Model studies of the flow in abdominal aortic aneurysms during resting and exercise conditions. J. Biomech. 32: 1319–1329.

Elger, D., Slippy, J., Budwig, R., Khraishi, T. and Johansen, K. (1995). A numerical study of the hemodynamics in a model abdominal aortic aneurysm AAA! Proc. ASME Symposium on Biomedical Fluids Engineering, R. A. Gerbsch and K. Ohba, eds., ASME FED 212: 15–22.

Ernst, C. (1993). Abdominal aortic aneurysm. N. Engl. J. Med. 328(16): 1167–1172.

Fukushima, T., Matsusawa, T. and Homma, T. (1989). Visualization and finite element analysis of pulsatile flow in models of the abdominal aortic aneurysm. Biorheology 26: 109–130.

Guzma´n, A. and Amon, C. (1996). Dynamical flow characterization of transitional and chaotic regimes in converging–diverging channels. J. Fluid Mech. 321: 25–57.

Guzma´n, A., Moraga, N. and Amon, C. (1997). Pulsatile non-newtonian flow in a double aneurysm. Advances in Bioengineering, ASME BED 36: 87–88.

Guzma´n, A., Moraga, N., Mun˜oz, G. and Amon, C. (1997). Pulsatile non-newtonian flow in a converging–diverging tube. AIChESymp. Series 93(314): 288–294.

Khraishi, T., Elger, D., Budwig, R. and Johansen, K. (1996). The effects of modeling parameters on the hemodynamics of an abdominal aortic aneurysm AAA! Proc. 1996 ASME Fluids Engineering Division Summer Meeting, ASME FED 237: 349–356.

Moraga, N., Guzma´n A. and Rosas, C. (1997). Meca´nica de Fluidos No Newtonianos de FlujoTransienteenTuberı´a con Seccio´n Transversal Variable en el Espacio. VI Congreso La Ingenierı´aen la Industria del Cobre, Universidad nde Antofagasta, Chile, pp. 167–175.

Peattie, R. and Bluth, E. (1998). Experimental study of pulsatile flows in models of abdominal aortic aneurysms. Proc. 20th Annual International Conference of the IEEE Engineering in Medicine and Biology Society 20(1): 367–370.

Peattie, R., Cooper, J. and Day, A. (1999). Computational investigation of pulsatile flows and wall stresses in models of abdominal aortic aneurysms. Proc. 1st Joint BMES/EMBS Conference, p. 305.

Perktold, K. (1987). On the paths of fluid particles in an axisymmetrical aneurysm. J. Biomech. 20(3): 311–417.

Perktold, K., Gruber, K., Kenner, T. and Florian, H. (1984). Calculation of pulsatile flow and particle paths in an aneurysm-model. Basic Res. Cardiol. 79: 253–261.

Rodkiewicz, C., Viswanath, N. and Zajac, S. (1995). On the abdominal aortic aneurysm: numerical and *in vitro* experimental study. Proc. 1st 1995 Regional Conference IEEE Engineering in Medicine and Biology Society and 14th Conference of the Biomedical Engineering Society of India, pp. 2.86–2.87.

Schoephoerster, R., Oynes, F., Nunez, G., Kapadvanjwala, M. and Dewanjee, M. (1993). Effects of local geometry and fluid dynamics on regional platelet deposition on artificial surfaces. Arterioscler. Thromb. 13(12): 1806–1813.

Taylor, T. and Yamaguchi, T. (1994). Three-dimensional simulation of blood flow in an abdominal aortic aneurysm—steady and unsteady flow cases. ASME J. Biomech. Eng. 116: 89–97.

Varghese, S.S. and Frankel, S.H. (2003). Journal of Biomechanics 125: 445–460.

Viswanath, N., Zajac, S. and Rodkiewicz, C. (1997). On the abdominal aortic aneurysms: pulsatile state considerations. Med. Eng. Phys. 19(4): 343–351.

Wille, S. (1981). Pulsatile pressure and flow in an arterial aneurysm simulated in a mathematical model. J. Biomed. Eng. 3: 153–158.

Yu, S. (2000). Steady and pulsatile flow studies in abdominal aortic aneurysm models using particle image velocimetry. Int. J. Heat Fluid Flow 21: 74–83.

# Chapter 6

# Finite Volume Simulation of Biomagnatic Fluid Flow in a Model Arterial Aneurysm

*Md. Mahadul Islam,*[1,2] *Md. Mamun Molla,*[1,2,*] *Most. Nasrin Akhter*[3] and *Suvash C. Saha*[4]

## 1. Introduction

An aneurysm is an abnormal dilation in the arterial wall. Arteries have thick walls to withstand normal blood pressure. However, certain medical problems, genetic conditions, and trauma can damage or injure arterial walls. The force of blood pushing against the weakened or damaged walls can cause an aneurysm. An Abdominal Aortic Aneurysm (AAA) is a localized abnormal dilatation of an artery caused by weakness in the arterial wall. An aneurysm usually has an aortic diameter of over 3 cm, whereas the average diameter of the aorta is around 2 cm. But if the outer diameter exceeds 5.5 cm, the aneurysm is considered to be large. The aorta is the biggest artery in the human body and is the primary supplier of oxygen-rich blood. Therefore, a ruptured abdominal aortic aneurysm can cause life-threatening bleeding and lead to severe pain and massive internal bleeding or hemorrhage.

Among various types of aneurysms, we are going to focus on abdominal arterial aneurysms (AAA), where abnormal growth or bulging of the aorta occurs. The aorta

[1] Department of Mathematics and Physics, North South University, Dhaka-1229, Bangladesh.

[2] Center for Applied and Computational Sciences (CACS), North South University, Dhaka-1229, Bangladesh.

[3] Department of Mathematics, Dhaka University of Engineering & Technology, Gazipur 1707, Bangladesh.

[4] School of Mechanical and Mechatronic Engineering, University of Technology Sydney, Sydney, New South Wales, 2007, Australia.

* Corresponding author: mamun.molla@northsouth.edu

DOI: 10.1201/9781003287636-6

has an elastic wall, which can stretch and then shrink back as needed to adapt to the flow of blood. However, an abdominal aneurysm occurs at the weak spot due to the pressure of the blood flow. The bulge usually starts small and grows as the pressure continues, which can result in rupture. Once an aneurysm has ruptured, it will have the classic symptoms of abdominal pain: severe, constant, and radiating to the back. Only 10 to 25% of patients survive rupture due to extensive pre- and post-operative mortality. Annual mortality from ruptured aneurysms in the United States is about 15,000.

Many studies have been performed regarding the expansion of the arterial wall. However, an experiment was conducted by Deplano et al. (2007) on the behavior of rigid and compliant asymmetric models of abdominal aortic aneurysms (AAA) under exercise and resting flow rate. They mainly evaluated the vortices inside the AAA, where they found them highly dependent on both the flow waveforms and the wall's performance. A similar experiment has also been studied by Egelhof et al. (1998) with different-sized model aneurysms. They revealed that asymmetric AAA creates a weaker vortex that expands very slowly; also, the absence of turbulence formation was observed. Yip and Yu (2001) investigated the hydrodynamic stability of cyclic flows within rigid abdominal aortic aneurysm (AAA) models and the conditions for transition to turbulence. They found that the flow remains laminar at the beginning and that instability grows gradually during the acceleration phase over the entire cross-section.

An experimental study was carried out by Fillinger et al. (2003) to determine the rupture risk of AAA by comparing wall shear stress with maximum aneurysm diameter. AAA wall stress allocation was computationally determined *in vivo* with CT data, three-dimensional stress distributions for maximum wall stress using finite element analysis, and blood pressure during observation. The study shows that 3D AAA wall stress is superior to the maximum aneurysm diameter in determining AAA rupture risk. Finol and Amon (2001) numerically predicted the blood flow pattern and hemodynamic stresses in two aneurysm models using the spectral element method under pulsatile conditions. Keeping the time-averaged Reynolds number constant, they concluded that—oscillating priorities are higher in a large aneurysm and lower in a small aneurysm. Pulsatile flow fields, velocity patterns, and flow-induced wall stresses of AAA were examined by Peattie et al. (2004) to find the mechanical factors that lead to the rupture of AAA. They found that—an increase in turbulence affects the wall stress magnitude and raises the rupture risk. On the other hand, Ma and Turan (2011) used the Carreau model and Womersley inlet velocity distribution to perform a numerical simulation of non-Newtonian blood flow in a 3D abdominal aortic aneurysm bifurcating model. They presented the flow behavior inside an aneurysm for equal and unequal outlet pressures at iliac bifurcations.

Bresson et al. (2012) experimented on an intracranial aneurysm animal model where they applied intravenous anesthesia on a relaxed muscle to determine the computational flow dynamics. The result shows a parabolic velocity pattern with a maximum velocity at the distal neck of an aneurysm. Computational comparison in modeling stent deployment in aneurysmatic cerebral vessels had been investigated by Bernardini et al. (2012), where the stent in the aneurysmatic vessel was modeled in four different initial positions. However, Xu and Lee (2009) used the Lattice Boltzmann Method (LBM) with an extrapolation boundary technique to develop a new object-oriented scientific programming strategy (Xu and Lee, 2009) to examine the flow phenomenon inside an aneurysm with ring-shaped stent obstacles. The result showed that the height of the stent obstacles effectively reduces the vorticity near the dome of an aneurysm.

Sheard (2009) computed the pulsatile flow through a model aneurysm of the sinusoidal bulge to determine the wall shear stress and flow dynamics. He used a high-order spectral element algorithm to determine the wall shear stress of the tube and aneurysm bulge and also found a plot for vorticity within an aneurysm. Swillens et al. (2008) concluded that AAA significantly alters wave reflection and hemodynamics in the aorta with apparently measurable effects in humans. Li (2010) established that as the stress decreases, it increases the diameter of the bulge at the shoulder region, eventually leading to the moving up of the shoulder region and a lengthening of an aneurysm.

The problem of biomagnatic fluid flow due to the presence of a steady magnatic field in a lid-driven cavity has been studied by Tzirtzilakis and Xenos (2013), who considered biomagnatic fluid as a homogeneous, Newtonian fluid and also as an electrically conducting magnetic fluid. The calculation was based on Ferrohydrodynamics (FHD) principles and Magnetohydrodynamics (MHD), and the numerical technique was based on a simple algorithm using a finite volume method on a spread-outstretched grid. They concluded that this increase in some vortices reduces the flow field's velocity close to the cavity's lower region where the magnatic field is strong. Tzirtzilskis (2008) concluded that even a negligible magnatic field strength could cause a considerable change in temperature and velocity field and conducted another study. However, other researchers (Tzirtzilakis et al., 2006; Loukopoulos and Tzirtzilakis, 2004; Misra et al., 2010; Sarma, 2012) have carried out some studies considering biomagnatic fluid flow in the presence of a magnatic field inside a variety of geometric conditions. Pia et al. (1996) numerically investigated the biological fluid dynamics in a high magnatic field. They considered the force arises due to the interaction between the applied magnatic field and the fluid magnatization vector using the Navier-Stokes model. They observed that an increased magnatic field reduces the blood flow rate and determines the non-dimensional parameters that direct it. They concluded that the presence of the magnatic field results in the formation of vortices about a magnetic axis.

In this paper, we have investigated the flow pattern of biomagnetic fluid through a model arterial aneurysm in the presence of different magnetic field induction where the finite volume numerical technique has been applied to illustrate the flow field.

## 2. Formulation of the Problem

The Biomgnatic fluid (blood) flows through a modeled arterial aneurysm is considered to be viscous, laminar, incompressible, homogeneous and Newtonian. The flow is exposed to a magnatic source at a place close to the center of the lower wall of the aneurysm at 0.2 h. The simulation is performed for different magnetic induction, $B_0$ = 1, 4 and 8 maintaining constant Reynolds number of 500 where, $B_0$ is define as,

$$\bar{B}_0 = \bar{\mu}_0 \bar{H}$$

Which is the product of magnatic permeability of vacuum $\mu_0$ and magnatic field strength $\bar{H}$. Blood is considered to be an electrically conducting Newtonian magnetic fluid that exhibits paramagnatic behavior with a density, $\rho$ = 1050 $gm/cm^3$ and a dynamic viscosity of $\mu = 3.45 \times 10^3$ $Pa.s$ for the flow field computation. We have considered negligible increment of the fluid viscosity due to the magnetic field.

## 2.1 Computational Geometry

The dilation of the aneurysm bulge and the rupture risk depend on the hemodynamic stresses on the wall and changes in wall mechanics. Blood flow patterns influence both of these factors through geometry. Thus, the geometry of AAA is better for understanding AAA and its physiological impacts before rupture.

In the present study, the steady flow of a Newtonian fluid in a dilated channel is considered where the dilation is defined by an axisymmetric-shaped aneurysm, as shown in Fig. 1. We have considered the aorta proximal and distal of the AAA as a straight, rigid channel excluding any branching arteries.

Due to the presence of the aneurysm, the height of the channel, $\delta$, is a variable in the streamwise direction (i.e., $\delta = \delta\ (x)$). Away from the aneurysm, the height of the channel is constant and is represented here using D (i.e., $\delta = h$ in the region either upstream or downstream of the bulge). For this model, the aneurysm is centered at 10 h downstream of the channel inlet (i.e., the inlet location is x/h = –10) and 10 h upstream from the channel outlet where, the length of the aneurysm is considered to be 8 h.

The axisymmetric model is formulated here having a cosine shaped bulge on both anterior and posterior wall using the following formula:

$$\frac{y}{h} = 1 + \frac{fc}{2}\left[1 + \cos\left(\frac{\pi x}{4h}\right)\right]; \qquad -4h \le x \le 4h$$

where $f_c = \delta/h$ is a parameter that controls the degree of dilation of the aneurysm. In the present study, $f_c$ is fixed to 1, which results in a 100% dilation of the cross-sectional area at the center of the aneurysm.

However, our investigation retains the flow criteria such as (1) the channel wall is rigid and (2) the flow is axisymmetric and time-independent. A Newtonian blood model is considered here since we have focused on aneurysms in the largest artery. Again, the relative change in the diameter of arteries is minimal, and therefore assumed that the error introduced by taking a fixed diameter is small.

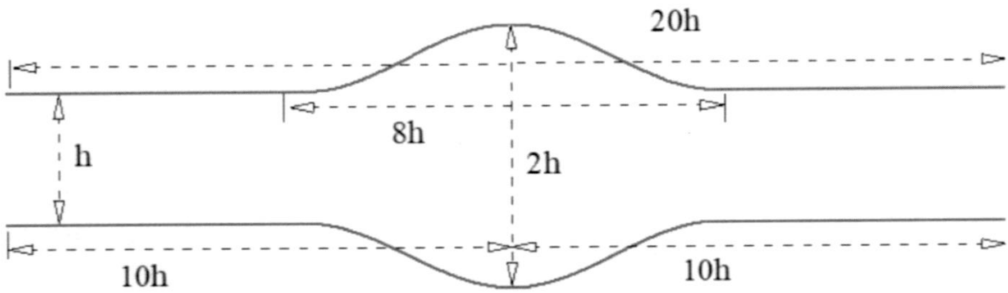


**Figure 1.** Schematic model of the computational geometry.

## *2.2 Governing Equations*

The equation of continuity:

$$\frac{\partial \bar{u}}{\partial \bar{x}}+\frac{\partial \bar{v}}{\partial \bar{y}}=0 \tag{1}$$

The Navier-Strokes equation of *u* and *v* momentum in dimensional form can be written as follows-

$$\frac{\partial \bar{u}}{\partial \bar{t}}+\bar{u}\frac{\partial \bar{u}}{\partial \bar{x}}+\bar{v}\frac{\partial \bar{u}}{\partial \bar{y}}=-\frac{1}{\rho}\frac{\partial \bar{p}}{\partial \bar{x}}+\frac{\mu}{\rho}\left(\frac{\partial^2 \bar{u}}{\partial \bar{x}^2}+\frac{\partial^2 \bar{u}}{\partial \bar{y}^2}\right)+\mu_0\bar{M}\frac{\partial \bar{H}}{\partial \bar{X}}-\bar{\sigma}\bar{B}_y^2\bar{u}+\sigma\bar{B}_x\bar{B}_y\bar{v} \tag{2}$$

$$\frac{\partial \bar{v}}{\partial \bar{t}}+\bar{u}\frac{\partial \bar{v}}{\partial \bar{x}}+\bar{v}\frac{\partial \bar{v}}{\partial \bar{y}}=-\frac{1}{\rho}\frac{\partial \bar{p}}{\partial \bar{y}}+\frac{\mu}{\rho}\left(\frac{\partial^2 \bar{v}}{\partial \bar{x}^2}+\frac{\partial^2 \bar{v}}{\partial \bar{y}^2}\right)+\mu_0\bar{M}\frac{\partial \bar{H}}{\partial \bar{y}}-\bar{\sigma}\bar{B}_x^2\bar{v}+\sigma\bar{B}_x\bar{B}_y\bar{u} \tag{3}$$

The Magnatization $\bar{M}$ is proportional to Magnatic field intensity $\bar{H}$, where $\bar{\chi}$ is a constant called magnatic susceptibility.

$$\bar{M}=\bar{\chi}\bar{H}$$

In the above relation, the components of magnatic field intensity $\bar{H}_x$ and $\bar{H}_y$ along the $\bar{x}$ and $\bar{y}$ coordinates ($\bar{H}=(\bar{H}_x, \bar{H}_y)$) are given respectively,

$$\bar{H}_x=\frac{\gamma}{2\pi}\frac{\bar{y}-\bar{b}}{(\bar{x}-\bar{a})^2+(\bar{y}-\bar{b})^2} \tag{4}$$

$$\bar{H}_y=-\frac{\gamma}{2\pi}\frac{\bar{x}-\bar{a}}{(\bar{x}-\bar{a})^2+(\bar{y}-\bar{b})^2} \tag{5}$$

where $(\bar{a}, \bar{b})$ is the point where the magnetic source is placed and $\gamma$ is the magnetic field strength at this point $(\bar{x}=\bar{a}, \bar{y}=\bar{b})$. The magnitude $\bar{H}$, of the magnetic field intensity, is given by

$$\bar{H}(\bar{x},\bar{y})=\left[\bar{H}_x^2+\bar{H}_y^2\right]^{\frac{1}{2}} \tag{6}$$

$$=\frac{\gamma}{2\pi}\frac{1}{\sqrt{(\bar{x}-\bar{a})^2+(\bar{y}-\bar{b})^2}} \tag{7}$$

**Non-dimensionalization**

To make non-dimensionalize, the equations we have used:

$$x = \frac{\overline{x}}{h}, \quad y = \frac{\overline{y}}{h}, \quad p = \frac{\overline{p}}{\rho u_r^2}$$

$$\tau = \frac{t\overline{u}_r}{h}, \quad u = \frac{\overline{u}}{\overline{u}_r}, \quad v = \frac{\overline{v}}{u_r}$$

$$\overline{B} = \mu_0 \overline{H}, \; H_x = \frac{\overline{H}_x}{\overline{H}_0}, \; H_y = \frac{\overline{H}_y}{\overline{H}_0}$$

Non-dimensional continuity equation,

$$\frac{\partial u}{\partial x} + \frac{\partial v}{\partial y} = 0 \tag{8}$$

Using the above dimensionless parameters we get non-dimensional u-momentum equation:

$$\frac{\partial u}{\partial \tau} + u\frac{\partial u}{\partial x} + v\frac{\partial u}{\partial y} = -\frac{\partial p}{\partial x} + \frac{1}{\mathrm{Re}}\left(\frac{\partial^2 u}{\partial x^2} + \frac{\partial^2 u}{\partial y^2}\right) + M_{nf} H\frac{\partial H}{\partial x} - N[H_y^2 u - H_x H_y v] \tag{9}$$

Similarly, the non-dimensional v-momentum equation:

$$\frac{\partial v}{\partial \tau} + u\frac{\partial v}{\partial x} + v\frac{\partial v}{\partial y} = -\frac{\partial p}{\partial x} + \frac{1}{\mathrm{Re}}\left(\frac{\partial^2 v}{\partial x^2} + \frac{\partial^2 v}{\partial y^2}\right) + M_{nf} H\frac{\partial H}{\partial y} - N[H_x^2 v - H_x H_y u] \tag{10}$$

The above equations contain some non-dimensional parameters such as:

$$\text{Reynolds number, } \mathrm{Re} = \frac{h u_r \rho}{\mu}$$

$$\text{Magnatic number, } M_{nf} = \frac{\mu_0 \overline{\chi} H_0^2}{u_r^2 \overline{\rho}}$$

$$\text{Stuart number, } N = \frac{\sigma \mu_0 H_0^2}{\rho u_r} = \frac{Ha^2}{\mathrm{Re}}$$

The parameter $M_{nf}$ is the magnatic number, the parameter $N$, is the Stuart number, which is the ratio of the square of Hartman number $Ha$, to the Reynolds number. Where,

$$Ha^2 = \frac{\sigma \mu_0 \overline{H}_0^2 h^2}{\mu} \tag{11}$$

For a specific Reynolds number, the increase in magnatic number is equivalent to increase in magnatic field strength $\overline{H}_0$. The magnitude $H$ equation of the magnatic field intensity from (Tzirtzilakis and Xenos, 2013) can be derived as,

$$\overline{H}(x, y) = \frac{|b|}{\sqrt{(x-a)^2 + (y-b)^2}} \tag{12}$$

In order to find the numerical results, it is necessary to assign values to the dimensionless parameters. The conditions adopted in (Tzirtzilakis and Xenos, 2013) is considered in which the fluid is blood (biomagnatic fluid density, $\rho = 1050\ kgm^{-3}$ and dynamic viscosity, $\mu = 3.2 \times 10^{-3}\ kgm^{-1}\ s^{-1}$). The width of the inlet and outlet was considered $h = 5.0 \times 10^{-2}\ m$, whereas the magnatic source is placed at below the lower wall of the aneurysm and the Reynolds number, Re is equal to 500. The dimensionless parameter $M_{nf}$ can be written as:

$$M_{nf} = \frac{\mu_0 \overline{\chi} \overline{H}_0^2}{u_r^2 \rho} = \frac{\mu_0 \overline{H}_0 \overline{\chi} \overline{H}_0}{u_r^2 \rho} \tag{13}$$

$$M_{nf} = \frac{\overline{B}_0 \overline{M}_0}{u_r^2 \rho} = \frac{\overline{B}_0 \overline{M}_0 h^2 \rho}{\mu^2 \operatorname{Re}^2} \tag{14}$$

where $\overline{B}_0$ and $\overline{M}_0$ are the magnetic induction and the magnatization at the point $(a, 0)$, respectively. The quantity $\overline{u}_r$ has been replaced using the definition of the Reynolds number $\overline{u}_r = \dfrac{\mu \operatorname{Re}}{h\rho}$.

Thus, by assuming that saturation magnatization $M_0 = 40\ Am^{-1}$ is attained, it is possible to produce a relation which gives a realistic range of values for the $M_{nf}$ related to the applied magnatic field $B_0$ and the Reynolds number (Tzirtzilakis and Xenos, 2013).

$$M_{nf} = \frac{1.0253906 \times 10^7 B_0}{\operatorname{Re}^2}$$

For the flowing blood the electrical conductivity is always greater than that of the stationary. In this case the electrical conductivity, $\sigma$ of blood is assumed to be temperature independent and equal to 0.8 $Sm^{-1}$ as in (Tzirtzilakis and Xenos, 2013).

$$N = \frac{\sigma h \mu_0^2 \overline{H}_0^2}{\rho u_r} = \frac{\mu}{h \rho u_r} \frac{h^2 \sigma \overline{B}_0^2}{\mu}$$

$$N = \frac{1}{\operatorname{Re}} Ha^2 = 0.625 \frac{B_0}{\operatorname{Re}}$$

### *2.3 Boundary Conditions*

In this study the boundary conditions for the velocity components $u$ and $v$ have fully developed time invariant parabolic profile at the inlet $u = 2u_r \left[1 - \left(\dfrac{r}{h/2}\right)^2\right]$ and no-slip condition is used at the walls, where velocity u and v are zero. At the outlet we have applied zero gradient outflow condition, where $\dfrac{\partial u}{\partial x} = 0$ and $\dfrac{\partial v}{\partial x} = 0$.

## 3. Result and Discussion

The numerical simulations are performed to clearly understand the flow pattern through a dilated channel after placing a magnatic source close to the lower wall of the bulge region. In this section, we have considered the fluid flow inside the aneurysm to be steady with a constant Reynolds number of 500 and magnetic field, $B_0 = 1$, 4 and 8. When the flow is steady, fluid characteristics like velocity, pressure, shear stress, etc., do not change over time. Numerical results are used to plot wall shear stress, wall pressure, centerline velocity, streamwise velocity, vorticity, and streamlines for different magnatic fields, which in detail explain the behavior of the flow pattern inside the arterial aneurysm.

### *3.1 Grid-independent Test*

Several grids are rectangular mashes used to split up the computational domain, defining the discrete points by which the numerical solutions are computed. In order to achieve greater accuracy and convergent results for different parameters, the grid size is varied. After the simulation is performed with varying grid sizes, we have found almost the same results for each grid combination, so the solution is considered to be grid-independent. We have used different grid combinations for determining more precise results that are given below:

$$\text{Case 1: } (240 \times 100) \approx (x \times y)$$

$$\text{Case 2: } (260 \times 110) \approx (x \times y)$$

$$\text{Case 3: } (280 \times 120) \approx (x \times y)$$

Grid independence test has been conducted using the above three cases to validate the code in terms of the streamwise velocity at different axial positions of the symmetrically shaped model arterial aneurysm, without the effect of the presence of magnatic source considering constant Reynolds number of 500. The result has been plotted in Fig. 2, and it is observed that all three grid combinations give the same velocity pattern at various arterial locations. Here, we noticed that the results overlap, and a few figures hold parabolic shapes which are symmetric concerning the central axis of the artery. From Figs. 2(d) to 2(g), slight negative velocity is observed near the upper and lower wall of the dilated region, which is caused due to the flow recirculation.

### *3.2 Streamwise Velocity*

Here, we are going to examine the velocity pattern at different positions of a model arterial aneurysm. The result of streamwise velocity at various arterial positions shows how the velocity develops gradually inside the artery in the presence of different magnatic fields. Each figure of Fig. 3 shows the streamwise velocity at a particular position of an arterial aneurysm for three distinct magnetic fields: $B_0 = 1, 4$ and 8 and $Re = 500$. The X-axis of each figure represents the velocity profile whereas the Y-axis represents the arterial location.

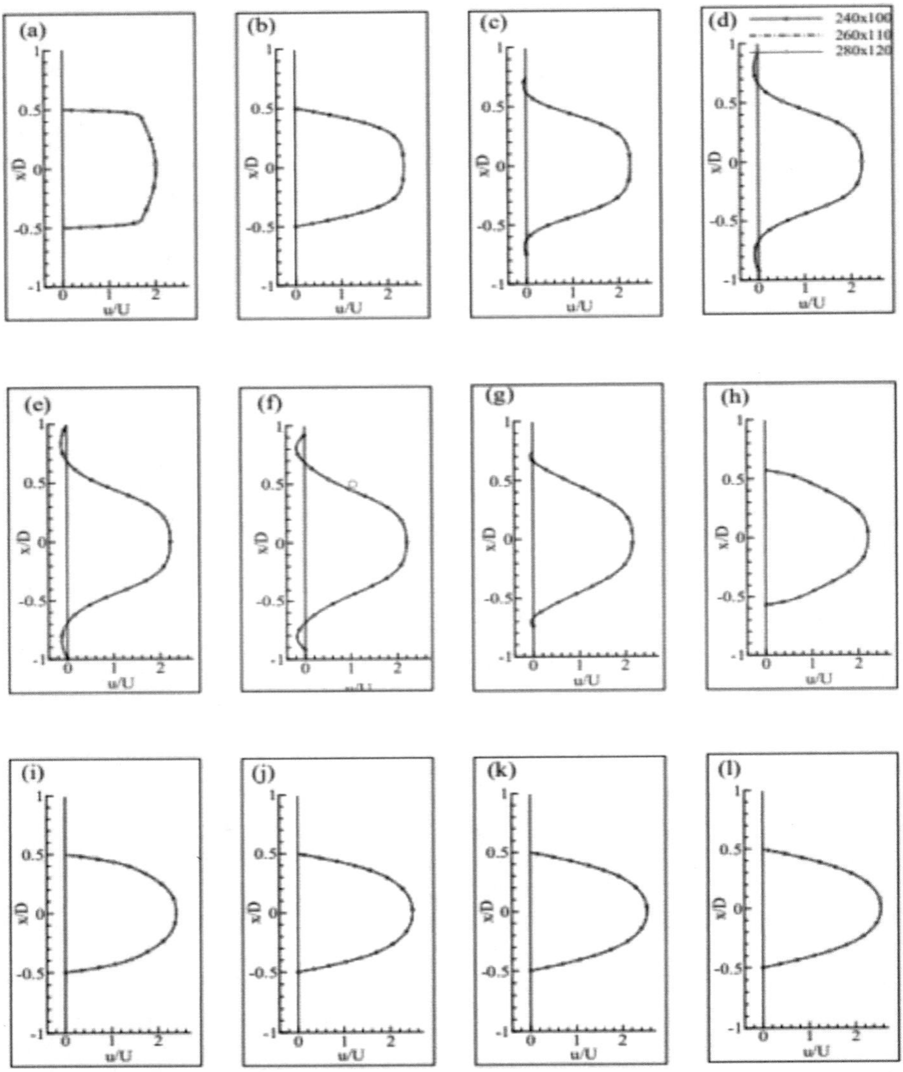

**Figure 2.** Grid Independent test for streamwise velocity in an axisymmetric shaped aneurysm at different axial (a) intel, x/h = –10.0, (b) x/h = –4.0, (c) x/h = –2.0, (d) x/h = –1.0, (e) x/h = 0.0, (f) x/h = 1.0, (g) x/h = 2.0, (h) x/h = 3.0, (i) x/h = 4.0, (j) x/h = 6.0, (k) x/h = 8.0, and (l) outlet, x/h = 10, while Re = 500 and $B_0 = 8$.

Figure 3(a) shows the velocity pattern at the inlet position where all the magnatic fields follow the same shape. The velocity pattern near the proximal neck is shown in Figs. 3(b) and 3(c), where all three magnatic fields follow the same parabolic shape, but the magnitude increases. Before the center of the dilation, we can observe slight negative velocity near the lower wall of the aneurysm, along with slight velocity deviation for three different magnatic fields, shown in Fig. 3(d).

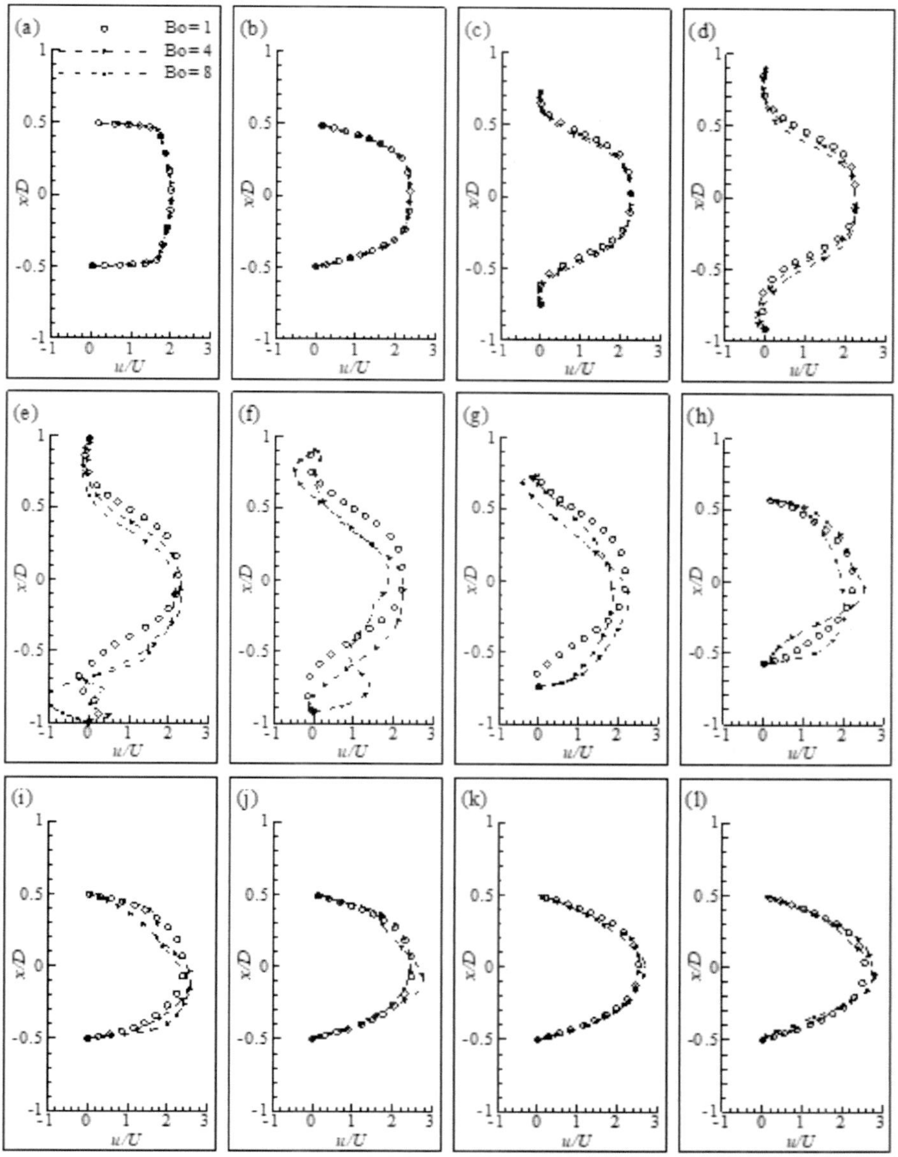


**Figure 3.** Streamwise velocity u/U, at different axial positions for $B_0 = 1$, 4 and 8 (a) intel, x/h = –10, (b) x/h = –4.0, (c) x/h = –2.0, (d) x/h = –1.0, (e) x/h = 0.0, (f) x/h = 1.0, (g) x/h = 2.0, (h) x/h = 3.0, (i) x/h = 4.0, (j) x/h = 6.0, (k) x/h = 8.0, and (l) outlet, x/h = 10.0 considering steady Newtonian flow for Re = 500.

Structures 3(e) to 3(h) show the streamwise velocity pattern from the center to the distal neck of the aneurysm, where we can observe a non-symmetrical velocity pattern concerning the central axis of the artery. From here, we can also follow that all three magnatic fields have different velocity variations near the lower wall of the aneurysm, but a higher magnatic field shows more velocity variation. Moreover, this bulge region also shows negative velocity near both the aneurysm's upper and lower walls due to the magnatic field's presence. However, entering the post-aneurysm area from Figs. 3(i) to 4(k), the velocity pattern for the lower magnatic field starts getting back to the parabolic shape, but the velocity for the higher magnatic field $B_0 = 8$ does not maintain the symmetrical shape as well as the curved shape. In the post-aneurysm zone, we observed no negative velocity, which indicates the absence of recirculation. The Fig. 3(l) represents the velocity at the outlet position, which is precisely similar to the inlet velocity.

The result of streamwise velocity is not uniform at the dilated region due to the presence of the magnatic field. Moreover, negative velocity is observed, which indicates the formation of a recirculation region inside the bulge.

### *3.3 Wall Shear Stress*

Wall Shear Stress, WSS arises due to the force per unit area exerted by a flowing fluid on the arterial wall, which occurs in the direction of the tangential plane. This frictional force resists the fluid to flow smoothly and exert an extra pressure on the wall. WSS, $\tau$ is define as-

$$\tau_W = \mu \left.\frac{\partial u}{\partial y}\right|_W$$

where $\mu$ is the dynamic viscosity, $u$ is the flow velocity parallel to the wall and y is the distance perpendicular to and away from the wall. So, we can define WSS as the product of dynamic viscosity and shear rate or velocity gradient $\frac{\partial u}{\partial y}$. It is an important parameter to determine stress inside an aneurysm. For steady state, we have plotted both upper and lower wall stress in an artery after applying different magnatic fields. Figure 4 shows the upper and lower wall shear stress of an axisymmetric shaped arterial aneurysm for $B_0 = 1, 4$ and 8 where, the upper and lower wall both experience almost constant shear stress up to pre-aneurysm zone.

Figure 4(a) shows three different magnatic fields' upper wall shear stress. From here, we can observe that WSS for all the magnatic fields has a slight drop at the proximal neck, which remains constant up to the center of the aneurysm. After the center of the dilation, a negative wall shear stress is observed for both $B_0 = 4$ and 8. They again provide the maximum WSS with a magnitude of 0.09 near the distal end of the aneurysm. From the center to the post-aneurysm region, continuous variation in stress value is observed for the higher magnatic field. But the stress pattern $B_0 =$ 1 remains constant up to the center of the bulge, and when it enters the distal end, WSS rises to 0.1 and remains almost endless. So, a lower magnatic field creates less variation in WSS near the upper wall of the aneurysm.

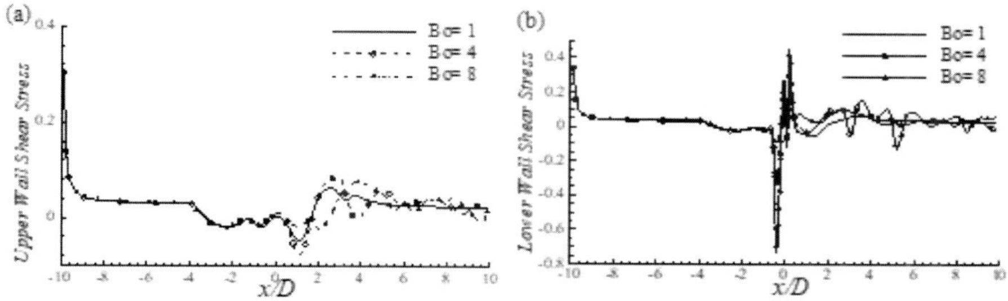


**Figure 4.** (a) Upper Wall Shear Stress, and (b) Lower Wall Shear Stress inside an aneurysm due to the effect of external magnatic field considering steady Newtonian flow for $Re = 500$ and $B_0 = 1$, 4 and 8.

The variation in lower wall shear stress at different arterial locations is shown in Fig. 4(b), where all the magnatic fields cause similar stress patterns like upper wall shear stress up to the proximal neck. After the proximal neck, there is a huge fluctuation of wall shear stress for $B_0 = 4$ and 8 and near the center, we can observe a range of maximum positive and negative WSS of (–0.62, 0.32) for $B_0 = 8$. The result for $B_0 = 1$ varies less compared to other results whereas, the result for $B_0 = 8$ shows highly oscillating behavior.

Since the magnatic source is placed close to the center of the lower wall of the aneurysm, we observe more variation in stress patterns near the lower border. Therefore, negative wall shear stress appears in the dilated region for a higher magnatic field due to decreased velocity. Thus, the fluid moves anticlockwise, creating recirculation in the bulged area.

### *3.4 Wall Pressure*

Wall pressure generally refers to the force applied per unit area on the arterial wall in the perpendicular direction. The main reason for the formation of AAA is the pressure exerted by the flowing fluid through the aorta. Therefore, the pressure inside the arterial aneurysm is essential in determining the flow behavior and rupture risk.

Figure 5 shows an axisymmetric-shaped arterial aneurysm's upper and lower wall pressure in different magnatic fields and a constant Reynolds number of 500. We observed that the upper and lower walls experience symmetrical pressure distribution patterns from here.

Figure 5(a) shows upper wall pressure distribution at different arterial positions. From here, we can observe that the pressure increases from the pre-aneurysm zone to the center of the dilation, and after that, it decreases at the same rate for all the magnatic fields, following a symmetric pattern. All the magnatic fields give maximum wall pressure at the center of the aneurysm. As a result, the upper wall experiences the highest pressure of +45.62 for $B_0 = 8$. However, the rate of increase in wall pressure is more for higher magnatic fields and vice versa.

The lower wall pressure at different arterial positions is shown in Figure 5(b), where pressure remains zero from the inlet to the proximal neck of the artery. After that, pressure rises sharply up to the center of the dilation and falls sharply by maintaining a symmetrical pattern. Here, the rate of increase in pressure is much

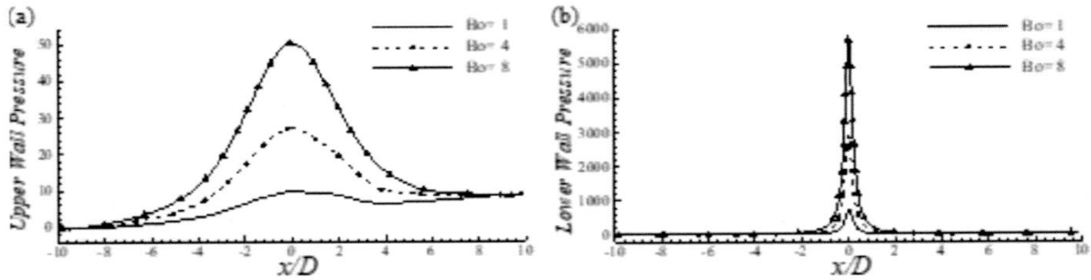


**Figure 5.** (a) Upper Wall Pressure, and (b) Lower Wall Pressure inside an aneurysm due to the effect of external magnatic field considering steady Newtonian flow for $Re = 500$ and $B_0 = 1$, 4 and 8.

higher compared to upper wall pressure. It is also observed that the lower arterial wall experiences maximum pressure of 5749 for $B_0 = 8$, and the lower magnatic field experience maximum pressure of 717 at the center, which is much higher than the upper wall pressure. Therefore, the effect of applying a magnatic field near the arteries is apparently evident in the case of arterial wall pressure. The magnatic source is placed close to the lower bulge, so this might be a reason behind getting higher pressure at the lower wall of the aneurysm. Therefore, higher wall pressure found inside the aneurysm can weaken the arterial wall and cause further elongation.

### 3.5 Centreline Velocity

To have complete knowledge of the fluid velocity at every instantaneous position of the arterial aneurysm, centreline velocity plays an important role. Figure 6 shows the variation in centerline velocity caused by the steady flow through arterial aneurysm in the presence of a magnatic field, $B_0 = 1$, 4 and 8.

From Fig. 6(a), we can observe the cross-stream centerline velocity for all the magnatic fields, which remain constant at zero from the inlet to the proximal neck. After the proximal neck to the outlet position, we can observe continuous fluctuation between positive and negative velocity for $B_0 = 4$ and 8, whereas the velocity pattern for $B_0 = 1$ remains almost constant near about zero. After the distal end, we get maximum and minimum centerline velocity of (+0.78, –0.78) for $B_0 = 8$. We can observe negative centerline velocity for all three magnatic fields, but it caused the least negative acceleration. Therefore, higher magnatic flux creates more oscillation in centerline velocity and vice versa. Moreover, the negative centerline velocity indicates that there is a backward flow, which results in the formation of recirculation inside the aneurysm.

Figure 6(b) shows the streamwise centerline velocity pattern of the fluid flow due to the presence of various magnatic fields in a modeled arterial aneurysm. The figure shows more variation in centerline velocity for a higher magnatic field along the horizontal direction. Throughout the aneurysm, we observe a higher fluctuation range for all the magnatic fields from the inlet to the outlet position. All the magnatic fields follow almost the same velocity pattern up to the center of the aneurysm with an initial velocity of 2 at the inlet position. But from the center to the outlet position of the aneurysm, speed continuously fluctuates. The minimum magnatic field rate is seen at the distal end. Moreover, higher magnatic flux gives more oscillating

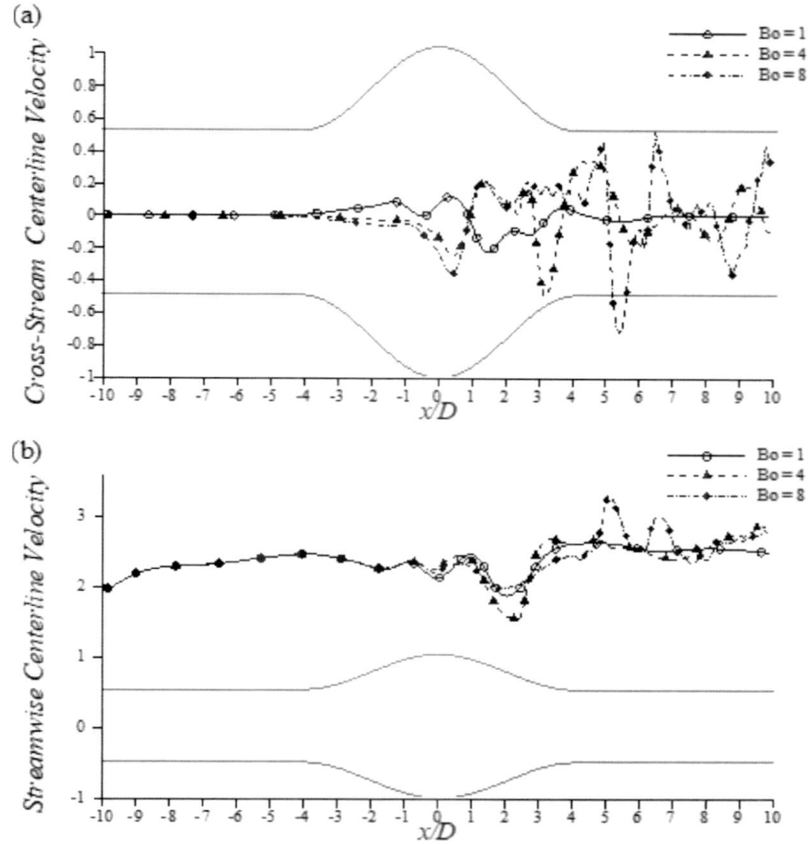


**Figure 6.** Centreline Velocity, $u/U$ inside an inside an aneurysm due to the effect of external magnatic field considering steady Newtonian flow for $Re$ = 500 and $B_0$ = 1, 4 and 8 (a) Cross stream centerline velocity, (b) Streamwise centerline velocity.

velocity where the maximum magnitude of the acceleration is 1.54 for $B_0 = 8$ and the minimum speed is 1.35 for $B_0 = 1$.

### 3.6 Streamline

A streamline is an imaginary line in the fluid that is tangent at every point to the local instantaneous velocity field. It also indicates the direction of the velocity of blood particles at a particular issue at a fixed instant of time. They are continuous lines that do not intersect at the exact moment because a fluid particle cannot have a different velocity at the same point. Figure 7 shows the streamlines in an axisymmetric shape arterial aneurysm using a constant Reynolds number of 500 and other magnatic fields, $B_0$ = 1, 4 and 8. The streamlined result for various magnatic fields shows no vortex formation at the artery's central axis. At the same time, vortices are formed on both sides of the dilated region for all the magnatic fields, which are not symmetric due to the presence of a magnatic source close to the lower wall of the aneurysm.

Figure 7(a) shows the streamlines for the magnatic field, $B_0 = 1$ where the flow is laminar with high-velocity intensity at the central axis of the artery. Still, recirculation is observed inside the dilated region with negative velocity intensity. From here, we can keep one large recirculation zone near the upper wall of the dilation and two large vortices near the lower division. The flow pattern inside the aneurysm for the higher magnatic field $B_0 = 1$ is represented by the streamlines shown in Fig. 7(b). The streamlines near the central axis of the artery are almost straight, with slight distortion near the distal end, where the velocity intensity decreases. We also observe that small vortices increase near the bulge, and the velocity intensity rises compared to the lower magnatic field. On further increase in the magnatic field to $B_0 = 8$, the velocity intensity at the middle arterial sections increases, and the streamline gets even more distorted at the distal neck, according to the result shown in Fig. 7(c). It is also observed that small vortices are formed from the proximal neck to the center region of the bulge, whereas no vortex is developed at the lower dilated region. So, with the increase in magnatic field strength, the recirculation area, along with the concentration of the vortex, decreases near both walls of the bulge region, and the velocity pattern also gets distorted near the lower border of the distal end regions.

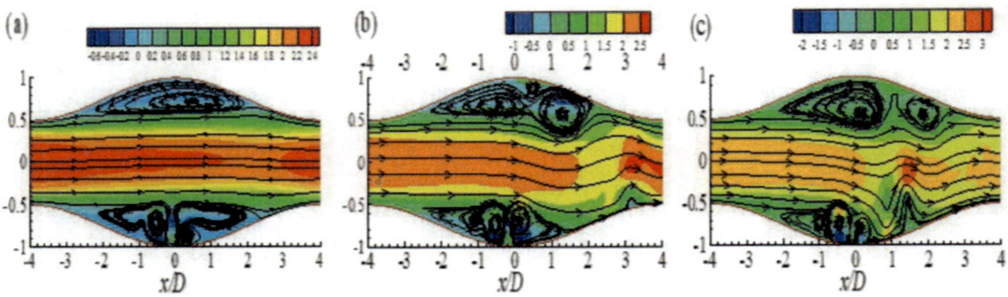


**Figure 7.** Streamline inside a symmetrical shaped aneurysm for $Re = 500$ (a) $M_{nf} = 41.02$, $N = 1.25 \times 10^{-3}$, $B_0 = 1$, (b) $M_{nf} = 164.06$, $N = 2.0 \times 10^{-2}$, $B_0 = 4$ and (c) $M_{nf} = 328.12$, $N = 1.25 \times 10^{-3}$, $B_0 = 8$.

### *3.7 Vorticity*

The vorticity describes the change in velocity vector when the fluid moves an extremely small distance in the direction perpendicular to it. However, the vorticity of the fluid dynamics explains the local circulatory motion of the fluid. The streamwise vorticity quantifies the curl of the vector field, $\bar{V}$.

$$\omega = \nabla \times \bar{V} = \frac{\partial v}{\partial x} - \frac{\partial u}{\partial y}$$

Figure 8 represents the vorticity of the fluid particles inside an axisymmetric-shaped arterial aneurysm in the presence of different magnetic fields $B_0 = 1, 4$ and 8, where we have used 15 unequal contour levels between their maximum and minimum value, shown in the legend color bar. We can observe solid and dotted streamlines from all the figures, representing positive clockwise rotation and negative

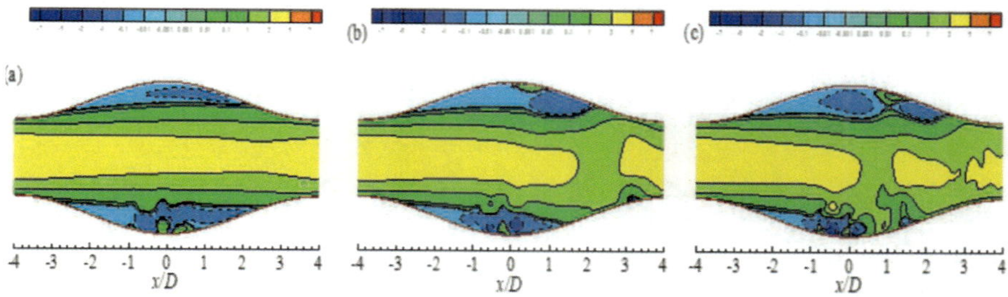

**Figure 8.** Velocity for three different magnatic field and $Re = 500$ (a) $B_0 = 1$ (b) $B_0 = 4$ (c) $B_0 = 8$.

anticlockwise rotation of fluid particles, respectively. Here, negative velocity is only observed close to the upper and lower wall of the dilated region.

Figure 8(a) represents the fluid vorticity for $B_0 = 1$ where, solid straight contour lines are observed with positive velocity intensity throughout the central axis of the artery. Near the upper and lower wall of the bulge that is from $-3 < \frac{x}{D} < +3$ we can observe large vortex formation with maximum magnitude of anticlockwise vortex which can be clearly seen from the color legend.

As the magnatic field increases to $B_0 = 4$, we can observe that the magnitude of positive vorticity decreases near the distal end, and from the center to the outlet, the contour lines get distorted, illustrating different velocity contours shown in Fig. 8(b). Vortexes are seen to propagate toward the distal end near the aneurysm's upper wall. Figure 8(c) represents the vorticity pattern for a magnetic field, $B_0 = 8$ where the magnitude of negative vorticity decreases further near the upper and lower border of the dilated region. Along the central axis of the artery, the contour lines get even more distorted, and the vorticity magnitude decreases. We also observe that the vortex gets smaller close to the lower wall of the aneurysm, whereas new vortices are formed close to the upper wall.

Therefore, as the magnatic field increases, a negative recirculating area near both the upper and lower wall of the bulge decreases, whereas the velocity intensity at the central axis of the artery decreases, and the contour line gets distorted.

## 4. Conclusion

The biomagnatic fluid flow through an axisymmetric-shaped arterial aneurysm in the presence of a magnatic source close to the lower wall of the aneurysm is studied based on finite volume numerical technique for steady Newtonian flow. From this investigation, we have observed the variation in the flow field, wall pressure, wall shear stress, and vortex dynamic by plotting the numerical result for different magnatic fields, i.e., $B_0 = 1, 4, 8$ and $Re = 500$, to have complete knowledge about the vortex dynamic and flow, pattern inside the arterial aneurysm.

The influence of the flow field and the presence of a magnatic field induce pressure on both the upper and lower walls of the bulge for the steady case where maximum pressure is observed at the center of the aneurysm. Tremendous pressure is monitored at the center of the lower wall of the aneurysm, which is about

100 times higher than upper wall pressure due to the placement of a magnatic source close to the lower wall. Cross-stream centerline velocity shows frequent velocity fluctuation in the downstream region of the aneurysm for a higher magnetic field. On the other hand, streamwise centerline velocity offers a higher instability range, but more variation is seen for the higher magnatic field. The result of wall shear stress for steady Newtonian flow shows more variation of WSS at the lower wall of the aneurysm compared to the upper wall for the higher magnatic field. However, more negativity in wall shear stress is observed for a higher magnatic field before the center of the lower border of the aneurysm.

On increasing the magnatic field strength, recirculation formation decreases, as seen from the streamlined diagrams. In the case of vorticity, we can observe that as the magnatic field increases negative recirculating area decreases and becomes scattered near both walls of the bulge for steady flow. From our investigation, the presence of a magnatic source significantly influences the flow pattern and the formation of the vortex.

## References

Bernardini, A., Larrabide, I., Petrini, L., Pennati, G., Flore, E., Kim, M. et al. (2012). Deployment of self-expandable stents in aneurysmatic cerebral vessels: comparison of different computational approaches for interventional planning. Comp. Meth. Biomech. Biomed. Eng. 15: 303–311.

Bresson, D., Kharboutly, Z., Mounayer, C., Couquet, C., Gory, B. and Legallais, C. (2012). Computational flow dynamics based on an intracranial aneurysm animal model. Comp. Meth. Biomech. Biomed. Eng. 15: 46–48.

Deplano, V., Knapp, Y., Bertrand, E. and Gaillard, E. (2007). Flow behaviour in an asymmetric compliant experimental model for abdominal aortic aneurysm. J. Biomed. 40: 2406–2413.

Egelhoff, C., Budwig, R., Elger, D., Khraishi, T. and K.H., J. (1998). Model studies of the flow in abdominal aortic aneurysms: During resting and exercise conditions. J. Biomech. 32: 1319–1329.

Fillinger, M.F., Marra, S.P., Raghavan, M.L. and Kennedy, F.E. (2003). Prediction of rupture risk in abdominal aortic aneurysm during observation: Wall stress versus diameter. J. Vasc. Surg. 37: 724–732.

Finol, E.A. and Amon, C.H. (2001). Blood flow in abdominal aortic aneurysms: pulsatile flow hemodynamics. J. Biomech. Eng. 123: 474–484.

Li, Z. (2010). Computed Wall Stress May Predict the Growth of Abdominal Aortic Aneurysm. 32nd Annual Int. Conf. IEEE EMBS Buenos Aires, Argentina pp. 2626–2629.

Loukopoulos, V.C. and Tzirtzilakis, E.E. (2004). Biomagnetic channel flow in spatially varying magnetic field. Int. J. Eng. Sci. 42: 571–590.

Ma, J. and Turan, A. (2011). Pulsatile non-newtonian haemodynamics in a 3D bifurcating abdominal aortic aneurysm model. Comp. Meth. Biomech. Biomed. Eng. 14: 683–694.

Misra, J.C., Sinha, A. and Shit, G.C. (2010). Flow of a biomagnetic viscoelastic fluid: application to estimation of blood flow in arteries during electromagnetic hyperthermia, a therapeutic procedure for cancer treatment. App. Math. Mech. 31: 1405–1420.

Peattie, R.A., Riehle, T.J. and Bluth, E.I. (2004). Pulsatile flow in fusiform models of abdominal aortic aneurysms: flow fields, velocity patterns and flow-induced wall stresses. J. Biomech. Eng. 126: 438–446.

Pia, V., Haik, Y. and Bio, C. (1996). Bio-magnetic fluid dynamics. ASME. Publi. FED 237: 761–774.

Sarma, M. (2012). Analysis of blood flow through stenosed vessel under effect of magnetic field. Int. J. Basic Sci. Social Sci. [IJBSS], ISSN: 2319-2968 (Online) 1.

Sheard, G.J. (2009). Flow dynamics and wall shear-stress variation in a fusiform aneurysm. J. Eng. Math. 64: 379–390.

Swillens, A., Lanoye, L., Backer, J.D., Stergiopulos, N., Verdonck, P.R., Vermassen, F. et al. (2008). Effect of an abdominal aortic aneurysm on wave reflection in the aorta. Biomed. Eng. IEEE Transactions on 55: 1602–1611.

Tzirtzilakis, E., Xenos, M., Loukopoulos, V. and Kafoussias, N. (2006). Turbulent biomagnetic fluid flow in a rectangular channel under the action of a localized magnetic field. Int. J. Eng. Sci. 44: 1205–1224.

Tzirtzilakis, E.E. and Xenos, M.A. (2013). Biomagnetic fluid flow in a driven cavity. Meccanica 48: 187–200.

Tzirtzilskis, E. (2008). Biomagnetic fluid flow in a channel with stenosis. Physica D: Nonlinear Phenomena 237: 66–81.

Valerie, D., Yannick, K., Eric, B. and Emmanuel, G. (2007). Flow behaviour in an asymmetric compliant experimental model for abdominal aortic aneurysm. J. Biomech. 40: 2406–2413.

Xu, X. and Lee, J.S. (2009). Application of the lattice Boltzmann method to flow in aneurysm with ring-shaped stent obstacles. Int. J. Num. Methods in Fluids 59: 691–710.

Yip, T.H. and Yu, S. (2001). Cyclic transition to turbulence in rigid abdominal aortic aneurysm models. Fluid Dyn. Res. 29: 81–113.

# Chapter 7

# Computational Fluid Dynamics of Cerebrospinal Fluid Flow

*Isabella Francis,*[1] *Akshoy Ranjan Paul,*[2] *Md. Mamun Molla*[3] and *Suvash C. Saha*[1,*]

## 1. Introduction

Cerebrospinal fluid (CSF) is a colourless transparent liquid protecting the brain and Spinal Cord (SC) from physical and chemical damage. It circulates through the brain's four ventricles, subarachnoid space, and central canal of the spinal cord. The four ventricles include two lateral ones and a third and fourth ventricle (Fig. 1a). The intracranial cavity consists of the brain, cranium, blood, arachnoid, and CSF (Fig. 1b). This CSF fluid is part of the extracellular fluid produced by ependymal cells lining a complex network of capillaries in the ventricles of the brain, called choroid plexuses (Fig. 1c). These contain various substances which are both organic and inorganic: Glucose, lactic acid, proteins, urea, positively charged ions (sodium, potassium, magnesium, calcium), and negatively charged ions (chloride and bicarbonate). The total volume of CSF produced by these choroid plexuses in a healthy adult is around 130 to 150 ml, with 25 to 30 ml in the lateral ventricles, 2 to 3 ml in the third and fourth ventricles, 100 to 125 ml in cranial subarachnoid spaces, and nearly 25 ml of the fluid in the Spinal Subarachnoid Space (SSS) (Condon

[1] School of Mechanical and Mechatronic Engineering, University of Technology Sydney (UTS), 15 Broadway, Ultimo, NSW-2007, Australia.

[2] Department of Applied Mechanics, Motilal Nehru National Institute of Technology Allahabad, Prayagraj-211004, Uttar Pradesh, India.

[3] Department of Mathematics & Physics, North South University, Dhaka-1229, Bangladesh.

* Corresponding author: Suvash.Saha@uts.edu.au

DOI: 10.1201/9781003287636-7

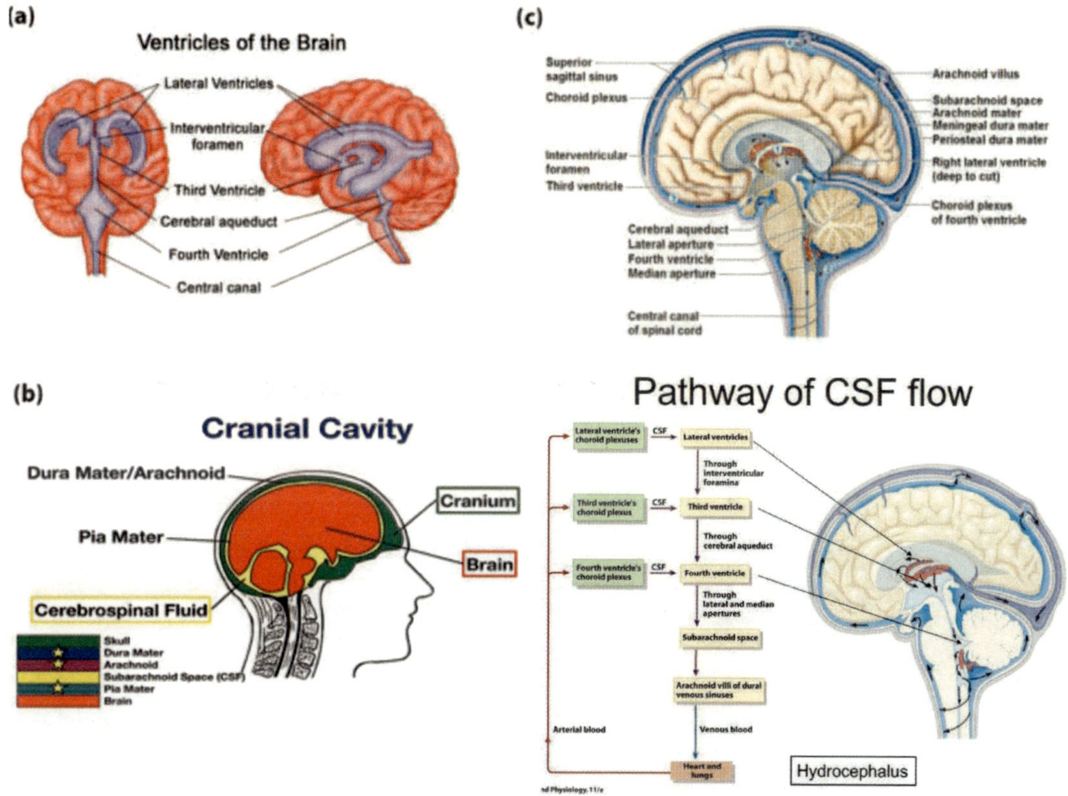


**Figure 1.** (a) Major parts of the brain, including the four ventricles, in a posterior view (left image) and lateral (right image) view, (b) Cranial cavity inside the brain, (c) Detailed parts of the brain, (d) Pathway of CSF flow leaving and entering the brain and the spinal cord, with a schematic displaying the detailed parts mentioned in (c).

et al., 1986; Sakka et al., 2011). CSF has many essential bodily functions. First, CSF ascertains a suitable chemical environment for neuronal signalling by balancing the ions that help facilitate action potentials (Johanson et al., 2008). Second, it provides physical protection by serving as a shock-absorbing medium, preventing the bounce of the brain and the spinal cord (cranial cavity, vertebral canal) (Hamilton et al., 2016). Third, it is involved in the exchange of nutrients between the blood and nerve tissue in the brain's deeper structures (Peate, 2008). The choroid plexus covers the floor of the lateral ventricle and the roof of the third ventricle. CSF flows from the choroid plexus through the body and tail of the lateral ventricle and into the third ventricle via the interventricular foramen (Fig. 1d). It then flows to the fourth ventricle via the cerebral aqueduct, exiting through the lateral and medial apertures and follows two paths: One rises back to the brain. The other flows down the sub-dural space surrounding the spinal cord and returns to the brain, joining CSF flows from the fourth ventricle (Levine, 1999). From there, CSF flows in the subarachnoid space, draining into the superior sagittal sinus. It then flows down the posterior surface and is absorbed into the dural venous sinuses via the arachnoid villi.

## 2. Magnetic Resonance Imaging in CSF Flows

The study of CSF flow has gained increased attention due to the relation between its pathological features and other diseases, such as multiple sclerosis, hydrocephalus, cryptococcal meningitis, and Alzheimer's disease (Silverberg et al., 2006). Another benefit is its capability to administer medicinal treatments directly into the CSF cavity for conditions in cases where they cannot pass the blood-brain barrier.

MRI of real-life patients used in some studies is a non-invasive measurement of the craniospinal dynamics. In other words, MRI accurately captures the *in vivo* brain anatomy (Masoumi et al., 2013) and measures CSF and brain tissue motion within the spine and cranium to analyze the CSF flow field (Alperin et al., 1996). Velocity-encoded/velocimetric MRI can be used to obtain CSF velocities and flow rates at specific locations. The technique that has received the most attention in recent years is single-slice two-dimensional phase-contrast MRI (2D PC MRI) flow measurements in one direction because it allows for *in vivo* measurement of variables like peak CSF flow velocity, pulse wave velocity in the SSS and timing relations between CSF and arterial pulsations (Wagshul et al., 2006). The 2D PC MRI measurements have the disadvantage of having a one-dimensional velocity encoding direction, such as in-plane or through-plane velocity encoding, and a limited spatial coverage that may not accurately reflect the whole CSF compartment of interest. Recent developments in the realm of medical imaging have completely opened up new MRI CSF measurement methodologies. Progressively, it has been believed that time-dependent three-dimensional velocity encoded PC MRI measuring methods (4D PC MRI) are well adapted to estimate complex flow phenomena in CSF-filled regions such as in the brain's ventricles (Stadlbauer et al., 2010) and hydrocephalus patients (Stadlbauer et al., 2012). Due to its three-dimensional information, wider field of view, and relatively tiny voxel size, 4D PC MRI has superior diagnostic usefulness than traditional 2D PC MRI techniques.

Specifically, MRI images are obtained by manual segmentation of the brain's sagittal, coronal, and transverse images using an MRI scanner and software (e.g., ITK Snap). The intricate anatomy of the brain is captured. The cerebellum, fourth ventricle and aqueduct of Sylvius, lateral ventricles and foramen of Monro, pontine cisterns, spinal cord, spinal canal, and the SAS are segmented separately (e.g., SURFdriver, Medviso, FreeSurfer), subsequently registered for improved reconstruction accuracy, and exported in STL (stereolithography) format. After this process, each MRI slice is smoothed and simplified using filtering and reconstruction software (e.g., Rhinoceros, CATIA, MeshLab, Mimics, VMTK, Blender) to remove segmentation errors and be converted into computational models ready to be used by meshing software for grid generation (e.g., ANSYS meshing, ANSYS ICEM, Gambit).

## 3. Computational Fluid Dynamics Modelling

The finite volume/element Computational Fluid Dynamics (CFD) modelling approach of cerebrospinal fluid hydrodynamics has been extensively applied in the past decade, providing more profound knowledge of the Human Ventricular System (HVS) inside the brain and the pathological processes related to CSF. This, in turn, helps diagnose brain diseases such as hydrocephalus, the build-up of fluid in the brain ventricles leading to increased ventricle size and brain pressure. In addition, studying circulatory CSF dynamics is essential as it affects the stem cell differentiation (Benitah, 2012) and gives rise to intracranial pressure (ICP), which varies in people suffering from severe infectious brain diseases. Early studies of CSF developed mathematical lumped parameter models to understand the complexities behind the ICP alterations (Chopp and Portnoy, 1980; Ursino and Lodi, 1997). Over the past decade, computational power has improved significantly, and software simulation has become more popular. Especially in biological flows where experimental results are challenging to obtain, CFD modelling enables the investigation of pressure, shear, flow rate, velocity, vortex, particle and even drug distributions in the respiratory (Francis and Saha, 2022a; Francis and Saha, 2022b; Saha et al., 2022), nervous (Pahlavian et al., 2015), and haemodynamic systems (Saha et al., 2022). Therefore, most CSF studies shifted their CSF flow mathematical modelling to CFD modelling (Jacobson et al., 1996), where accurate simulation results provide valuable information about the flow dynamics in healthy and diseased cases. This, in turn, provides the chance to understand brain diseases by modelling abnormal CSF flows.

### *3.1 CFD Studies in Regular CSF Flows (Healthy Brain)*

During the 1950s, two different theories described the nature of the CSF flow. The first theory was observed and discussed by John Pappenheimer, who considered the flow to follow the bulk flow theory, where blocks at various locations define the path of the CSF. The bulk flow of CSF in the cerebral fluid space was first described as the "third circulation", with the first two being the blood and lymphatic circulation (Cushing, 1925). The second theory by Edgar Bering stated that the CSF flow is a pulsatile flow oscillating with each cardiac cycle, changing direction at various times throughout the domain. For a few decades, the bulk flow theory dominated the pulsatile one until around two decades ago, when results of flow-sensitive Magnetic-Resonance-Imaging (MRI) proved the pulsatile nature of CSF flows (Kurtcuoglu et al., 2005; Linninger et al., 2005).

Nowadays, most studies consider the CSF flow as a pulsatile flow, following that the choroid plexus expands in synchronization with the heartbeat (Ursino, 1988). Early CFD patient-specific models have neglected the cranial subarachnoid space (SAS) and investigated CSF flow dynamics only in the ventricular space (Fin and Grebe, 2003; Jacobson et al., 1999; Kurtcuoglu et al., 2005). This is mainly due to the need for clinical data on the flow field in the SAS and its anatomical complexity, making it challenging to model. The first study tackling the flow field in the SAS in a very simplified and idealized geometry was conducted by Jacobson et al. (1999). Loth et al. (2001) created a physiologically accurate spinal SAS model and elaborately explained the characteristics and features of the CSF flow. Advancing even further to the study of the cranial SAS was Linninger et al. (2005), with a pulsatile CSF flow and a 2D non-realistic model. Some studies employed one-way Fluid-Structure Interaction (FSI) to investigate the effect of the pulsatile CSF flow on brain tissue displacement and stresses (Masoumi et al., 2013). The ventricles are considered as elastic walls for deformability, where the diastolic backflow can be observed during the diastolic time. However, some studies ignored the pulsatile flow and the choroid plexus expansion to simplify the CFD model and assumed constant velocity flow and rigid ventricles. However, two-way FSI is preferred as it represents more realistic phenomenon for many biological flows surrounded by tissues and muscles (Shukla et al., 2020). In addition, some created a 2D model to represent the CSF flow inside the brain, while others employed a 3D model.

For example, Gupta (2009b) used high-resolution MRI images and CFD to investigate patient-specific CSF dynamics in different brain areas, including the fourth ventricle, inferior cranial space, and the superior subarachnoid space (SAS) (Fig. 2). An entire healthy 3D realistic SAS model with porous media of anisotropic permeability was reconstructed from MRI data of a healthy subject. The boundary conditions employed in CFD include a pulsatile flow initiating from three distinct places: Cerebellomedullary Cistern, Pontine Cistern, and SSS. A constant pressure boundary condition was also set at the interface between the aqueduct and the fourth ventricle. Using velocimetric MRI and CFD, the velocity and pressure variation were computed. They reported an 80 Pa pressure increase in one cardiac cycle and a maximum CSF velocity of 15 cm/s in the inferior of the aqueduct. The flow Reynolds number was significant in the pontine cistern indicating a convective flow, and small in the cerebellomedullary cistern showing a relatively diffusive flow.

A more recent study by Fillingham et al. (2022) simulated CSF flow in a whole brain 3D geometry extracted and segmented from MRI brain scans of a healthy individual. Using the commercial CFD software STARCCM+, they revealed that the flow in and out of the ventricles had a little phase delay (about 20% of the cardiac cycle) compared to the flow in the subarachnoid region. They have also shown that the pressure in the subarachnoid area has more substantial transient effects than the pressure in the lateral ventricles.

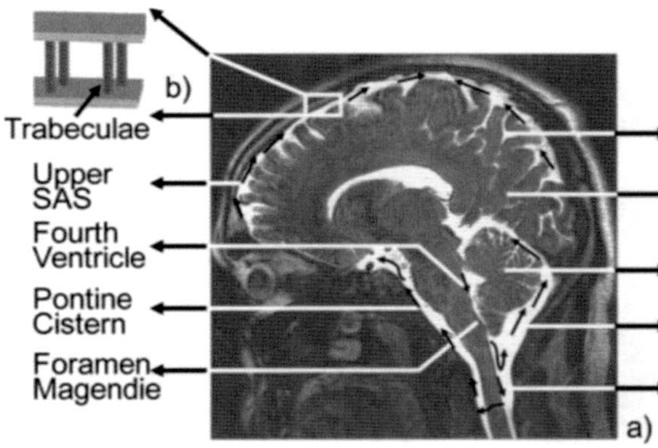


**Figure 2.** Image showing the CSF flow path inside the brain (Gupta et al., 2009b). (a) detailed parts of the brain and upper part of the spine, including the ventricles, brain parenchyma, and spinal SAS, (b) trabeculae, a strand of collagen tissue between the brain and the skull.

## 3.2 CFD Studies in Abnormal CSF Flows (Diseased Brain)

Computational studies with diseased or unhealthy organs of the human body have now become standard practice (Singh et al., 2020; Larpruenrudee et al., 2022) to understand the change in the biological flows and the root cause of the diseases. Hydrocephalus (1 in 2000 births) (Wiswell et al., 1990), Chiari malformations (1 in 3000), and syringomyelia (1 in 8000) (Speer et al., 2003) are devastating neurological diseases caused by abnormal CSF dynamics. Additionally, abnormal CSF flow and altered brain tissue mechanics have become essential factors in the aetiology of Alzheimer's Disease (AD) (Blennow and Zetterberg, 2018; Rasmussen et al., 2018). Researchers are looking into the relationship between CSF dynamics and disease states in terms of objective measures such as CSF velocities, integrated flow waveforms, and bi-directional flow jets. Combining CFD with subject-specific medical imaging of *in vivo* geometry and flow is a potent technique to aid in understanding the potential relationship between CSF dynamics and disease. This approach has been widely employed in vascular haemodynamics (Xie et al., 2022) and is appealing since it can give subject-specific, non-invasive information regarding disease states.

### 3.2.1 Hydrocephalus

Some CFD studies focused on a brain disease related to CSF flows, hydrocephalus, a neurological disorder where the CSF flow is disturbed, and the ICP is increased. The abnormal build-up of CSF within the brain (Fig. 3) resulting from an imbalance between its production and circulation causes hydrocephalus (Vardakis et al., 2013). Therefore, proper movement of CSF is essential for regular production and absorption as the CSF flow passes through the brain passages and its four ventricles. The abnormal and disturbed movement of CSF caused by stenosis of the cerebral aqueduct leads to edema (an increase of fluid in the extracellular space of the brain) and the distention, displacement, and distortion of the cerebral ventricles and the

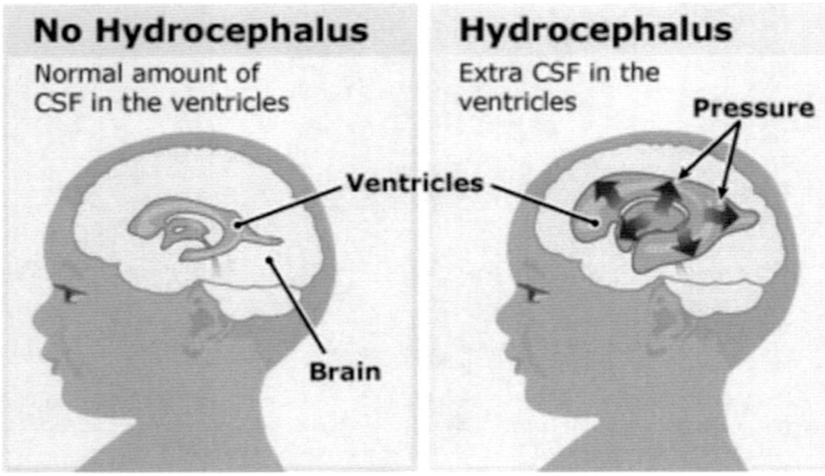


**Figure 3.** Comparison between a healthy brain and a hydrocephalic brain in the amount of CSF in the ventricles and its resulting increased pressure on the brain tissue.

brain tissue. Experimental/clinical methods studying hydrocephalus are complicated and lack accurate measurements of biomechanical loadings and local flow patterns (Vinchon et al., 2012). Therefore, computer simulations/CFD are required to gain more insight into the nature of the disease. For example, Tully and Ventikos (2009) coupled the poroelasticity of the brain tissue with CFD to study the hydrodynamics of CSF. This study uses a simplified spherically symmetric brain geometry with physiologically accurate dimensions of the ventricles, skull, subarachnoid space, and stenosed aqueduct. Using the finite volume Multiphysics software CFD-ACE+ (ESI group), the poroelasticity equations were coupled with flow solvers. Results proved the importance of this coupling as the ventricular enlargement in acute hydrocephalus cases due to the local aqueduct stenosis patterns is significantly smaller than estimated.

Treatment of hydrocephalus patients can be done in several ways. One way is called Endoscopic Third Ventriculostomy (ETV), which involves introducing an endoscope through the skull and creating a small opening in the bottom of the third ventricle. This hole in the brain allows CSF to bypass any obstruction to its outflow. Another treatment is a procedure called shunting. This process entails the placement of two catheters and a flow control valve system to transfer extra cerebrospinal fluid (CSF) from the subarachnoid space or the ventricles of the brain to an area of the body where it can be absorbed. Several CFD studies conducted by Gholampour et al. investigated the relationship between clinical symptoms of individuals with hydrocephalus before and after shunting (Gholampour, 2018b; Gholampour et al., 2017). Recently, Gholampour and Fatouraee (2021) aimed to evaluate the difference in CSF dynamic simulation in hydrocephalus patients when applying three different inlet and outlet boundary conditions. Using MRI data from 8 healthy people and 11 hydrocephalus patients, three-dimensional geometrical models were constructed to evaluate the intracranial compliance of the hydrocephalus patients. The maximum CSF pressure and the ventricular system volume were the most effective and accurate parameters in evaluating the patients' conditions.

### *3.2.2 Chiari I Malformation (CMI)*

CMI is a congenital craniospinal disorder in which a downward displacement of brain tissues into the spinal canal occurs. In particular, the cerebellar tonsils, the lower portion of the cerebellum, block the Foramen Magnum (FM), the aperture at the base of the skull (Fig. 4), causing spinal fluid to back up into the spinal canal. In some circumstances, the CSF may accumulate inside the spinal cord due to the spinal canal being blocked by CMI. The SC develops a syrinx-like cavity (fluid-filled cyst) when the CSF concentration rises, which causes the nerve fibres to stretch and finally become damaged. Patients with CMI may experience a range of mild to severe neurological symptoms and pain. These symptoms are hypothesized to be caused by aberrant pressure on the spinal cord, brain stem, or cerebellum due to tissue crowding at the craniovertebral junction (Tubbs et al., 2009).

A static anatomic MRI of the brain and spine can assess whether a patient has CMI and how chronic it is. Researchers discovered variations in CSF pressure, flow resistance, and velocities between pre-surgery CMI patients, healthy subjects, and post-surgery CMI patients. Dynamic elements of the CSF biomechanical environment, such as resistance to flow and compliance of the craniospinal system, are not observable by the conventional, straightforward anatomic MRI diagnostic approaches. Hydrodynamic measurements can help doctors better predict which patients may experience increased symptoms and respond best to treatment. Therefore, a potent tool to better reflect the pathophysiology of the disease and function simultaneously as a prognostic predictor of CMI is the combination of thorough measurement with modern *in vivo* techniques and engineering study of CSF hydrodynamics with the use of three-dimensional CFD.

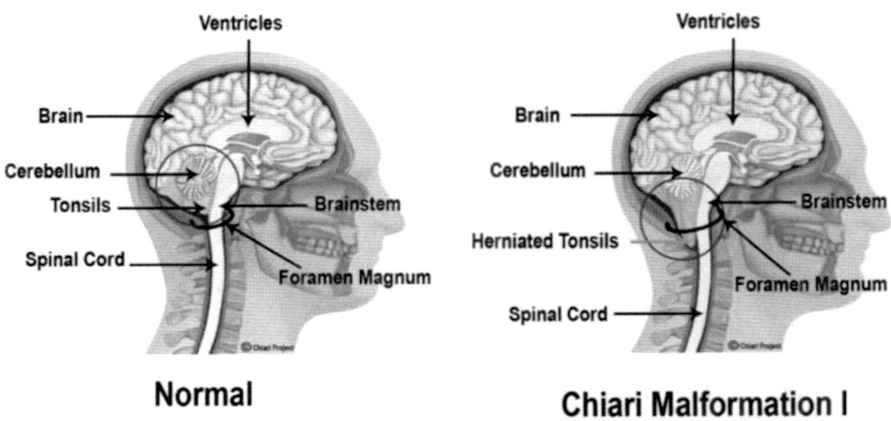


**Figure 4.** Comparison between a healthy brain and a brain with Chiari Malformation I with the tonsils blocking the foramen magnum.

For example, Clarke et al. (2013) applied CFD to analyze the CSF pressure in a diseased state associated with CMI called syringomyelia, where fluid-filled cysts develop in the spinal cord. This disease has proven to cause alterations in the CSF flow velocity leading to brain enlargement. This study developed CFD geometry models from MRI data for a healthy participant, a CM patient with syringomyelia and a CM patient without syringomyelia. Results showed that the diseased cases have much higher peak CSF pressure values than the standard healthy case. Masoumi et al. (2013) used MRI images of a patient to compare between two models, using rigid and flexible walls through fluid-solid interaction to obtain the ventricular displacement

**Table 1.** CFD-CSF studies in literature for diseased/injured brain: Hydrocephalus (HP), Chiari Malformation I (CMI), Subarachnoid haemorrhage (SAH).

| **Studies** | **2D/3D** | **Constant/ Pulsatile flow** | **Rigid/ Deformable tissue** | **Main results** | **CFD software** |
|---|---|---|---|---|---|
| (Tully and Ventikos, 2009) | 3D | Pulsatile | Rigid and Deformable | Ventricular enlargement significantly smaller than estimated in HP | CFD-ACE+ |
| (Clarke et al., 2013) | 3D | Constant | Rigid | Much higher peak CSF pressure values in diseased CMI case. | ANSYS Fluent |
| (Masoumi et al., 2013) | 2D | Pulsatile | Rigid and Deformable | Velocity and pressure drop decrease by two-fold in FSI model in HP | ADINA |
| (Pahlavian et al., 2015) | 3D | Pulsatile | Rigid with moving boundary conditions | Central nervous system tissue motion increased CSF pressure of a CMI patient | ANSYS Fluent |
| (Farnoush et al., 2016) | 3D | Pulsatile | Rigid | ETV treatment for HP reduces the ventricular pressure by nine-fold | ANSYS CFX |
| (Thalakotunage and Sethaput, 2016) | 3D | Pulsatile | Rigid | Peak velocity is much higher in HP patients at the narrowest point of the aqueduct | ANSYS CFX |
| (Lloyd et al., 2017) | 3D | Pulsatile | Rigid | Peak pressures in CMI patients without a syrinx were higher than in healthy subjects | ANSYS CFX |
| (Gholampour, 2018a) | 3D | Pulsatile | Deformable | Increase in a hydrocephalic patient's intracranial compliance after shunting | ADINA |
| (Khani et al., 2020) | 3D | Pulsatile | Rigid | Neurapheresis system for SAH patients increases average steady streaming CSF velocity | ANSYS Fluent |
| (Khani et al., 2020) | 3D | Pulsatile | Rigid | Neurapheresis therapy for SAH patients is most efficient for blood removal from CSF flow | ANSYS Fluent |
| (Gholampour and Fatouraee, 2021) | 3D | Pulsatile | Deformable | Maximum CSF pressure is most important for evaluating HP patients' conditions | ADINA |

and stress and analyze the effect of the brain parenchyma on the diastolic backflow, CSF velocities and pressure drop, and ICP gradients. They considered a realistic pulsatile laminar flow to accurately compare the CSF flow velocity and pressure to experimental results. Their simulation results matched the experimental MRI results of Hoffman et al. (2000). A more recent study by Lloyd et al. (2017) has shown that patients with a syrinx have pressure versus time profiles that are more similar to healthy controls. However, patients with a Chiari Malformation I (CMI) have peak pressures that occur earlier and are much higher than those from healthy subjects.

Some brain injuries lead to brain diseases. For example, a Subarachnoid Haemorrhage (SAH) is a severe life-threatening stroke caused by bleeding into the space surrounding the brain. After SAH, hydrocephalus may occur where CSF accumulates, increasing the pressure on the brain. Table 1 lists some studies in literature tackling injured/diseased brain cases and mentions the methods/simplifications used along with the leading results found.

## 4. 4D PC MRI versus CFD of CSF in Cervical SSS

Examining the brain and spinal cord morphology alone cannot explain patient symptoms and surgical results. Thus, diagnosing and evaluating craniospinal diseases such as CMI requires understanding CSF dynamics in the cervical Spinal Subarachnoid Space (SSS). There has been a growing interest in 4D PC MRI for its ability to analyze CSF flow dynamics quantitatively and qualitatively and shed light on complex flow phenomenon, such as secondary flow and vortex strength associated with craniospinal disorders. When examining CSF flow in CMI patients with and without a syrinx, for instance, Bunck et al. (2012) discovered that 4D PC MRI allowed for the detection of higher CSF peak velocities than single-plane 2D PC MRI data.

On the other hand, several CFD studies have tackled CSF motion in the cervical SSS with or without craniospinal disorders. Some of these studies have used CFD with anisotropic porous media for the tissue (Gupta et al., 2009a), a combination of CFD and Lattice Boltzmann method (Stockman, 2007), numerical model/wave propagation technique (Bertram, 2009), with/without tissue motion, arachnoid trabeculae (strands of connective tissue between the brain and skull), and nerve roots. The first CFD simulation of the CSF motion in a stiff wall SSS was performed by Loth et al. (2001). The simulated geometry did not incorporate small anatomical components like the SC nerve roots, denticulate ligaments (anchoring ligaments of the spinal cord), or arachnoid trabeculae. Stockman (2007) looked at how minute the anatomical small structures affected the CSF motion and behaviour and discovered that when the spacing was symmetrical around the SC, the delicate structures' existence had no impact on the velocity profiles. Following this research, it has been widely accepted that tiny structures in the SSS have no discernible influence on CSF velocity profiles at the macroscale. Based on MRI results, Roldan et al. (2009) created rigid physiologically realistic SSS models without tiny features and simulated the motion of the CSF flow. The findings showed anterolateral flow jets in the CSF that were diverse around the SC.

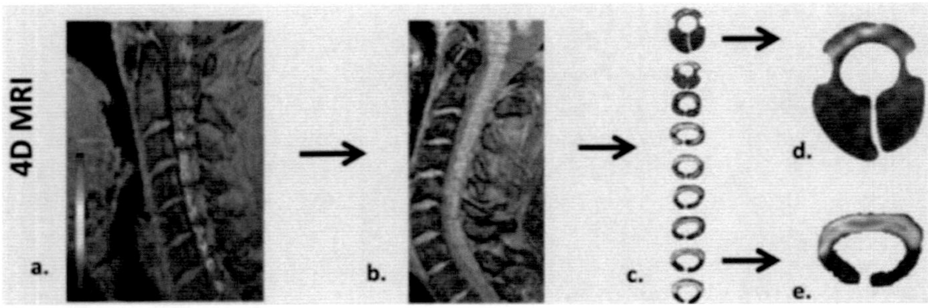


**Figure 5.** 4D MRI of cervical SSS in a healthy subject (Yiallourou et al., 2012). (a) velocity vectors, (b) axial planes orthogonal to the cervical SSS, (c) velocity contours on the planes, and (d) zoomed-in velocity contours.

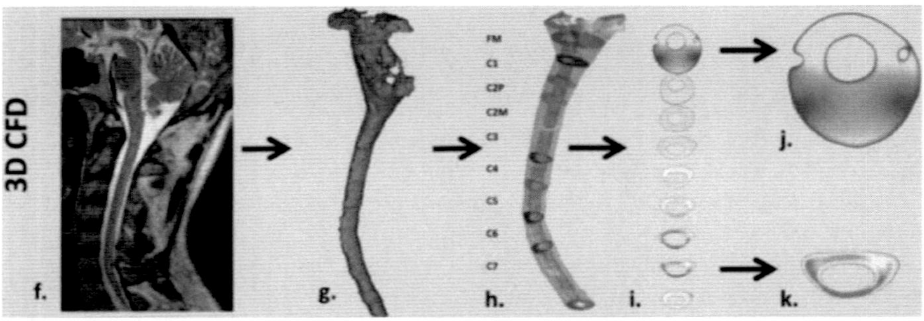


**Figure 6.** 3D CFD analysis of cervical SSS in a healthy subject (Yiallourou et al., 2012). (f) MRI scan of high resolution for CFD geometry generation, (g) three-dimensional geometry of cervical SSS with geometric smoothing, (h) axial planes orthogonal to the cervical SSS, (i) velocity contours on the axial planes, (j) (k) zoomed-in view of velocity contours on two axial planes.

Yiallourou et al. (2012) utilized 4D MRI with the GTflow software to capture the accurate anatomical structure of a patient-specific rigid cervical SSS (Fig. 5a). Numerous orthogonal axial planes were manually defined on the spinal axis (Fig. 5b). According to the findings (Fig. 5c, d, and e), the axial planes showing high velocity magnitudes caused by the vascular flow were frequently seen near the FM region close to the left and right vertebral arteries. This study compared the results and measurements of the 4D PC MRI with CFD results of velocity contours on the same axial planes of a 3D cervical SSS (Fig. 6f to k). The average peak velocity magnitudes obtained by 4D PC MRI for healthy and CM participants at the FM were consistently more significant than those predicted by CFD simulations. Concerning the peak velocity magnitudes or peak pulsatile velocity profiles across a wide span of CSF flow parameters in relation to disease, sex, and age, the 4D PC MRI measurements did not show comparable CSF dynamics with the CFD results. They proposed that the disparities could be explained by the noise in the 4D PC MRI measurements and the omission of tiny structures and tissue motion in the cervical SSS in the CFD simulation.

## 5. *In Vitro* Experimental Models

*In vitro* flow models based on 2D PC MRI readings and anatomical data have been created to replace *in vivo* complicated techniques that might adversely affect patients' health. The computational methods and approaches used to understand CSF flows paved the way for *in vitro* model fabrication, analysis, and CFD validation. For many biological flows around and inside organs, *in vitro* models are being designed and created by fabricating microchips called organ-on-a-chip models (Francis et al., 2022), such as lung-on-a-chip, liver-on-a-chip, heart-on-a-chip, stomach-on-a-chip, muscle-on-a-chip, blood vessels-on-a-chip, kidney-on-a-chip. Primary and cancer cells from patients are used to mimic exact *in vivo* flow and cell conditions. *In vitro* experiments for CSF flows were performed by Martin et al. (Martin and Loth, 2009; Martin et al., 2010) to investigate the effects of spinal stenosis on CSF flow, which happens when the space inside the backbone is too small. This can pressure the spinal cord and nerves that travel through the spine. They have also considered the presence of a syrinx and analyzed its effects on spinal CSF dynamics. Bottan et al. (2012) created a 3D non-realistic model of intracranial pressure and CSF dynamics. A more recent study by Khani et al. (2020) performed *in vitro* experiments with fluoresce in tracers on a developed subject-specific CFD model from MRI scans to evaluate the efficiency of neurapheresis therapy, rapid extraction and removal of pathogens and Red Blood Cells (RBCs), thus filtering the CSF. The *in vitro* studies highlighted the significance of the mechanical characteristics of the neural tissue, including compliance and permeability, as well as the intricate FSI between the neural tissue and the obstruction to CSF flow.

## 6. Types of CFD Software Used for Simulation

Several CFD programs simulate the CSF flow inside the brain by numerically solving the Navier-Stokes equations. These programs follow the Finite Element Method (FEM) or the Finite Volume Method (FVM). These methods are highly adapted to handle geometrically challenging domains, such as the SAS, since they break the problem domain into more manageable/accessible portions, known as finite elements/volumes. Then, they are utilized for the spatial discretization of the studied geometries. The main difference between the two methods is that the FVM calculates accurate expressions for the solution's average value over a volume and then uses this information to build cell-based approximations of the answer.

On the other hand, the FEM calculates local approximations of a solution using local data and then combines them to form a global approximation. One of the most popular computer programs includes a combined meshing-CFD finite element software, Automatic Dynamic Incremental Nonlinear Analysis (ADINA), which can be used to mesh the required brain model and numerically solve the mathematical Navier-Stokes equations for viscous and incompressible CSF fluid flow to obtain desired results. The software STARCCM+, CFD-ACE+, and ANSYS finite volume FLUENT/CFX software are also popular in modelling CSF fluid flow. In the case of 1-way FSI between the brain parenchyma and the CSF, ANSYS finite element Transient Structural is used to couple the fluid and solid domains. Another

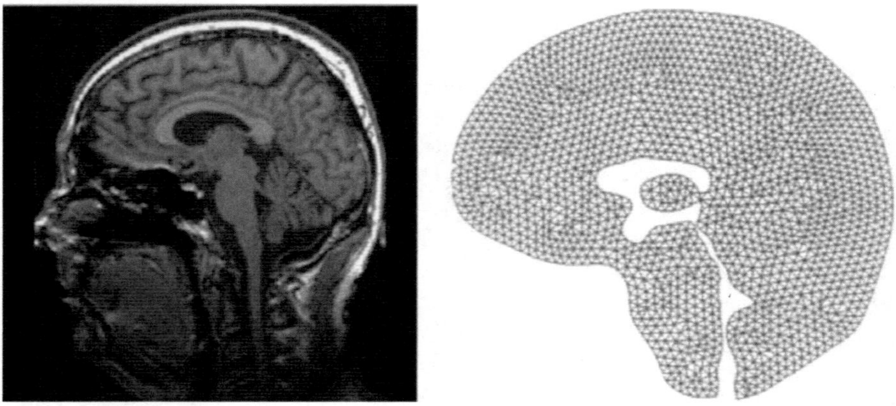

**Figure 7.** MRI brain scan of a volunteer (left), triangular meshing of the brain tissue of a 2D model extracted from MRI scan (right) (Masoumi et al., 2013).

software is GTflow, which is used for data processing of 4D PC MR data sets, flow visualization, and quantification. For the meshing part, triangular (or quadrilateral or hexagonal) face elements are used for 2D models for the fluid, and solid components, as can be seen from the right image of Fig. 7 (Masoumi et al., 2013), and tetrahedral (or polyhedral) volume elements with inflation layers are used for 3D models. Due to intricate geometrical features, fine mesh near the walls are important to capture wall turbulence in case of biological flows (Srivastav et al., 2019b). Hence, grid refinement near walls and use of suitable wall functions are required.

## 7. Numerical Simulation

The CSF flow environment can be non-invasively examined in healthy and diseased patients cases using CFD simulations of the flow field. The FEM or FVM are employed to analyze hydrodynamic variables such as pressure and velocity. The CSF fluid is generally considered and simulated as an incompressible fluid with a Newtonian behaviour, having the physical properties of water at average body temperature with a constant dynamic viscosity (around 0.001 Pa.s) and constant density (about 1000 kg/m$^3$). The CSF flow is considered laminar due to a small Reynolds number $Re = \frac{\rho D_h V}{\mu}$, where V is the velocity (taken either as a constant of about 0.1 m/s or as a pulsatile flow with changing values throughout the cardiac cycle), and $D_h$ is the hydraulic diameter, $D_h = \frac{4Ac}{Pwet}$, with $Ac$ being the cross-sectional area and $Pwet$ being the wetted perimeter where the CSF fluid is wetting the surface. For the cranial CSF flow models, three velocity inlets should be imposed at three distinct places: cerebellomedullary cistern, pontine cistern, and SSS. For the cervical spinal canal CSF flow models, one velocity inlet should be imposed at the bottom of the model (caudal end), directing CSF upwards to the cranial end (outlet) as the CSF returns from the spine to the brain. A no-slip boundary condition should be imposed at the walls with a pressure outlet boundary condition having a gauge pressure of 0 Pa, thus modelling at atmospheric pressure. A transient simulation runs for at least one cycle with small time step sizes of about 0.001 to 0.005 s.

The efficacy of the various numerical schemes/algorithms available in a CFD package eventually affects the solution of CFD simulation (Kumar et al., 2019). Hence, careful selection of numerical schemes/algorithms is essential. The Navier-Stokes equations for the incompressible CSF flow are solved with a coupled pressure-velocity solver. For example, the SIMPLE and the SIMPLE-Consistent (SIMPLEC) algorithms ensure mass conservation by solving velocity and pressure correction equations separately. The difference between the two algorithms lies in the expression for the face flux correction. Another commonly used solver is the pressure implicit with the Splitting of Operators Scheme, PISO, which improves the efficiency of the calculation by performing mesh neighbour and skewness corrections. In addition, the COUPLED method provides a replacement for the pressure-based segregated scheme with the pressure-velocity coupling of the SIMPLE type. When the mesh quality is poor or huge time steps are applied to transient flows, the COUPLED algorithm must be used where the pressure-based continuity and momentum equations are solved together. Concerning the spatial and time schemes, the most commonly used are the second-order upwind for momentum discretization and PRESTO (Pressure Staggering Options) for pressure discretization. The convergence criteria for velocity, continuity, and momentum are usually set to 1e-06 s.

For the studies with MRI velocity and pressure monitoring points manually placed in the model, validation is performed by comparing the magnitudes and time of peak velocity between the simulated and the MRI-measured profiles. Similarly, CFD models are validated by *in vitro* models by obtaining quantitative data of the results and performing linear regression analysis. P-tests and t-tests that indicate the significant differences between the two compared data can also be used. Most studies computationally observed and evaluated CSF dynamics in the brain by generating qualitative data of flow streamlines representing CSF motion, coloured by CSF velocity magnitude (cm/s) (Fig. 8) (Gholampour, 2018a). The CSF's maximum velocity reaches 3.1 cm/s at the narrowest cross-section. Pressure contours at different planes in the model and the whole brain are also obtained where higher pressure values are present in the middle where the ventricles reside, and the CSF flow is produced (Fig. 9) (Gholampour and Fatouraee, 2021).

In addition, quantitative data of average and peak values have been collected, and dimensionless numbers (e.g., Womersley number) are calculated in pulsatile fluid flow studies to describe the unsteady nature of fluid flow in response to an unsteady pressure gradient. The Womersley number ($\alpha$) is an expression relating the pulsatile flow frequency to viscous effects. It is denoted by $\alpha = \dfrac{\rho\omega U}{\mu U L^{-2}} = \dfrac{\omega L^2}{v}$, where L is an appropriate length scale, $\omega$ is the angular frequency of the oscillations, and $\rho$, $v$, $\mu$ are the CSF density, kinematic viscosity, and dynamic viscosity, respectively.

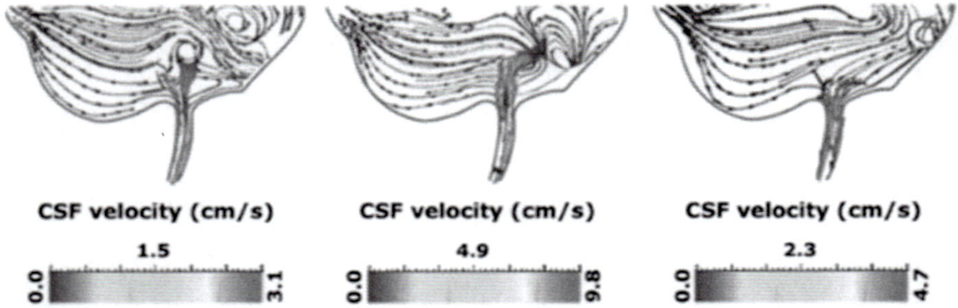


**Figure 8.** Streamlines of CSF velocity magnitude in the inferior section of the third ventricle at the mid-systole (Gholampour, 2018a).

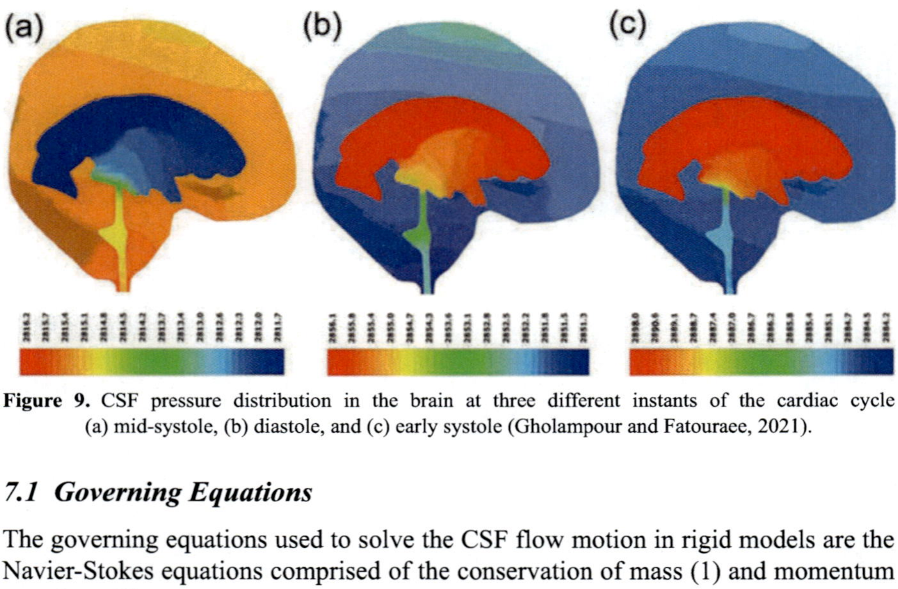


**Figure 9.** CSF pressure distribution in the brain at three different instants of the cardiac cycle (a) mid-systole, (b) diastole, and (c) early systole (Gholampour and Fatouraee, 2021).

## *7.1 Governing Equations*

The governing equations used to solve the CSF flow motion in rigid models are the Navier-Stokes equations comprised of the conservation of mass (1) and momentum equations (2).

$$\rho(\nabla \cdot \boldsymbol{u}) = \dot{S}_m \tag{1}$$

$$\rho\left(\frac{\partial \boldsymbol{u}}{dt} + \boldsymbol{u} \cdot \nabla \boldsymbol{u}\right) = -\nabla p + \mu \nabla^2 \boldsymbol{u} + \boldsymbol{S_P} \tag{2}$$

where $\boldsymbol{u}$ is the flow velocity vector, $\rho$ is the density, $p$ is the pressure, $\mu$ is the viscosity, $\dot{S}_m$ is the rate of CSF generation per unit volume in the fourth ventricle, and $\boldsymbol{S_P}$ is the pressure gradient due to the morphology of the SAS.

## *7.2 Fluid-Structure Interaction Equations*

The term "Fluid-Structure Interaction" (FSI) refers to how a fluid flow interacts with a moving or deformable structure. By applying FSI between the CSF fluid and the solid parts of the brain, some researchers have looked at and reproduced the CSF space displacement and spinal tissue displacement. These models emphasized the significance of the mechanical characteristics of brain tissue, including compliance and permeability, as well as the intricate FSI involving the neural tissue and CSF flow blockage. The FSI between the CSF and the brain tissue is a 1-way coupling where a portion of the fluid's kinetic energy is consumed by the elastic ventricular walls, leading to the deformation of the brain tissue and the storage of the energy as potential energy. This is the main reason behind the decrease in velocity in FSI models compared to rigid models.

The continuity equation (3) for the fluid flow with a moving mesh interface is the following:

$$\rho_F \frac{\partial \boldsymbol{u_F}}{\partial t} + \rho_F \left( (\boldsymbol{u_F} - \boldsymbol{W}) \cdot \nabla \right) u_F = -\nabla p + \mu \nabla^2 \boldsymbol{u_F} + \boldsymbol{f_F^B} \tag{3}$$

The terms $W$, $p$, $f_F^B$, and $\mu$ are the moving mesh velocity vector, CSF pressure, body force per unit volume, and CSF dynamic viscosity, respectively. The terms $u_F$, $\rho_F$, and $(u_F - W)$ are the CSF velocity vector, density, and relative fluid velocity concerning the moving coordinate velocity.

### *7.2.1 Solid (Brain Tissue) Formulation*

In most studies, the brain tissue was modelled as a linear material, specifying Young's modulus ($E$) and Poisson ratio (ν). The arbitrary Lagrangian-Eulerian formulation is applied having the following equation (4):

$$\frac{\partial \tau_{ij}^s}{\partial x_j} = \rho^s \frac{\partial^2 d_i^s}{\partial t^2} \tag{4}$$

where $\tau_{ij}^s$ represents the Cauchy stress tensor, $d_i^s$ represents the solid displacement component, and $\rho^s$ represents the solid tissue density.

The kinetic coupling equations (5) and (6), which denote the no-slip situations at the boundary, are the following:

$$d_i^s = d_i^f \tag{5}$$

$$\frac{d_i^s}{\partial t} = \frac{d_i^f}{\partial t} \tag{6}$$

where $d$ is the displacement, and the superscripts s and f represent the brain tissue and CSF, respectively.

The kinetic coupling situation comes from the equilibrium of forces between the CSF (fluid) and the brain tissue (solid) at the border interface wall (7):

$$\sigma_{ij}^s \cdot \boldsymbol{n}_j = \sigma_{ij}^f \cdot \boldsymbol{n}_j \tag{7}$$

where $\sigma_{ij}$ is the stress tensor, $\boldsymbol{n}$ is the unit vector ordinary to the boundary of the CSF solution and solid regions. Considering a no-slip condition at the brain wall, the flow velocity is $\boldsymbol{u_F} = 0$.

## 8. Limitations/Simplifications of Few CSF CFD Studies

Most CSF models created to date have either excluded the entire brain or relied on assumptions about the anatomy, flow boundary conditions, characteristics and material properties of brain tissue. The primary limitations/simplifications of various studies other than the ones already mentioned in Table 1 (constant velocity, 2D model, and rigid model) are the following:

1. Not modelling the effect of CSF absorption into arachnoid villi
2. Considering the CSF as a Newtonian solution (water) with constant density and viscosity
3. Modelling the brain tissue as linear elastic material rather than viscoelastic or non-linear hyperelastic
4. Deformation of the spinal cord due to the pressure of the solid boundary is considered negligible
5. Lack of a large study population as studies consider patient-specific or idealized models
6. Not including secondary motion caused by tissue-to-tissue contact.

These limitations/simplifications are related to some drawbacks that CFD simulations for CSF flows have. While CFD has proved its ability to represent and mimic real-life physical problems and solutions, there are generally known disadvantages of CFD. These include numerical errors such as round-off and interpolation errors, difficulty applying correct initial and boundary conditions, and requiring simplifications and assumptions. Specifically, applying CFD to simulate CSF flows inside the brain and spinal cord has some drawbacks and challenges. A critical consideration in CFD FSI modelling of a patient-specific geometry is the complex and computationally expensive simulations that require new moving geometry for the walls. In addition, the numerical 3D approximation of the CSF flow field depends on the boundary conditions obtained from the *in vivo* anatomical and flow data. However, *in vivo*, measurements have only supplied a limited amount of information on the effect of the delicate structures inside the SSS on the CSF flow field due to the spatial resolution restriction of the current PC MRI techniques. Even with the most advanced MRI techniques now available, the anatomical size of these structures is too tiny to estimate. Especially in diseased brain cases, although there has been much progress in estimating brain mechanics and understanding the physiology of CSF flow, it is still challenging to accurately characterize the compliance, deformation, and resultant CSF flow in a given patient. The lack of *in vivo* accurate information continues to be a barrier to prospective therapeutic approaches and diagnostic evaluations for hydrocephalus, AD, and associated dementias. This presents a glaring gap in CFD modelling and highlights the necessity to collect reliable *in vivo* measurements and create a patient-specific methodology to evaluate the viscoelastic characteristics of brain tissue and the spatiotemporal evolution of CSF velocities in both healthy and pathological situations.

## 9. CFD Future Direction of CSF Flow

CFD simulations of CSF flow have demonstrated tremendous potential in its use by giving a chance to study and understand CSF motion and behaviour in normal healthy brains. In addition, it paves the way for developing novel therapeutic approaches/interventions by predicting CSF flow behaviour in an injured or diseased brain. The extraction of a CAD model of any biological organ from CT-scan/MRI images is a herculean task as it involves the removal of insignificant tissues while considering the intricacies of the geometrical features of the flow boundaries (Srivastav et al., 2013). Several CFD limitations and drawbacks (section 8) still force the application of simplifications regarding the brain and spinal cord model material, geometry, and boundary conditions to this date. Moreover, in general, the spatial and temporal discretization of the flow domain, the iterative solver, the level of convergence criteria, the mesh quality, the initial and boundary conditions, and other features of the computational setup might all have problems that may affect obtaining accurate, realistic CSF flow results. Therefore, the significant advance in computationally simulating accurate quantitative and qualitative behaviour of CSF flows will only be possible with technological advancements in software and hardware that will allow for the quick calculation of large systems of equations and the reconstruction of anatomically precise patient data from high-resolution imaging. Furthermore, doctors and scientists should give initial and boundary conditions in CSF physiological and pathological states precise and trustworthy values. As a final step, mathematicians and engineers should execute strict validation and verification procedures of proper mathematical models to increase the reliability of the CSF CFD results.

## Acknowledgments

The authors would like to acknowledge the University of Technology Sydney (UTS) library for providing access to journal articles that helped shape this chapter.

## Nomenclature

| | |
|---|---|
| CSF | Cerebrospinal fluid |
| HVS | Human Ventricular System |
| ICP | Intracranial pressure |
| CFD | Computational fluid dynamics |
| MRI | Magnetic resonance imaging |
| PC | Phase Contrast |
| ADINA | Automatic Dynamic Incremental Nonlinear Analysis |
| CMI | Chiari I malformation |
| AD | Alzheimer's disease |
| SAS | Cranial Subarachnoid space |
| SC | Spinal Cord |
| SSS | Spinal Subarachnoid space |
| FM | Foramen Magnum |

| | |
|---|---|
| FSI | Fluid-structure interaction |
| CNS | Central Nervous System |
| CATIA | Computer-Aided Three-Dimensional Interactive Application |
| ICEM | Integrated Computer-aided Engineering and Manufacturing |
| GAMBIT | Geometry and Mesh Building Intelligent Tool |
| VMTK | Vascular Modeling Toolkit |
| FVM | Finite volume method |
| FEM | Finite element method |

## References

Alperin, N., Vikingstad, E.M., Gomez-Anson, B. and Levin, D.N. (1996). Hemodynamically independent analysis of cerebrospinal fluid and brain motion observed with dynamic phase contrast MRI. Magnetic Resonance in Medicine 35(5): 741–754.

Benitah, S.A. (2012). Defining an epidermal stem cell epigenetic network. Nature Cell Biology 14(7): 652–653.

Bertram, C.D. (2009). A numerical investigation of waves propagating in the spinal cord and subarachnoid space in the presence of a syrinx. Journal of Fluids and Structures 25(7): 1189–1205.

Blennow, K. and Zetterberg, H. (2018). Biomarkers for Alzheimer's disease: current status and prospects for the future. Journal of Internal Medicine 284(6): 643–663.

Bottan, S., Poulikakos, D. and Kurtcuoglu, V. (2012). Phantom model of physiologic intracranial pressure and cerebrospinal fluid dynamics. IEEE Transactions on Biomedical Engineering 59(6): 1532–1538.

Bunck, A.C., Kroeger, J.R., Juettner, A., Brentrup, A., Fiedler, B., Crelier, G.R. et al. (2012). Magnetic resonance 4D flow analysis of cerebrospinal fluid dynamics in Chiari I malformation with and without syringomyelia. European Radiology 22(9): 1860–1870.

Chopp, M. and Portnoy, H.D. (1980). Systems analysis of intracranial pressure: comparison with volume-pressure test and CSF-pulse amplitude analysis. Journal of Neurosurgery 53(4): 516–527.

Clarke, E.C., Fletcher, D.F., Stoodley, M.A. and Bilston, L.E. (2013). Computational fluid dynamics modelling of cerebrospinal fluid pressure in Chiari malformation and syringomyelia. Journal of Biomechanics 46(11): 1801–1809.

Condon, B., Wyper, D., Grant, R., Patterson, J., Hadley, D., Teasdale, G. et al. (1986). Use of magnetic resonance imaging to measure intracranial cerebrospinal fluid volume. The Lancet 327(8494): 1355–1357.

Cushing, H. (1925). The third circulation and its channels. Lancet 2(851-857): 547.

Farnoush, A., Tan, K., Juge, L., Bilston, L.E. and Cheng, S. (2016). Effect of endoscopic third ventriculostomy on cerebrospinal fluid pressure in the cerebral ventricles. Journal of Clinical Neuroscience 23: 63–67.

Fillingham, P., Levendovszky, S.R., Andre, J., Parsey, C., Bindschadler, M., Friedman, S. et al. (2022). Patient-specific computational fluid dynamic simulation of cerebrospinal fluid flow in the intracranial space. Brain Research 147962.

Fin, L. and Grebe, R. (2003). Three dimensional modeling of the cerebrospinal fluid dynamics and brain interactions in the aqueduct of sylvius. Computer Methods in Biomechanics and Biomedical Engineering 6(3): 163–170.

Francis, I. and Saha, S.C. (2022a). Surface tension effects on flow dynamics and alveolar mechanics in the acinar region of human lung. Heliyon e11026.

Francis, I. and Saha, S.C. (2022b). Computational fluid dynamics and machine learning algorithms analysis of striking particle velocity magnitude, particle diameter, and impact time inside an acinar region of human lung. Physics of Fluids.

Francis, I., Shrestha, J., Paudel, K.R., Hansbro, P.M., Warkiani, M.E. and Saha, S.C. (2022). Recent advances in lung-on-a-chip models. Drug Discovery Today.

Gholampour, S. (2018a). FSI simulation of CSF hydrodynamic changes in a large population of non-communicating hydrocephalus patients during treatment process with regard to their clinical symptoms. PloS One 13(4): e0196216.

Gholampour, S. (2018b). FSI simulation of CSF hydrodynamic changes in a large population of non-communicating hydrocephalus patients during treatment process with regard to their clinical symptoms. PloS One 13(4): e0196216.

Gholampour, S. and Fatouraee, N. (2021). Boundary conditions investigation to improve computer simulation of cerebrospinal fluid dynamics in hydrocephalus patients. Communications Biology 4(1): 1–15.

Gholampour, S., Fatouraee, N., Seddighi, A.S. and Seddighi, A. (2017). Evaluating the effect of hydrocephalus cause on the manner of changes in the effective parameters and clinical symptoms of the disease. Journal of Clinical Neuroscience 35: 50–55.

Gupta, S., Soellinger, M., Boesiger, P., Poulikakos, D. and Kurtcuoglu, V. (2009a). Three-dimensional computational modeling of subject-specific cerebrospinal fluid flow in the subarachnoid space. Journal of Biomechanical Engineering 131(2).

Gupta, S., Soellinger, M., Boesiger, P., Poulikakos, D. and Kurtcuoglu, V. (2009b). Three-dimensional computational modeling of subject-specific cerebrospinal fluid flow in the subarachnoid space. Journal of Biomechanical Engineering 131(2).

Hamilton, M., Gruen, J.P. and Luciano, M.G. (2016). Introduction: adult hydrocephalus. Neurosurgical Focus 41(3): E1.

Hofmann, E., Warmuth-Metz, M., Bendszus, M. and Solymosi, L. (2000). Phase-contrast MR imaging of the cervical CSF and spinal cord: volumetric motion analysis in patients with Chiari I malformation. American Journal of Neuroradiology 21(1): 151–158.

Jacobson, E.E., Fletcher, D.F., Morgan, M.K. and Johnston, I.H. (1996). Fluid dynamics of the cerebral aqueduct. Pediatric Neurosurgery 24(5): 229–236.

Jacobson, E.E., Fletcher, D.F., Johnston, I.H. and Morgan, M.K. (1999). Computer modelling of CSF flow in the subarachnoid space. Journal of Clinical Neuroscience 6(6): 498–500.

Jacobson, E.E., Fletcher, D.F., Morgan, M.K. and Johnston, I.H. (1999). Computer modelling of the cerebrospinal fluid flow dynamics of aqueduct stenosis. Medical & Biological Engineering & Computing 37(1): 59–63.

Johanson, C.E., Duncan, J.A., Klinge, P.M., Brinker, T., Stopa, E.G. and Silverberg, G.D. (2008). Multiplicity of cerebrospinal fluid functions: new challenges in health and disease. Cerebrospinal Fluid Research 5(1): 1–32.

Khani, M., Sass, L.R., McCabe, A.R., Zitella Verbick, L.M., Lad, S.P., Sharp, M.K. et al. (2020). Impact of Neurapheresis system on intrathecal cerebrospinal fluid dynamics: a computational fluid dynamics study. Journal of Biomechanical Engineering 142(2).

Khani, M., Sass, L.R., Sharp, M.K., McCabe, A.R., Zitella Verbick, L.M., Lad, S.P. et al. (2020). *In vitro* and numerical simulation of blood removal from cerebrospinal fluid: comparison of lumbar drain to Neurapheresis therapy. Fluids and Barriers of the CNS 17(1): 1–17.

Kumar, B., Srivastav, V.K., Jain, A. and Paul, A.R. (2019). Study of numerical schemes for the CFD simulation of human airways. International Journal of Integrated Engineering 11(8): 32–40.

Kurtcuoglu, V., Poulikakos, D. and Ventikos, Y. (2005). Computational modeling of the mechanical behavior of the cerebrospinal fluid system.

Kurtcuoglu, V., Soellinger, M., Summers, P., Boomsma, K., Poulikakos, D., Boesiger, P. et al. (2005). Reconstruction of cerebrospinal fluid flow in the third ventricle based on MRI data. International Conference on Medical Image Computing and Computer-Assisted Intervention, Springer.

Larpruenrudee, P., Islam, M.S., Paul, G., Paul, A.R. and Saha, S.C. (2022). Model for pharmaceutical aerosol transport through stenosis airway. *In*: Yashwant Pathak and Nazrul Islam (eds.). Handbook of Lung Targeted Drug Delivery Systems: Recent Trends and Clinical Evidences, Online ISBN: 9781003046547, Chapter-8, CRC Press, Florida (USA).

Levine, D.N. (1999). The pathogenesis of normal pressure hydrocephalus: a theoretical analysis. Bulletin of Mathematical Biology 61(5): 875–916.

Linninger, A.A., Tsakiris, C., Zhu, D.C., Xenos, M., Roycewicz, P., Danziger, Z. et al. (2005). Pulsatile cerebrospinal fluid dynamics in the human brain. IEEE Transactions on Biomedical Engineering 52(4): 557–565.

Lloyd, R.A., Fletcher, D.F., Clarke, E.C. and Bilston, L.E. (2017). Chiari malformation may increase perivascular cerebrospinal fluid flow into the spinal cord: a subject-specific computational modelling study. Journal of Biomechanics 65: 185–193.

Loth, F., Yardimci, M.A. and Alperin, N. (2001). Hydrodynamic modeling of cerebrospinal fluid motion within the spinal cavity. Journal of Biomechanical Engineering 123(1): 71–79.

Martin, B.A. and Loth, F. (2009). The influence of coughing on cerebrospinal fluid pressure in an in vitro syringomyelia model with spinal subarachnoid space stenosis. Cerebrospinal Fluid Research 6(1): 1–18.

Martin, B.A., Labuda, R., Royston, T.J., Oshinski, J.N., Iskandar, B. and Loth, F. (2010). Spinal subarachnoid space pressure measurements in an in vitro spinal stenosis model: implications on syringomyelia theories. Journal of Biomechanical Engineering 132(11).

Masoumi, N., Framanzad, F., Zamanian, B., Seddighi, A., Moosavi, M., Najarian, S. et al. (2013). 2D computational fluid dynamic modeling of human ventricle system based on fluid-solid interaction and pulsatile flow. Basic and Clinical Neuroscience 4(1): 64.

Pahlavian, S.H., Loth, F., Luciano, M., Oshinski, J. and Martin, B.A. (2015). Neural tissue motion impacts cerebrospinal fluid dynamics at the cervical medullary junction: a patient-specific moving-boundary computational model. Annals of Biomedical Engineering 43(12): 2911–2923.

Peate, I. (2008). Body fluids: components and disorders of cerebrospinal fluid. British Journal of Healthcare Assistants 2(9): 434–436.

Rasmussen, M.K., Mestre, H. and Nedergaard, M. (2018). The glymphatic pathway in neurological disorders. The Lancet Neurology 17(11): 1016–1024.

Roldan, A., Wieben, O., Haughton, V., Osswald, T. and Chesler, N. (2009). Characterization of CSF hydrodynamics in the presence and absence of tonsillar ectopia by means of computational flow analysis. American Journal of Neuroradiology 30(5): 941–946.

Saha, S.C., Francis, I. and Nassir, T. (2022). Computational inertial microfluidics: optimal design for particle separation. Fluids 7(9): 308.

Saha, S.C., Francis, I., Huang, X. and Paul, A.R. (2022). Heat transfer and fluid flow analysis in a realistic 16-generation lung. Physics of Fluids 34(6): 061906.

Sakka, L., Coll, G. and Chazal, J. (2011). Anatomy and physiology of cerebrospinal fluid. European Annals of Otorhinolaryngology, Head and Neck Diseases 128(6): 309–316.

Shukla, R.K., Srivastav, V.K., Paul, A.R. and Jain, A. (2020). Fluid structure interaction studies of human airways. Sādhanā 45, article number 229: 1–6, https://doi.org/10.1007/s12046-020-01460-9.

Silverberg, G., Mayo, M., Saul, T., Fellmann, J. and McGuire, D. (2006). Elevated cerebrospinal fluid pressure in patients with Alzheimer's disease. Cerebrospinal Fluid Research 3(1): 1–6.

Singh, D., Jain, A. and Paul, A.R. (2020). Numerical study on particle deposition in healthy human airways and airways with glomus tumor. *In*: Rizvanov, Albert A., Singh, Bikesh Kumar and Ganasala, Padma (eds.). Lecture Notes in Bioengineering, Albert A. Rizvanov et al. (eds.). Advances in Biomedical Engineering and Technology, Online ISBN: 978-981-15-6329-4, Chapter 31. Springer, Singapore. https://doi.org/10.1007/978-981-15-6329-4_31.

Speer, M.C., Enterline, D.S., Mehltretter, L., Hammock, P., Joseph, J., Dickerson, M. et al. (2003). Chiari type I malformation with or without syringomyelia: prevalence and genetics. Journal of Genetic Counseling 12(4): 297–311.

Srivastav, V.K., Paul, A.R. and Jain, A. (2013). Effects of cartilaginous rings on airflow and particle transport through simplified and realistic human upper respiratory tracts. Acta Mechanica Sinica 29(6): 883–892. https://doi.org/10.1007/10.1007/s10409-013-0086-2.

Srivastav, V.K., Paul, A.R. and Anuj Jain, A. (2019b). Capturing the wall turbulence in CFD simulation of human respiratory tract. Mathematics and Computers in Simulation 160: 23–38, https://doi.org/10.1016/j.matcom.2018.11.019.

Stadlbauer, A., Salomonowitz, E., Van Der Riet, W., Buchfelder, M. and Ganslandt, O. (2010). Insight into the patterns of cerebrospinal fluid flow in the human ventricular system using MR velocity mapping. Neuroimage 51(1): 42–52.

Stadlbauer, A., Salomonowitz, E., Brenneis, C., Ungersböck, K., van der Riet, W., Buchfelder, M. et al. (2012). Magnetic resonance velocity mapping of 3D cerebrospinal fluid flow dynamics in hydrocephalus: preliminary results. European Radiology 22(1): 232–242.

Stockman, H.W. (2007). Effect of anatomical fine structure on the dispersion of solutes in the spinal subarachnoid space.

Thalakotunage, A. and Sethaput, T. (2016). Quantification of CSF velocity through the narrowest point in aqueduct of sylvia for normal and normal pressure hydrocephalus patient by CFD analysis. International Journal of Pharmacy and Pharmaceutical Sciences 8(2).

Tubbs, R.S., Bailey, M., Barrow, W.C., Loukas, M., Shoja, M.M. and Oakes, W.J. (2009). Morphometric analysis of the craniocervical juncture in children with Chiari I malformation and concomitant syringobulbia. Child's Nervous System 25(6): 689–692.

Tully, B. and Ventikos, Y. (2009). Coupling poroelasticity and CFD for cerebrospinal fluid hydrodynamics. IEEE Transactions on Biomedical Engineering 56(6): 1644–1651.

Ursino, M. (1988). A mathematical study of human intracranial hydrodynamics part 1—the cerebrospinal fluid pulse pressure. Annals of Biomedical Engineering 16(4): 379–401.

Ursino, M. and Lodi, C.A. (1997). A simple mathematical model of the interaction between intracranial pressure and cerebral hemodynamics. Journal of Applied Physiology 82(4): 1256–1269.

Vardakis, J.C., Tully, B.J. and Ventikos, Y. (2013). Exploring the efficacy of endoscopic ventriculostomy for hydrocephalus treatment via a multicompartmental poroelastic model of CSF transport: a computational perspective. PloS One 8(12): e84577.

Vinchon, M., Rekate, H. and Kulkarni, A.V. (2012). Pediatric hydrocephalus outcomes: a review. Fluids and Barriers of the CNS 9(1): 1–10.

Wagshul, M.E., Chen, J.J., Egnor, M.R., McCormack, E.J. and Roche, P.E. (2006). Amplitude and phase of cerebrospinal fluid pulsations: experimental studies and review of the literature. Journal of Neurosurgery 104(5): 810–819.

Wiswell, T.E., Tuttle, D.J., Northam, R.S. and Simonds, G.R. (1990). Major congenital neurologic malformations: a 17-year survey. American Journal of Diseases of Children 144(1): 61–67.

Xie, J., Cheng, Z., Gu, L., Wu, B., Zhang, G., Shiu, W. et al. (2022). Evaluation of cerebrovascular hemodynamics in vascular dementia patients with a new individual computational fluid dynamics algorithm. Computer Methods and Programs in Biomedicine 213: 106497.

Yiallourou, T.I., Kröger, J.R., Stergiopulos, N., Maintz, D., Martin, B.A. and Bunck, A.C. (2012). Comparison of 4D phase-contrast MRI flow measurements to computational fluid dynamics simulations of cerebrospinal fluid motion in the cervical spine. PloS One 7(12): e52284.

# Chapter 8
# Ocular Flow Modelling

*Vivek Kumar Srivastav,*[1,*] *Aman Raj Anand*[2] and *Akshoy Ranjan Paul*[3]

## 1. Introduction

The human eye plays a crucial role in visual perception, capturing light from the surrounding environment and transmitting visual information to the brain. Due to its function as a visual system, it is often referred to as the organ of visualization. Eyes exhibit variations in shape, size, and color, and they possess a slightly asymmetrical spherical structure with a diameter of approximately 2.5 cm (Cheng et al., 2017). In modern times, Computational Fluid Dynamics (CFD) has emerged as a powerful technique for investigating fluid flow within complex systems. Within the field of ophthalmology, CFD has found extensive application in studying the fluid dynamics of the human eye and associated disorders.

## 2. Anatomy and Physiology of Ocular System

The human eye is a muddled and noteworthy device that is liable for seeing and permits us to see our environmental elements. The eye is the essential organ of the visual framework, and it is comprised of a few interrelated parts, each with its own physical design and physiological capability. The schematic diagram of the human eye and its different parts is shown in Fig. 1. In this chapter, we will take a gander at the life structures and physiology of the visual framework, with an accentuation on the eye. The eye is a tangible organ that transforms light into electrical signs that the mind can decipher. It is an empty, round structure with a measurement of around 2.5 centimeters (Villamarin et al., 2012; Silva et al., 2014; Cheng et al., 2017).

[1] Applied Science & Humanities, Government Engineering College Siwan, Bihar, India.
[2] Department of Mechanical Engineering, IIT-BHU, UP, India.
[3] Department of Applied Mechanics, Motilal Nehru National Institute of Technology Allahabad, Prayagraj-211004, India.
* Corresponding author: vivekapril12@gmail.com

DOI: 10.1201/9781003287636-8

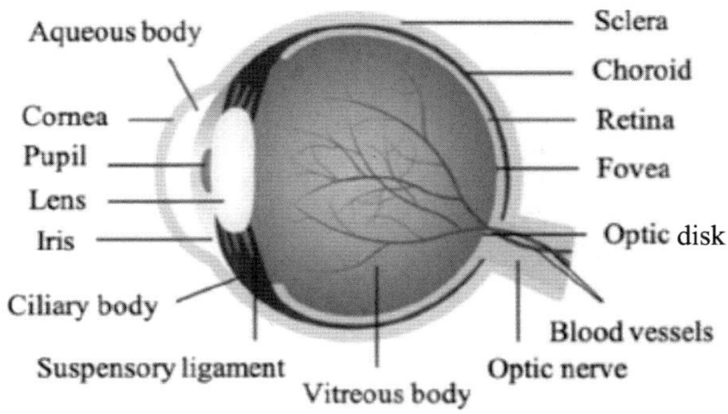


**Figure 1.** Human eye anatomy. Source: Villamarin et al., 2012.

The solid, misty sclera, frequently known as the white of the eye, is the eye's furthest covering. The cornea is framed from the front part of the sclera, which is straightforward and permits light to enter the eye.

The conjunctiva is a slender, clear layer that covers the cornea and safeguards the eye from unfamiliar particles. The conjunctiva likewise covers the eyelids. The eyelids, specifically, carry out a significant role in safeguarding the eye and distributing tears all through its surface to hold dampness and wipe out soil. The eyelashes and temples additionally help to protect the eye from unfamiliar things (Safdel et al., 2021).

Moving further inside the eye, we go over the choroid, a vascular layer that conveys nourishment to the retina's external layers. The choroid contains dim color cells that assist in lessening inside reflections inside the eye and consequently work on visual sharpness. The ciliary body is close to the choroid and comprises of ciliary muscles that alter the state of the focal point during convenience (zeroing in on objects at various distances) and ciliary cycles that make fluid humor.

The focal point is a straightforward, biconvex design upheld from the ciliary body by suspensory tendons. It refracts light and helps in zeroing in it on the retina. Convenience is achieved by contracting or loosening up the ciliary muscles, which changes the pressure on the suspensory tendons, permitting the focal point to move shape and change its central length. The foremost depression of the eye is separated into two chambers: the front chamber, which lies between the cornea and the iris, and the back chamber, which lies between the iris and the focal point. The two chambers are loaded up with watery humor, a watery liquid that feeds the connective cornea and focal point, and keeps up with intraocular pressure. The fluid humor is delivered by the ciliary cycles and is continually emptied out through the trabecular meshwork and Schlemm's channel to keep up with the appropriate strain.

The iris, a pigmented solid stomach (Ortiz et al., 2012), controls how much light enters the eye through the focal opening called the student. The size of the pupil is regulated by the contraction and relaxation of the iris muscles, which are controlled by the autonomic nervous system. In bright light, the iris contracts to decrease the pupil size and limit the amount of light entering the eye. Conversely, in dim light, the iris expands to allow more light to enter. The deepest layer of the eye is the retina, a fragile, light-touchy tissue made out of specific photoreceptor cells called bars and cones. Poles are answerable for fringe and faint light vision, while cones are liable for focal and variety vision. The retina additionally contains different layers, including the bipolar cells, ganglion cells, and the deepest layer called the retinal shade epithelium (RPE). At point when light enters the eye, it goes through the cornea, fluid humor, and focal point and arrives at the retina. The photoreceptor cells in the retina convert the light into electrical signs, which are then sent through the bipolar cells to the ganglion cells. The axons of the ganglion cells unite at the optic plate, shaping the optic nerve, which sends these electrical signs to the cerebrum for additional handling and translation (Crowder et al., 2013).

In summary, the visual system, primarily composed of the eye, serves as a crucial framework responsible for our ability to see. Its intricate structure and complex physiological processes work together to capture light, focus it onto the retina, and convert it into electrical signals that the brain can interpret. A comprehensive understanding of the detailed anatomy and physiology of the human eye is essential for comprehending the mechanisms underlying vision and the various disorders that can impact its functionality.

## 3. Fluid Mechanics of Ocular Diseases

There are several diseases associated with human eye. Major eye diseases which are analyzed by CFD are:

3.1 Glaucoma

3.2 Diabetic Retinopathy

3.3 Pink Eyes (Conjunctivitis)

### *3.1 Glaucoma*

Glaucoma affects the optic nerve and further results in vision loss. Glaucoma is one of the main causes of blindness. In glaucoma, Aqueous Humor (AH) outflow decreases which further increases the intraocular pressure (IOP). This increased pressure further damage retina ganglion cells (Ortiz et al., 2012). CFD is used to study the eye fluid dynamics of glaucoma sufferers. The flow of aqueous humor, the clear fluid that fills the front of the eye, has been specifically studied by many researchers. Researchers want to better understand the causes of glaucoma and create more efficient treatments by investigating the fluid dynamics of the eye. Girad et al. (2016) investigated the fluid flow in both healthy and glaucoma-affected eyes using CFD. They created a mathematical model of the eye that included all of the eye's tissues and fluids. It was found that glaucoma patients had higher ocular pressure than healthy patients and that the increased pressure was brought on by modifications in the aqueous humor flow.

The study of glaucoma, a condition that damages the optic nerve and can cause visual loss, is one area of research that has made use of CFD. Increased Intraocular Pressure (IOP) in glaucoma is thought to be caused by an accumulation of aqueous fluid inside the eye. Researchers have used CFD to model aqueous humor flow and

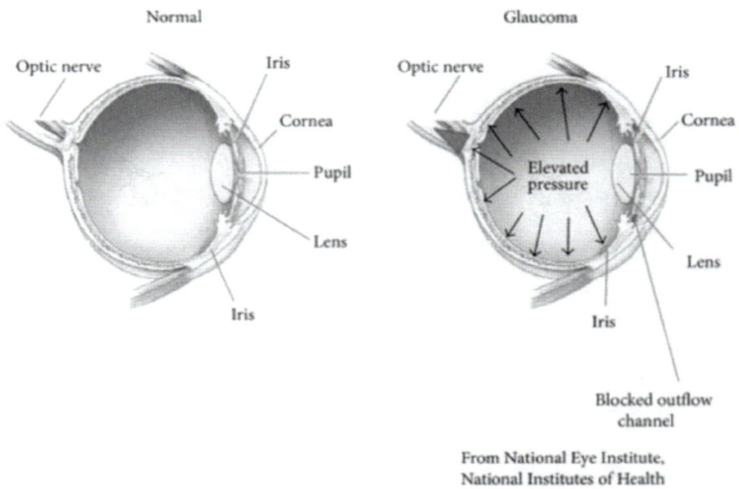


**Figure 2.** Human eyes anatomy (Normal eye and Glaucoma suffered eye). Source: Wirtz et al., 2016.

investigate how various elements, such as the positioning of drainage devices or the geometry of the eye, affect IOP. The examination of contact lens use is another field of inquiry that has made use of CFD. Changes in the tear film that covers the surface of the eye brought on by contact lenses may affect the comfort and clarity of vision. Researchers have used CFD to analyze the flow of tears and investigate how contact lenses affect the dynamics of the tear film. Beyond these particular uses, CFD has the potential to be an important tool for researching a variety of eye-related issues. CFD might be utilized, for instance, to investigate the movement of blood within the eye or the effect that changes in eye shape have on visual acuity.

## *3.2 Diabetic Retinopathy*

Diabetic retinopathy is a complication of diabetes that affects the blood vessels in the retina. CFD has been used to study the blood flow in the retina of diabetic patients.

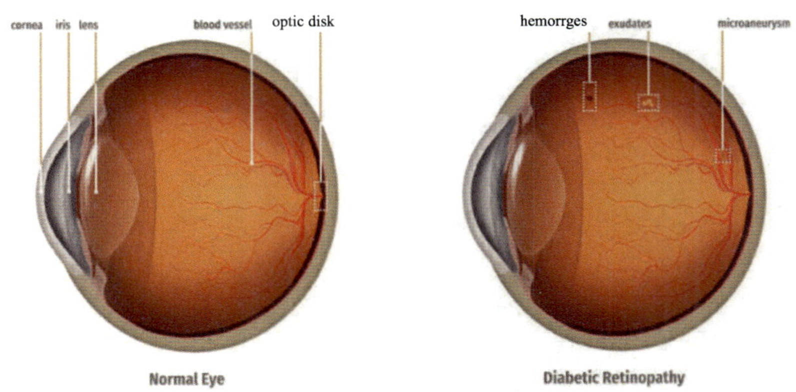


**Figure 3.** Human eye anatomy (Normal eye and diabetic retinopathy). Source: Tan et al., 2017.

By studying the blood flow in the retina, researchers hope to better understand the causes of diabetic retinopathy and develop more effective treatments.

## *3.3 Pink Eyes (Conjunctivitis)*

Pink eye, otherwise called conjunctivitis, is a typical eye contamination that causes irritation and redness in the conjunctiva, the dainty layer that covers the white piece of the eye. It can influence individuals, everything being equal, yet it is more common in kids. Pink eye can be brought about by different variables, including infections, microbes, allergens, and aggravations. Viral conjunctivitis is the most well-known sort of pink eye and is profoundly infectious. It is frequently connected with side effects like watery release, redness, and tingling in the impacted eye. Viral pink eye regularly begins in a single eye and can spread to the next eye within a couple of days. It, for the most part, settles all alone with next to no particular treatment, yet side effect help can be accomplished using counterfeit tears and cold packs. Bacterial conjunctivitis is one more type of pink eye, brought about by microbes, for example, Staphylococcus aureus or Streptococcus pneumoniae. It is portrayed by a thick, yellow, or greenish release that can make the eyelids stay together after awakening. Bacterial pink eye is likewise infectious and requires anti-microbial eye drops or treatment recommended by a medical service proficient to clear the contamination (Sun et al., 2014). Hypersensitive conjunctivitis happens when the conjunctiva is exposed to allergens like dust, pet dander, or residue parasites. It ordinarily influences the two eyes and is related to side effects like tingling, redness, and watery release. Keeping away from allergens and utilizing allergy medicine eye drops or oral prescriptions can give help from hypersensitive pink eye. To forestall the spread of pink eye, it is essential to rehearse great cleanliness, like washing hands routinely, abstaining from contacting the eyes, and not imparting towels or individual things to tainted people. Contact focal point wearers ought to briefly stop focal point use until the contamination settles. All in all, pink eye is a typical eye contamination that can be brought about by infections, microbes, allergens, or aggravations. It

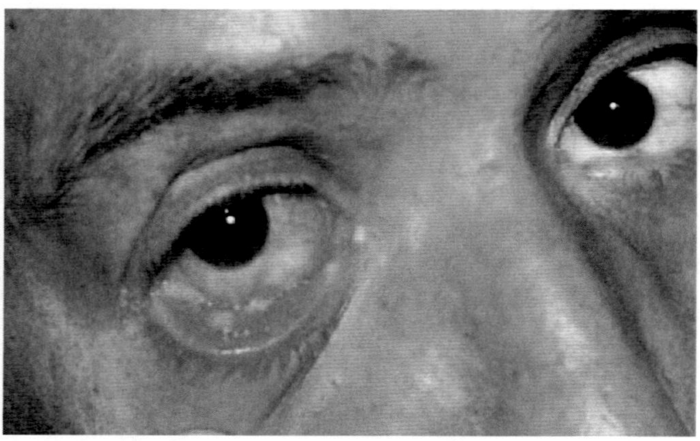

**Figure 4.** Human eye anatomy (Normal eye and Conjunctivitis eye). Source: Freitas et al., 2011.

appears as redness, tingling, and release in the impacted eye and can be exceptionally infectious. Treatment choices shift depending upon the basic reason, and it is crucial to train great cleanliness to forestall its spread. On the off chance that you suspect you have pink eye, counselling medical care proficient for an exact finding and suitable treatment is prudent.

## 4. Blood Flow Dynamics in Ocular System

The eye, similar to some other organs in the human body, requires a steady and satisfactory blood supply to work ideally. The many-sided organization of veins inside the eye guarantees the conveyance of oxygen and supplements while eliminating side effects. Understanding the elements of blood stream inside the visual framework is urgent for appreciating different eye conditions and sicknesses. In this paper, we will investigate the blood stream elements in the visual framework exhaustively.

The visual framework is provided with blood by two primary sources: the ophthalmic course and the ciliary supply routes (Farhadi et al., 2014; Silva et al., 2014). The ophthalmic course diverges from the inner carotid vein, while the ciliary corridors emerge from the ophthalmic supply route or its branches. These courses lead to an intricate organization of veins that supply various designs inside the eye.

Beginning from the foremost fragment of the eye, the bloodstream elements assume a crucial part in keeping up with the well-being of the cornea. The cornea is connective, and its sustenance essentially comes from the tear film on its surface. In any case, the limbal vessels, situated at the boundary between the cornea and the sclera, give some blood supply to the cornea.

Moving to the back fragment of the eye, the choroid, a profoundly vascular layer situated between the retina and the sclera, assumes a significant part in providing blood to the external layers of the retina (Abouali et al., 2014). The choroidal bloodstream is managed by different elements, including autoregulation, neighborhood metabolites, brain impacts, and visual perfusion pressure. Changes in the choroidal bloodstream have been related to different eye sicknesses, including age-related macular degeneration.

The retina, the deepest layer of the eye, has a high metabolic interest and requires very much directed blood supply (Joukar et al., 2015; Nejad et al., 2015; Tamm et al., 2015). The retinal bloodstream is basically controlled by the focal retinal corridor, which enters the eye through the optic nerve and branches into more modest arterioles and vessels inside the retina. The retinal bloodstream is fundamental for the conveyance of oxygen and supplements to the retinal cells, and any disturbance in this bloodstream can prompt vision debilitation (Joukar, 2015; Bhandari et al., 2021).

The guideline of the bloodstream in the visual framework is a perplexing cycle, including different systems (Girad et al., 2016). Autoregulation is one such system that keeps a somewhat consistent bloodstream in spite of changes in perfusion pressure. This autoregulation is intervened by the myogenic reaction of veins, metabolic variables, and brain control.

Moreover, the autonomic sensory system assumes a part in managing blood stream elements inside the visual framework. Thoughtful innervation causes vasoconstriction of the veins, while parasympathetic innervation prompts vasodilation (Rafiq et al.,

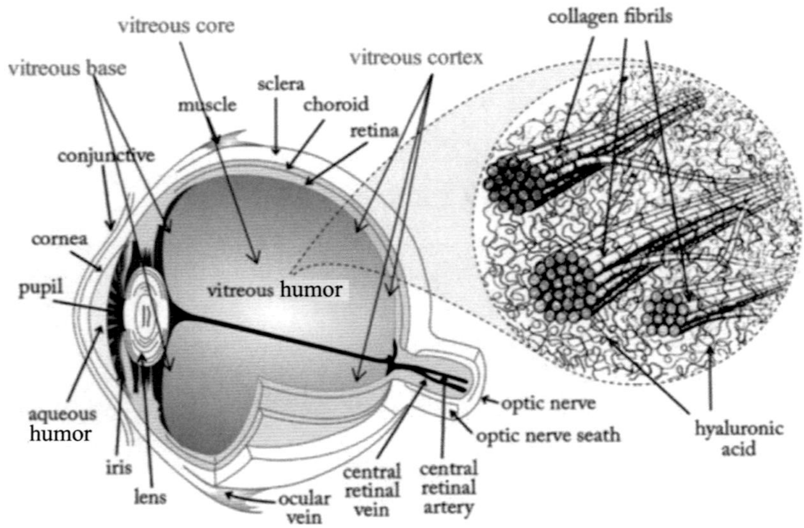


**Figure 5.** Fluid flow in the eye. Source: Monteiro et al., 2015.

2016). These instruments assist with controlling the bloodstream and keeping up with the proper perfusion pressure (Carreon et al., 2017; Karacali et al., 2019). A few methods can be utilized to concentrate on bloodstream elements inside the visual framework. Doppler ultrasound, laser Doppler flowmetry, fluorescein angiography, and Optomography angiography (OCTA) are among the ordinarily utilized strategies (Colciago et al., 2020). These procedures permit perception and estimation of bloodstream speed, volume, and perfusion in various designs of the eye.

Taking everything into account, understanding the elements of bloodstream inside the visual framework is vital for keeping up with visual well-being and forestalling vision-related messes. The intricate organization of veins, alongside autoregulatory systems and brain control (Bhandari et al., 2021), guarantees the conveyance of oxygen and supplements to the different designs of the eye. Progresses in imaging procedures have given important bits of knowledge to concentrating on bloodstream elements and their suggestions for visual illnesses (Wang et al., 2022). Further examination in this field will add to how we might interpret visual physiology and help in the advancement of novel remedial methodologies for visual circumstances.

## 5. Computational Fluid Dynamics of Eye: CFD Modeling of Ocular Flow

Computational fluid dynamics is one of the potent tools for simulating fluid flows and forecasting fluid behavior in a variety of biomedical problems. Engineering, biology, and medicine are just a few of the industries that frequently use the effective technology known as CFD. CFD has been used more and more in recent years to investigate the dynamics of flow in the human eye, particularly in relation to ocular diseases and disorders.

The proper representation of the geometry of the problem is one of the main difficulties in employing CFD (Goud, 2018). This problem is more apparent when analyzing the fluid behavior of intricate systems like the human eye. In this reference, the author(s) will talk about some of the problems with human eye analysis in CFD, as well as some of the methods that have been suggested to fix them.

## *5.1 Mechanism of the Human Eye and CFD*

The study of fluid flow in the human eye using CFD has gained popularity recently. A computational method called CFD enables the simulation of fluid flow using mathematical models. CFD can be used to simulate the movement of fluids, such as tears or aqueous humor, inside the eye. The human eye is a sophisticated and complicated organ that enables us to see the outside world. It is a sensory system that picks up light and transforms it into brain-transmitted neural signals that are then translated into visual representations. The cornea, lens, retina, and optic nerve are just a few of the structures that make up the eye, and they all work together to maintain clear vision.

The human eye's capacity to focus on objects at various distances is one of its most important features. This is accomplished via the lens's ability to alter shape and change the focus of incoming light. A vital component in focusing incoming light is also played by the cornea, a transparent outer layer. Also, photoreceptors, which are specialized cells in the retina at the back of the eye, transform light into electrical signals that are transmitted to the brain.

Light capture, transduction, and brain processing are the three phases that make up the human eye's visual mechanism. Light enters the eye at the first stage and is focused onto the retina via the cornea and lens. The photoreceptors in the retina convert light into electrical signals in the second step. The third stage culminates in the transmission of these electrical signals via the optic nerve to the brain, where they are translated into visual images.

## *5.2 CFD Modeling of Different Eye Diseases*

Computational Fluid Dynamics modeling can be used to study and analyze various eye diseases that involve fluid flow and fluid-structure interactions. While CFD cannot replace clinical evaluations and medical diagnoses, it can provide insights into the fluid dynamics within the eye and aid in understanding the underlying mechanisms of certain eye conditions. Here are a few examples: Diabetic retinopathy, pink eye, glaucoma, dry eye syndrome, ocular drug delivery, etc., it is important to note that CFD modeling of eye diseases is a complex task that requires accurate representation of eye geometry, appropriate boundary conditions, and reliable fluid and tissue properties. Collaboration between biomedical engineers, ophthalmologists, and other healthcare professionals is crucial to ensure the accuracy and validity of such models and their translation to clinical practice.

### *5.2.1 CFD Modeling for Diabetic Retinopathy*

Computational Fluid Dynamics modeling can be utilized to study certain aspects of diabetic retinopathy, a complication of diabetes that affects the blood vessels in the retina. Diabetic retinopathy involves changes in blood flow and vascular abnormalities within the retina, which can lead to vision loss if left untreated. While CFD cannot replace clinical examinations and diagnostics, it can provide insights into the blood flow dynamics and help understand the underlying mechanisms of the disease. Many researchers have used to analyze diabetic retinopathy which are summarised below:

Ding et al. (2020) indicate the use of CFD to analyze the hemodynamic of blood flow in diabetic retinopathy patients' retinal arteries. It sheds light on the changes in blood flow patterns and velocity in retinal arteries caused by diabetic retinopathy. Souza et al. (2021) present an overview of CFD in ophthalmology, including its usage in diabetic retinopathy research. It highlights a number of CFD-based research that have analyzed the hemodynamics of blood flow in the retinal vasculature and their relevance to diabetic retinopathy. Using CFD, Sanchez et al. (2015) explore the hemodynamics of blood flow in retinal vessels of diabetic retinopathy patients. It investigates the changes in flow patterns, velocity, and wall shear stress in the retinal vasculature caused by diabetic retinopathy. Tan et al. (2009) address the use of quantitative confocal imaging methods to analyze the human retinal microvasculature. While it does not address CFD explicitly, it does give insights into the imaging modalities used to analyze the retinal vasculature, which can supplement CFD research in diabetic retinopathy. Zhang et al. (2018) examine advancements in Optical Coherence Tomography (OCT) monitoring in diabetic retinopathy. While it does not directly address CFD, OCT imaging can give crucial anatomical and structural information that can be integrated with CFD analysis to acquire full knowledge of diabetic retinopathy etiology.

*Age-related Macular Degeneration (AMD)*: AMD is just a condition that affects the retina's center, which is where sharp, detailed vision is produced. The flow of blood in the retina of AMD patients has been examined using CFD. Researchers want to better understand the causes of AMD and create more efficient treatments by examining the blood flow in the retina. Popescu et al., 2011, carried out one study in which AMD was investigated using CFD. The blood flow in the retina of healthy patients and patients with AMD was simulated by the researchers using CFD (Popeşcu et al., 2011). They discovered that people with AMD had considerably lower retinal blood flow than healthy individuals. It was also found that the reason this reduced blood flow was the changes in the cellular organization of the blood vessels in the retina.

Gardner et al. (2017) provide a strategy for identifying Diabetic Macular Edoema (DME) from Proliferative Diabetic Retinopathy (PDR) in the context of diabetic retinopathy. While it does not focus on CFD analysis, it does emphasize the need of understanding the root causes and molecular pathways in these two disorders, which can possibly be supplemented by CFD investigations. Tawfik et al. (2019) investigate the uses of CFD and Optical Coherence Tomography Angiography (OCTA) in

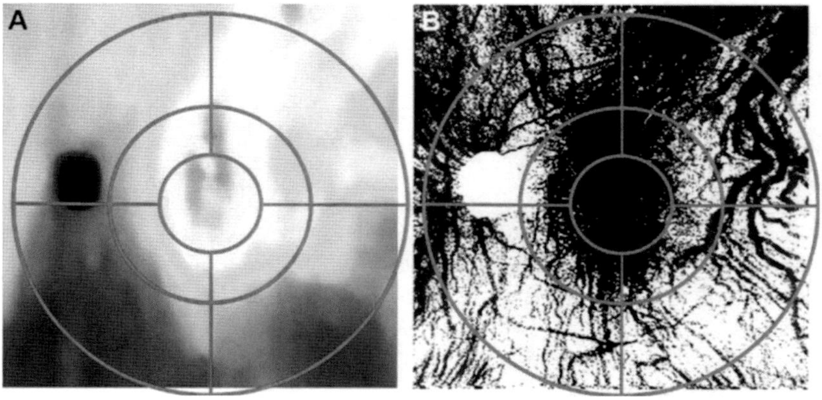


**Figure 6.** CFD modeling for Diabetic Retinopathy. Source: Zheng et al., 2016.

diabetic retinopathy research. It explains how CFD may provide information into the blood circulation of blood flow in the retinal circulation, and how OCTA can provide comprehensive imaging of the retina microvasculature, supplementing CFD research in understanding the pathophysiology of diabetic retinopathy.

### *5.2.2 CFD Modeling for Pink Eye*

In the case of pink eye, also known as conjunctivitis, CFD might not be directly applicable to predict about pink eye. Pink eye is an inflammation of the conjunctiva, the thin clear tissue that covers the white part of the eye and lines the inner surface of the eyelids. It is commonly caused by viral or bacterial infections, allergies, or irritants.

CFD simulations are typically utilized to study the behavior of fluids, such as air or liquids, in different environments. Although the eye contains fluids like tears, the nature of pink eye involves inflammation and infection rather than fluid dynamics. The diagnosis and treatment of pink eye are usually performed through clinical evaluation by a healthcare professional. They may examine the eye, inquire about symptoms, and perform tests, such as swabs or cultures, to identify the cause of the condition. Many researchers have used CFD studies for the prediction of pink eye disease which are summarized below.

Wang et al., 2022, created a CFD model to mimic ocular allergen deposition on the human conjunctiva. The model considered blinking, tear film dynamics, and fluid-particle interactions. To anticipate allergen deposition patterns on the ocular surface, the scientists used data from experiments on particle size distribution and blinking patterns. Rebuli et al., 2021 used CFD to predict the fate of allergens that are airborne deposited on the ocular surface. To simulate allergen deposition and transit on the ocular surface, the authors used a CFD model. The model took blinking, tear film dynamics, and particle adherence to the ocular surface into account. The researchers wanted to know what variables influence allergen spreading and retention on the ocular surface.

Alkozi and Pintor (2015), investigated the influence of tears on the ocular surface using numerical modeling. The scientists employed CFD modeling to investigate the fluid dynamics of tears on the ocular surface, taking into account aspects such as tear film thickness and distribution. The researchers wanted to know how tear dynamics influence the stability and dispersion of the tear film on the ocular surface.

The CFD investigation of tear film breakdown on an eye-shaped domain was the focus of Feng et al. (2016), To simulate tear film behavior and breakage on an eye-shaped surface, the scientists created a CFD model. The study sought to better understand the variables that influence tear film breakage and the consequences for ocular health. His colleagues studied tears film lipid layer thickness and ocular comfort following therapies for meibomian gland dysfunction. While it did not employ CFD directly, it did entail the measurement of tear film lipid layer thickness with a commercial tool called LipiView®. The researchers wanted to know if there was a link between tear film lipid layer thickness, meibomian gland dysfunction, and ocular comfort.

Ji et al. (2018), concentrated on creating a finite element model in three dimensions of the human eyeball. Although it did not directly employ CFD, it did entail numerical modeling of the mechanical behavior of the eyeball using the finite element approach. The goal of the research was to create a complete model for analyzing the biomechanical reaction of the eyeball under various stress circumstances.

### *5.2.3 CFD Modeling for Glaucoma*

Glaucoma is a group of eye diseases depicted by increased intraocular pressure (IOP) that can damage the optic nerve and lead to vision loss. CFD modeling can help simulate the fluid flow and pressure distribution within the eye to understand factors that contribute to increased IOP and assess the effectiveness of different surgical interventions or medication delivery systems. While CFD cannot diagnose glaucoma, it can provide insights into the flow patterns and pressure distributions within the eye, aiding in understanding the disease progression and potentially guiding treatment strategies.

CFD was used by Tamm et al. (2015), to examine the aqueous humor outflow routes in glaucoma. The researchers got insight into the mechanisms driving glaucoma and potential therapies by modeling fluid dynamics within the eye. Safdel et al. (2021), investigated the effect of intraocular pressure fluctuations on the optic nerve heads in glaucoma. The study provided insight on the biomechanical processes impacting the optic nerve in glaucoma by taking fluid dynamics and structural behavior into account. Karacali (2019) used CFD methods to evaluate the performance of a titanium outflow pipe implant used to treat glaucoma. The researchers evaluated the implant's efficacy and possible effects on intraocular pressure management by simulating fluid flow through it. Cereda et al. (2020), contrasted CFD and Fluid-Structure Interaction (FSI) techniques to analyze the ocular fluid dynamics impacts on an intraocular magnesium-based device. The researchers analyzed the device's performance and fluid behavior using CFD simulations, offering vital

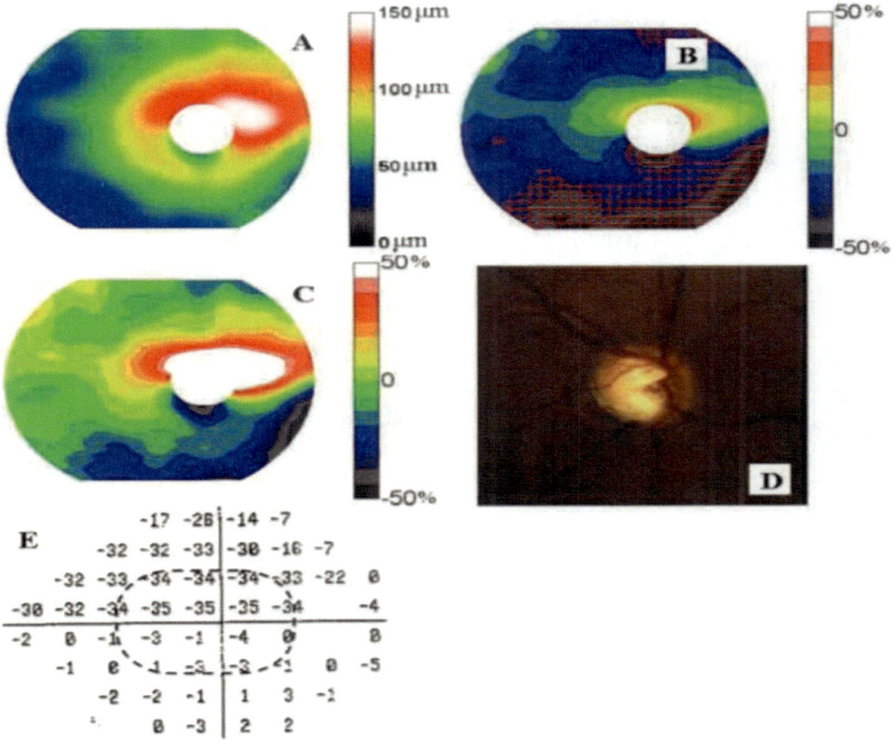


**Figure 7.** CFD modelling for Glaucoma. Source: Tan et al., 2009.

insights into its functioning for glaucoma care. Acott et al. (2021), performed an in-depth assessment of normal and glaucomatous outflow control. While not mainly concerned with CFD, it highlighted a number of CFD research and computational models that helped to improve knowledge of aqueous humor dynamics and outflow resistance pathways in glaucoma. Sun and Zhang (2012) studied the hydrodynamics of aqueous humor in the human eye using CFD models. The researchers got insight into fluid behavior and its implications for glaucoma pathogenesis by studying fluid flow patterns and pressure distribution. Harris et al. (2013) investigated aqueous flow in eyes using an Ahmed glaucoma valve using CFD models. The researchers investigated the fluid dynamics and pressure distribution within the eye, gaining vital insight on the performance and usefulness of the glaucoma valve. Hosseini et al. (2019) used CFD simulation to examine aqueous humor outflow in a realistic 3D model of the human eye. The researchers got insight into the complicated mechanisms controlling aqueous humor outflow and its possible consequences for glaucoma by analyzing fluid dynamics and pressure distribution.

## 6. Challenges Associated with Human Eye Analysis

One of the biggest changes in biomedical problems is to validate the computational results. The following challenges are encountered to investigate the accuracy of results:

A frequent method in many experimental researches, including those in psychology, neurology, and medicine, is human eye analysis. In order to improve the understanding of human behavior and cognitive processes, this method entails watching and documenting eye movements. Unfortunately, there are a number of difficulties with this approach that could compromise the accuracy and dependability of the findings (Macedo-de-Araújo et al., 2016).

1. Individual variations in eye movements pose a significant problem for human eye analysis. Everybody has distinctive eye movements, which might change depending on things like age, gender, and cultural background. These individual variations can make it challenging to compare data across participants and can cause results to vary. To overcome this difficulty, researchers frequently need to recruit a sizable sample size for their studies to guarantee that the findings are representative of the population being investigated.
2. Controlling for unimportant variables that may influence eye movements is another difficulty. Examples of environmental elements that can affect how people perceive visual stimuli include variations in illumination, noise levels, and other environmental conditions. These factors might cause discrepancies in visual processing data, which makes it challenging to infer reliable conclusions from the findings. By adopting standardized testing circumstances and making sure that all participants are assessed in the same manner, researchers need to properly control for these variables (Sánchez-Ferrer et al., 2017).
3. Choosing the right stimuli carefully in order to cause regular and significant eye movements is a related difficulty. The selection of visual stimuli has a significant impact on the outcomes since not all visual stimuli are equally effective at causing eye movements. For instance, more rich and meaningful stimuli, such faces, or settings, may be more successful at causing eye movements than plain geometric objects.
4. The problem of measurement and analysis is another difficulty. There are several different metrics that may be used to measure eye movements, including fixation time, saccade amplitude, and gaze duration. Eye movement data can be complex and challenging to analyze. It is important for researchers to use standardized and trustworthy measurements that are relevant for their study question. Moreover, analyzing eye movement data can take a lot of time and requires specific knowledge and tools S (Vidyadharan et al., 2015; Goele et al., 2019).
5. Finally, when using human eye analysis in experimental research, ethical issues must also be taken into mind. Researchers must take care to gain participants' informed permission and avoid subjecting them to unwarranted risk or injury. Researchers must also make sure that they respect the participants' right to privacy and confidentiality and that they do not utilize the data they collect in any way that would be damaging or discriminatory.

## 7. Conclusions

It is important to note that CFD modeling is a complex process that requires accurate representation of eye geometry, appropriate boundary conditions, and reliable fluid and tissue properties. Collaborations between biomedical engineers, ophthalmologists, and other healthcare professionals are crucial to ensure the validity and applicability of CFD models in clinical practice. CFD studies of ocular flow have enhanced our understanding of eye diseases, treatment interventions, and ocular drug delivery systems, contributing to advancements in ophthalmology and improving patient care in the field of ocular health. Based on the aforementioned studies, the following conclusions can be drawn:

1. To explore the fluid dynamics of the human eye in various illnesses, CFD has been employed extensively. Researchers want to better understand the underlying causes of these conditions and create more potent treatments by investigating the fluid dynamics of the eye. The studies mentioned above are just a handful of the numerous investigations that have been carried out in ophthalmology utilizing CFD.
2. Although being a beneficial method for experimental inquiry, human eye analysis has some drawbacks. Researchers must carefully take into account for individual differences, irrelevant variables, and measurement challenges in addition to making sure they use appropriate stimuli and ethical practises. Despite these challenges, studying the human eye has the potential to shed light on human mind and behavior, and it most likely will remain an important method in experimental study for some time to come.
3. The analysis of the human eye in CFD is a difficult topic requiring correct modeling of the intricate geometry of the eye as well as precise boundary conditions. The use of high-resolution imaging techniques, the simplification of the shape of the eye, and hybrid strategies that integrate experimental data with CFD simulations have all been suggested as solutions to these problems. Although if each strategy has its own advantages and disadvantages, they all have the potential to advance our knowledge of how fluids behave in the human eye and help in the research and development of novel treatments for eye illnesses.
4. Fluid flow in the human eye can be studied using CFD simulations, which can provide important details about the physiological and pathological circumstances that influence the eye. However, using experimental data for validation comes with a number of difficulties, such as the difficulty in obtaining accurate and reliable data, the absence of a standardized protocol for obtaining data, and difficulties in accurately simulating the intricate geometry of the eye in CFD simulations. Researchers in the fields of fluid dynamics, ophthalmology, and imaging technologies must work together to address these issues.
5. The acquisition and segmentation of the eye geometry, the determination of the computational domain, and the specification of boundary conditions are all steps in the crucial CFD pre-processing phase of human eye flow analysis.

Our understanding of the flow dynamics in the human eye and its role in ocular diseases and disorders has shown considerable promise thanks to CFD. More study is required to fully explore the potential of the rapidly expanding disciplines of OCT and CFD in human eye flow studies.

6. Understanding fluid flow in the intricate human ocular system can be done with the help of CFD. To simulate fluid dynamics in the eye, scientists utilize a range of approximations, such as the Navier-Stokes equations, the lattice Boltzmann method, particle image velocimetry, and analytical models. These models have the potential to increase our knowledge of ocular fluid dynamics and have shed light on the mechanisms underlying eye disorders.
7. For studying the flow characteristics of the human eye, CFD is a useful technique. Turbulence models like RANS, LES, and DNS are widely used to simulate these flows. These models, which have been used in numerous studies to shed light on the underlying fluid dynamics, provide these simplified versions of the complicated turbulence processes that occur in the human eye.
8. Our knowledge of the intricate fluid dynamics that take place in the eye might be substantially improved with the use of CFD in flow study of the human eye. Researchers can learn more about the causes of eye-related issues and create innovative treatments to enhance the health and function of the eye by accurately modeling these processes.
9. An intricate and sophisticated organ, the human eye is essential to vision. Its visual system involves the neuronal processing, transduction, and capture of light. A variety of eye-related phenomena, such as glaucoma and contact lens use, have been studied using CFD as a useful technique for analyzing fluid flow within the eye. It is anticipated that CFD techniques will take on a greater significance in eye research as they advance.

## References

Abouali, O., Neitz, M. and Luo, J. (2014). Numerical simulation for unsteady motions of the human vitreous humor as a viscoelastic substance in linear and non-linear regimes. Journal of Non-Newtonian Fluid Mechanics 204: 22–31.

Acott, T.S., Kelley, M.J. and Keller, K.E. (2021). Normal and glaucomatous outflow regulation. Progress in Retinal and Eye Research 82: 100897.

Acott, T.S., Kelley, M.J., Keller, K.E., Vranka, J.A. and Abu-Hassan, D.W. (2021). Normal and glaucomatous outflow regulation. Progress in Retinal and Eye Research 82: 100897.

Alkozi, H.A. and Pintor, J. (2015). TRPV4 activation triggers the release of melatonin from human non-pigmented ciliary epithelial cells. Experimental Eye Research 136: 34–37.

Bhandari, R., Li, Y. and Nguyen, T.H. (2021). Ocular fluid mechanics and drug delivery: a review of mathematical and computational models. Pharmaceutical Research 38: 2003–2033. https://doi.org/10.1007/s11095-021-03141-6.

Callico, G.M., Armengol, J. and Sanchez, C.I. (2017). Challenges and methodologies in retinal image analysis for diabetic expert systems with applications. 81–100. DOI: 10.1016/j.eswa.2017.07.020.

Carreon, T., van der Merwe, E. and Fellman, R.L. (2017). Aqueous outflow—A continuum from trabecular meshwork to episcleral veins. Progress in Retinal and Eye Research 57: 108e133.

Cereda, A., D'Agostino, C., Deparis, S., Forti, F. and Quarteroni, A. (2020). Evaluation of the ocular fluid dynamic effects on intraocular magnesium-based device: A comparison between CFD and FSI approaches. Medical Engineering and Physics 86: 20–28.

Cheng, Y.H., Li, S.M., Li, S.M., Zhang, Y., Wang, N. and Wang, Q.M. (2017). Computational fluid dynamics simulation of airflow in human eyes under forced convection. Journal of Biomechanical Engineering 139(3): 031011. doi: 10.1115/1.4035449.

Colciago, C.M., Deparis, S., Domanin, M., Riccobene, C., Schenone, E. and Quarteroni, A. (2020). Analysis of morphological and hemodynamical indexes in abdominal aortic aneurysms as preliminary indicators of intraluminal thrombus deposition. Biomechanics and Modeling in Mechanobiology 19: 1035–1053.

Crowder, C., Helton, A., Williams, P., Bartolo, D. and Lin, C. (2013). Numerical simulations of fluid pressure in the human eye. Applied Mathematics and Computation 219: 11119–11133.

Ding, J., Guo, J., Shu, C., Liu, Y. and Wang, J. (2020). Hemodynamics analysis of blood flow in retinal arteries of diabetic retinopathy patients using computational fluid dynamics. BMC Medical Imaging 20(1): 1–14. doi:10.1186/s12880-020-00519-3.

Farhadi, H.F., Quinones, R. Jr., Ryu, K., Li, Y. and Weinreb, R.N. (2014). Computational modeling of drug distribution in the posterior segment of the eye: Effects of device variables and positions. Mathematical Biosciences 255: 11–20.

Feng, Y., Simpson, M.J. and Yu, D. (2016). Wall shear stress in a dynamic model of the ocular surface tear film. Journal of Fluids Engineering 138(10): 101106. doi:10.1115/1.4033272.

Freitas, Dayvison& Hoagland, Brenda & Valle, Antonio & Fraga, Beatriz & Barros, Mônica&Schubach, Armando &Paes, Rodrigo &Cuzzi, Tullia&Valete-Rosalino, Cláudia&Zancope-Oliveira, Rosely&Galhardo and Maria Clara. (2011). Sporotrichosis in HIV-infected patients: Report of 21 cases of endemic sporotrichosis in Rio de Janeiro, Brazil. Medical Mycology : Official Publication of the International Society for Human and Animal Mycology 50: 170–8. 10.3109/13693786.2011.596288.

Fu, L., Hu, L. and Yang, H. (2019). A close-packed imprinted colloidal array for naked-eye detection of glycoproteins under physiological pH. Biosensors and Bioelectronics 142: 111499. https://doi.org/10.1016/j.bios.2019.111499.

Gardner, T.W. and Antonetti, D.A. (2017). A new vision for diabetic retinopathy: Distinguishing diabetic macular edema and proliferative diabetic retinopathy. Investigative Ophthalmology & Visual Science 58(13): 696–702. doi:10.1167/iovs.17-22846.

Girad, F., Cloutier, G., Ethier, C.R. and Simmons, C.A. (2016). Factors influencing lamina cribrosa microcapillary hemodynamics and oxygen concentrations. Investigative Ophthalmology & Visual Science 57. ISSN: 1552-5783.

Goele, T.A., Jainta, S., Himmelbach, M. and Karnath, H.O. (2019). Challenges and techniques in eye gaze tracking for human computer interaction. Proceedings of the 4th International Conference on Human Computer Interaction and User Experience in Indonesia (CHIuXiD '15). DOI: 10.1145/2829794.2829831.

Goud, D. V. (2018). Anterior Segment of Eye: Imaging and Image Analysis (Doctoral dissertation, Indian Institute of Technology Hyderabad).

Harris, A., Siesky, B., Zarfati, D., Anwaruddin, R. and Arciero, J. (2013). Computational fluid dynamics simulation of aqueous flow in eyes with an Ahmed glaucoma valve. Medical Engineering and Physics 35(9): 1179–1185.

Hosseini, H., Nassiri, N., Nassiri, N., Schoen, J. and Lo, R. (2019). Computational fluid dynamics simulation of aqueous humor outflow in a realistic 3D model of the human eye. Journal of Biomechanics 82: 315–322.

Ji, Y.W., Li, E.W., Kim, H.J., Kang, M.K., &Seo, K.Y. (2018). Effect of artificial tears on tear film in a mouse dry eye model induced by Scopolamine. Korean Journal of Ophthalmology 32(1): 21–28. doi:10.3341/kjo.2017.0028.

Joukar, A., Mahmoudi, R. and Shakeri, M. (2015). A comparative study of thermal effects of 3 types of laser in eye: 3D simulation with bioheat equation. Journal of Thermal Biology 49-50: 74–81.

Karacali, A., Ceylan, M. and Sahin, M. (2019). Analysis of titanium outflow pipe implant for glaucoma treatment in biomechanical engineering. special issue of the 8th international advances in applied physics and materials science congress (APMAS 2018). Computational Fluid Dynamics. DOI: 10.12693/APhysPolA.135.561.

Li, Y., Sun, M., Wang, Z., Yu, Y., Yao, X. and Liu, Z. (2020). Microfluidic device to study tear film dynamics and its dysfunction in dry eye disease. Lab on a Chip 20(12): 2195–2207. doi:10.1039/c9lc01210g.

Lopes, D.M., Gonçalves, R. and Ferreira, D. (2016). Challenges in automatic analysis of fundus images in retinal diseases. Journal of Ophthalmology, 2016, Article ID 1395492. DOI: 10.1155/2016/1395492.

Macedo-de-Araújo, R., Ferreira-Neves, H., Rico-del-Viejo, L., Peixoto-de-Matos, S.C. and González-Méijome, J.M. (2016). Light distortion and spherical aberration for the accommodating and nonaccommodating eye. Journal of Biomedical Optics 21(7): 075003–075003.

Montalbán, R., Beltrán, B., Pintor, J. and Mateos, J.M. (2015). Numerical simulation of the impact of tears on the ocular surface. Computers in Biology and Medicine 59: 141–149. doi:10.1016/j.compbiomed.2015.02.002.

Monteiro, J.P., Santos, F.M., Rocha, A.S., Castro-de-Sousa, J.P., Queiroz, J.A., Passarinha, LA. et al. (2015). Vitreous humor in the pathologic scope: insights from proteomic approaches. Proteomics Clin Appl. 2015 Feb; 9(1-2): 187–202. doi: 10.1002/prca.201400133. Epub 2015 Jan 19. PMID: 25523418.

Nejad, H.R., Karimi, N., Jafari, R. and Gholizadeh, H. (2015). Effect of tissue and atmosphere's parameters on human eye temperature distribution. Journal of Thermal Biology 47: 51–58.

Ortiz, S., Pérez-Merino, P., Gambra, E., de Castro, A. and Marcos, S. (2012). *In vivo* human crystalline lens topography. Biomed. Opt. Express 3: 2471–2488. 10.1364/BOE.3.002471.

Pak, K.Y., Tran, P.L. and Mergen, S. (2018). Computational modeling of drug transport across the *in vitro* cornea. Computers in Biology and Medicine 92: 139–146.

Popeşcu, A., Savencu, A., Antohe, I., Cucu, R.G., Constantin, C. and Păduraru, D. (2011). Age-related eye disease and mobility limitations in older adults. Investigative Ophthalmology & Visual Science 52(10): 7168–7174. doi: https://doi.org/10.1167/iovs.11-7564.

Rafiq, M.Q., Chen, Z. and Teng, J.T.C. (2016). Thermal behavior of human eye in relation with change in blood perfusion, porosity, evaporation and ambient temperature. Journal of Thermal Biology 62: 138–142.

Rebuli, N.P., Du, K., Ravi, N. and Sobolewski, M. (2021). Predicting the fate of airborne allergens deposited on the ocular surface: A CFD study. Journal of Aerosol Science 157: 105999. doi:10.1016/j.jaerosci.2021.105999.

Safdel, M.H., Asadnia, M., Alijanpour, S. and Shamsi, M. (2021). Fluid–structure interaction analysis of alteration of the intraocular pressure on the optic nerve head in glaucoma. Journal of the Iranian Chemical Society 18: 1–15. https://doi.org/10.1007/s12596-021-00755-2.

Sanchez, C., Asin, K.E. and Artigas, F. (2015). Vortioxetine, a novel antidepressant with multimodal activity: review of preclinical and clinical data. Pharmacology & Therapeutics 145: 43–57.

Sánchez-Ferrer, M.L., Grima-Murcia, M.D., Sánchez-Ferrer, F., Hernández-Peñalver, A.I., Fernández-Jover, E. and Del Campo, F.S. (2017). Use of eye tracking as an innovative instructional method in surgical human anatomy. Journal of Surgical Education 74(4): 668–673.

Silva, A.C., Ribeiro, R.M. and Ferreira, J.G. (2020). Flow dynamics of vitreous humour during saccadic eye movements. Journal of the Mechanical Behavior of Biomedical Materials 110: 103860.

Silva, F., Ambrósio, R. and Oliveira, L. (2014). Numerical simulation of aqueous humor flow: From healthy to pathologic situations. Applied Mathematics and Computation 226: 777–792.

Soleri, P., Boccardi, S., Barbani, N., Cascone, M.G. and Cristofani, E. (2017). Aqueous flow in the presence of a perforated iris-fixated intraocular lens. Meccanica 52: 577–586. DOI: 10.1007/s11012-015-0313-y.

Souza, N.M., Barbosa, J.M. and Machado, R.S. (2021). Computational fluid dynamics applied to ophthalmology: A review. Computers in Biology and Medicine 132: 104348. doi:10.1016/j.compbiomed.2021.104348.

Sun, W. and Zhang, Y. (2012). CFD simulation of aqueous humor hydrodynamics in human eye. Computers in Biology and Medicine 42(4): 379–385.

Sun, J.K., Lin, M.M., Lammer, J., Prager, S., Sarangi, R., Silva, P.S. et al. (2014). Disorganization of the retinal inner layers as a predictor of visual acuity in eyes with center-involved diabetic macular edema. JAMA Ophthalmology 132(11): 1309–1316.

Tamm, E.R., Braunger, B.M. and Fuchshofer, R. (2015). The aqueous humor outflow pathways in glaucoma: A unifying concept of disease mechanisms and causative treatment. European Journal of Pharmaceutics and Biopharmaceutics 95: 173–181.

Tan, Jen Hong & Fujita, Hamido & Sivaprasad, Sobha & Bhandary, Sulatha & Rao, A. & Chua, Kuang and Acharya, U. (2017). Automated segmentation of exudates, haemorrhages, microaneurysms using single convolutional neural network. Information Sciences 420. 10.1016/j.ins.2017.08.050.

Tan, O., Chopra, V., Lu, A.T., Schuman, J.S., Ishikawa, H., Wollstein, G. et al. (2009). Detection of macular ganglion cell loss in glaucoma by Fourier-domain optical coherence tomography. Ophthalmology 2009 Dec; 116(12): 2305–14.e1-2. doi: 10.1016/j.ophtha.2009.05.025. Epub 2009 Sep 10. PMID: 19744726; PMCID: PMC2787911.

Tan, P.E., Yu, P.K. and Balaratnasingam, C. (2016). Quantitative confocal imaging of the retinal microvasculature in the human retina. pp. 55–68. *In*: The Ocular Circulation. Springer.

Tawfik, A.M., Rival, D.E. and Sun, J.K. (2019). Computational fluid dynamics and optical coherence tomography angiography in diabetic retinopathy. pp. 29–42. *In*: Retinal Imaging and Diagnostics. Springer.

Tiang, J.J., Tan, B. and Acharya, U.R. (2016). Effects of aqueous humor hydrodynamics on human eye heat transfer under external heat sources. Medical Engineering and Physics 38: 776–784.

Varikooty, J., Keir, N.J. and Jones, L. (2016). Effect of tear film components on the evaporation rate of tears in healthy and dry eyes. Investigative Ophthalmology & Visual Science 57(5): 2210–2217. doi:10.1167/iovs.15-18977.

Venkatavaradhan, P., R, S., Goud, N.V. and kumar, M.P. (2014). Challenges in automated analysis and interpretation of human retinal imaging. International Journal of Engineering Research and Applications 4(6): 265–270.

Vidyadharan, A., Kumar, V. and Reddy, M.C. (2015). A survey on challenges and techniques of eye tracking systems. International Journal of Advanced Research in Computer Science and Software Engineering 5(9): 519–523.

Villamarin, A., Navarro, P., Ortiz, S. and Castellanos, A. (2012). 3D simulation of the aqueous flow in the human eye. Medical Engineering & Physics 34: 1462–1470.

Wang, W., Tian, X. and Chen, X. (2022). Evaluation of flow field in the anterior segment during irrigation and aspiration in cataract surgery with three-dimensional computational simulation. Medical Engineering and Physics 99: 103741.

Wang, Z. and Meng, Y. (2021). Computational fluid dynamics modeling of ocular allergen deposition on human conjunctiva. Journal of Aerosol Science 157: 105959. doi:10.1016/j.jaerosci.2021.105959.

Wirtz, Mary and Keller, Kate. (2016). The role of the IL-20 subfamily in Glaucoma. Mediators of Inflammation 1–8. 10.1155/2016/4083735.

Zhang, X. and Balaratnasingam, C. (2018). Advances in OCT imaging for diabetic retinopathy. pp. 45–65. *In*: Diabetes and Retinopathy. Academic Press.

Zheng, F., Gregori, G., Schaal, K.B., Legarreta, A.D., Miller, A.R., Roisman, L. et al. (2016). Choroidal thickness and choroidal vessel density in nonexudative age-related macular degeneration using swept-source optical coherence tomography imaging. Invest Ophthalmol Vis Sci. 2016 Nov 1; 57(14): 6256–6264. doi: 10.1167/iovs.16-20161. PMID: 27849311; PMCID: PMC5114036.

# Chapter 9

# Modeling of Miscellaneous Bio-Fluid Flows

*Akshoy Ranjan Paul,*[1,*] *Anuj Jain*[1] and *Suvash C. Saha*[2]

## 1. Introduction

Miscellaneous bio-fluid flows are a wide range of fluid flow phenomena that occur in the body, including lymphatic flow, and cerebrospinal fluid flow. These flows can be classified according to their origin, purpose, or dynamics. For example, blood flow is primarily driven by the heart, providing oxygen and nutrients to the body's cells and carrying away metabolic wastes. Lymphatic flow is a one-way, low-pressure flow of fluid that helps maintain tissue fluid balance and transports immune cells, lipids, and other substances throughout the body. Cerebrospinal fluid flow is a low-pressure flow of fluid that circulates around the brain and spinal cord, providing a protective cushion and helping to regulate intracranial and spinal pressure. Other miscellaneous bio-fluid flows include mucus flow, sweat flow, and urine flow.

Modeling of miscellaneous bio-fluid flows involves the use of mathematical equations to describe and predict the behavior of various bio-fluids such as blood, urine, saliva, etc. It requires the use of sophisticated mathematical tools such as Computational Fluid Dynamics (CFD) and Finite Element Analysis (FEA). These tools are used to model the flow of the fluid and its interaction with the surrounding geometry. The modeling process also considers the effects of pressure, temperature, viscosity, and other fluid properties. By using these tools, researchers can simulate the behavior of bio-fluids and gain insight into how they behave under different

[1] Department of Applied Mechanics, Motilal Nehru National Institute of Technology Allahabad, Prayagraj-211004, India.

[2] School of Mechanical and Mechatronic Engineering, University of Technology Sydney (UTS), 15 Broadway, Ultimo, NSW-2007, Australia.

* Corresponding author: arpaul@mnnit.ac.in

DOI: 10.1201/9781003287636-9

conditions. This knowledge can be used to better understand disease processes, improve medical treatments, and develop new medical devices.

## 2. Lymphatic Flow

Lymphatic flow is the movement of lymph through the lymphatic system. It occurs when lymph moves through the lymphatic vessels toward the heart, carrying waste products and proteins away from tissues and organs (Fig. 1). Lymphatic flow is powered by the contraction of skeletal muscles, lymphatic pump techniques, and breathing. It is important for maintaining fluid balance and transporting immune cells and other substances throughout the body.

### *2.1 Anatomy and Physiology of Lymphatic System Flows*

The lymphatic system is a network of vessels, organs, and tissues that help to filter and transport lymph fluid throughout the body (Fig. 1). This fluid is composed of white

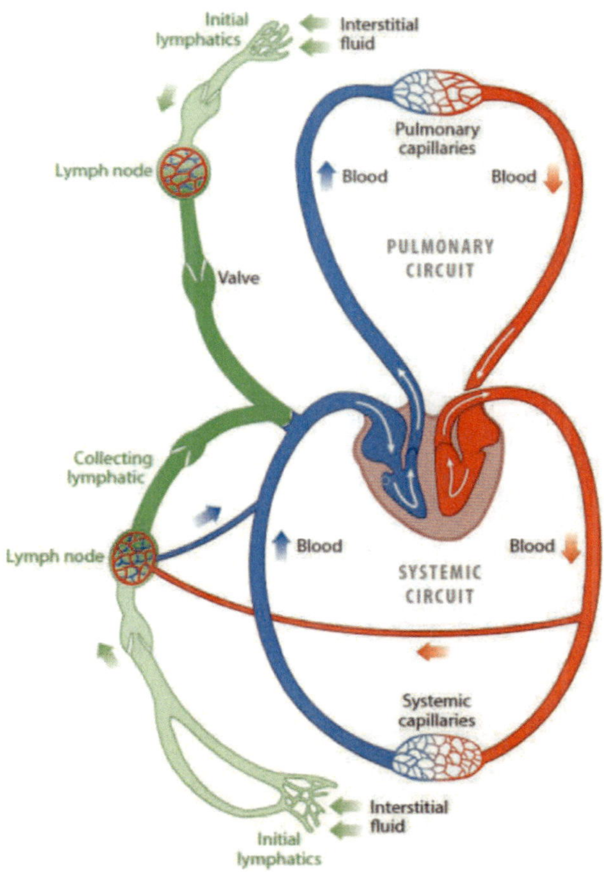


**Figure 1.** Blood circulation and Lymphatic vascular system [Lubopitko Encycl. Corp., 2016].

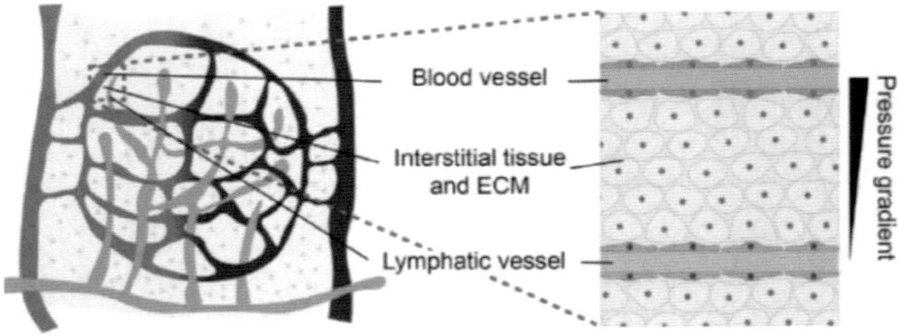


**Figure 2.** Transport unit of solid tissues, showing a blood vessel, interstitial tissue and extracellular matrix (ECM), and a lymphatic vessel [Lee et al., 2022].

blood cells, proteins, and other substances that help to fight infection and disease. The lymphatic system is responsible for removing excess fluid, waste, and toxins from the body. It also helps to transport nutrients and oxygen to cells and tissues. The lymphatic system is composed of lymphatic vessels, lymph nodes, and lymphoid organs. Lymphatic vessels are thin tubes that transport lymph fluid throughout the body (Fig. 2). Lymph nodes are small, bean-shaped organs that filter the lymph fluid and help to fight infection. Lymphoid organs are larger organs that produce and store lymphocytes, which are white blood cells that help to fight infection. The lymphatic system flows in one direction, from the tissues to the lymph nodes and then to the lymphatic vessels. The lymphatic vessels then transport the lymph fluid to the thoracic duct, which empties into the bloodstream. The lymphatic system is important for maintaining fluid balance in the body and for fighting infection. Lymph tissues are located throughout the body and include tonsils, adenoids, spleen, thymus, lymph nodes, Mucosa-Associated Lymphoid Tissue (MALT), and Peyer's patches.

## *2.2 Lymph Properties*

Lymph is a clear, slightly yellowish fluid that is composed of white blood cells, proteins, and other substances. It is found in the lymphatic system, which is a network of vessels and nodes that helps to transport fluid, nutrients, and waste products throughout the body. Lymph is important for maintaining the body's fluid balance, fighting infection, and removing waste products from the body. It also helps to transport fat and fat-soluble vitamins throughout the body. The density of lymph is about 1000 kg/m$^3$ (Mazzoni et al., 1987) and the dynamic viscosity is between $1.5 \times 10^{-6}$ and $2.2 \times 10^{-6}$ Ns/m$^2$ (Dixon et al., 2006; Burton-Opitz and Nemser, 1917).

## *2.3 Lymphatic Vessels*

The largest lymphatic vessel in the human body is the thoracic duct. It is a large, tube-like structure that runs from the abdomen to the chest and is responsible for carrying lymphatic fluid from the lower body to the heart. The diameter of the thoracic duct in the human body is typically between 5 and 8 mm (Knipe et al.,

2013). It is estimated that the lymphatic system produces approximately 3 liters of lymph fluid per day (Kumar et al., 2019). The type of lymphatic flow is pulsatile and hence is characterized by a Womersley number of 0.1–0.2 (Kumar and Saini, 2018). The flow in the lymphatic vascular system is always laminar.

## *2.4 Lymphatic Valves*

Lymphatic valves are one-way valves that are located in the lymphatic vessels. They are responsible for regulating the flow of lymph fluid and preventing it from flowing backward. The valves are made up of two leaflets of connective tissue that form a semilunar shape along the lymphatic vessels. They open and close in response to pressure changes in the lymphatic system, allowing lymph fluid to flow in one direction only. This helps to ensure that the lymphatic system is able to efficiently transport lymph fluid throughout the body. The smallest lymphatic vessels contain primary lymphatic valves, while the larger collecting vessels feature bi-leaflet secondary lymphatic valves, which actively pump the lymphatic fluid (lymph). Extrinsic pumping can be caused by adjacent tissue movement, while intrinsic contractility of specialized muscle cells in lymphatic vessel walls can result in intrinsic pumping.

Fluid is taken up from the interstitial by the initial lymphatics. The fluid in the lymphatic vessel is known as lymph. This lymph is transported from the initial lymphatics to larger vessels which have walls that are able to contract and propel the lymph forward, as well as valves to reduce backward flow (Fig. 3). The lymph is transported through the body by these collecting lymphatic vessels and is returned to the bloodstream.

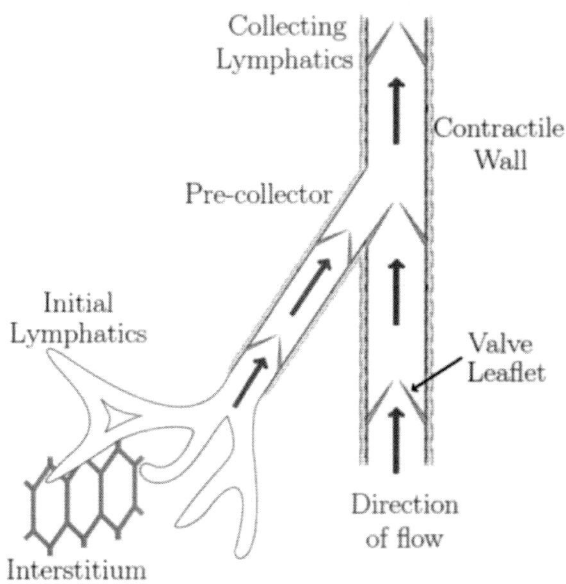


**Figure 3.** Exaggerated view of a Schematic of a section of the lymphatic system.

## *2.5 Role of Primary Valves in Interstitial Fluid Take-Up*

Interstitial fluid take-up is the process by which lymphatic vessels absorb fluid from the interstitial space (the space between cells) in the body. This fluid is then carried to the lymph nodes, where it is filtered and returned to the bloodstream. This process is an important part of the lymphatic system's role in maintaining fluid balance and immunity in the body. Primary valves exist in the confluence of primary lymphatic vessels and in the larger vessels of the lymphatic system (called primary lymphatics) play an important role in interstitial fluid take-up by regulating the amount of fluid that enters and exits the tissue. They act as one-way valves that allow the fluid to move in and out of the tissue while preventing any backflow of fluid. This ensures that the tissue remains properly hydrated while helping to maintain the balance of fluid in the body. Secondary lymphatic valves are valves found in the larger collecting lymphatic vessels (called collecting lymphatics), which are more distal than the primary lymphatic vessels. These valves too are one-way valves that are designed to prevent the backflow of lymph and help to regulate the rate of flow of lymph. They are important for maintaining the correct pressure and flow of lymph in the body, which is critical for the proper functioning of the lymphatic system.

## *2.6 Intrinsic Pumping in the Lymphatic Vascular System*

Intrinsic pumping in the lymphatic vascular system is a process whereby the walls of the lymphatic vessels contract rhythmically to propel lymph throughout the body. This pumping action helps to maintain the pressure of the lymphatic system and facilitates the transport of nutrients and wastes. Intrinsic pumping is carried out by the smooth muscle cells that line the walls of the lymphatic vessels. The contractions of these cells create a pulsating motion that moves along the walls of the vessels and pushes the lymph forward.

## *2.7 Mass Transport of Vasoactive Substances in Lymphatic Vessels*

The lymphatic system is a network of vessels and organs that transport lymph, a clear fluid containing white blood cells, throughout the body. This system also plays an important role in the transport of vasoactive substances, such as hormones, cytokines, and growth factors, which are essential for maintaining homeostasis. The lymphatic vessels are lined with a single layer of endothelial cells, which are responsible for the transport of vasoactive substances. These cells are able to absorb and transport molecules, such as hormones, cytokines, and growth factors, from the interstitial fluid into the lymphatic vessels. The molecules are then transported to the lymph nodes, where they are further processed and released into the bloodstream.

The vasoactive substances can be transported through the lymphatic vessels by various means, including diffusion, active transport, and bulk flow. Diffusion is the process by which molecules move from an area of higher concentration to an area of lower concentration. Active transport is the process by which molecules are moved against a concentration gradient, usually requiring energy. Bulk flow is the movement of molecules in a large group, usually due to pressure or gravity. Diffusion is the most common way for vasoactive substances to be transported.

The transport of vasoactive substances in the lymphatic vessels is regulated by several factors, including the size of the molecules, the concentration of the molecules in the interstitial fluid, and the presence of specific receptors on the endothelial cells. Additionally, the transport of these molecules is affected by the contractile activity of the lymphatic vessels, which is regulated by the autonomic nervous system. The transport of vasoactive substances in the lymphatic vessels is essential for maintaining homeostasis and for the proper functioning of the immune system. It is also important for the delivery of nutrients and hormones to the cells and tissues of the body.

### *2.8 Fluid Mechanics of Lymphatic Diseases*

Lymphatic diseases are a group of disorders that affect the lymphatic system, which is a network of vessels and nodes that help to transport lymph fluid throughout the body. The lymphatic system is also responsible for filtering out bacteria, viruses, and other toxic and foreign substances from the body. This fluid is thus important for maintaining the body's immune system. When the lymphatic system is not functioning properly, it can lead to a variety of lymphatic diseases including lymphedema, lymphoma, and Hodgkin's disease. Primary lymphedema is a condition in which the limbs swell due to an accumulation of fluid. It is caused by a disruption in the lymphatic system due to a genetic defect, which is responsible for draining fluid from the body. Lymphedema can affect any part of the body but is most commonly seen in the arms and legs. Symptoms of lymphedema include swelling, a feeling of heaviness, and tightness in the affected area. Secondary lymphedema is a type of lymphedema caused by damage to the lymphatic system due to surgery, radiation therapy, trauma, infection, or other medical conditions. Lymphoma is a type of cancer that affects the lymphatic system, which is part of the immune system. It is a cancer of the lymphocytes, which are white blood cells that help fight infection. The normal range for lymphocyte concentration in the blood is between 20–45% of the total White Blood Cell (WBC) count. In cases of lymphoma, the lymphocyte concentration typically increases significantly. This is because lymphoma cells are a type of lymphocyte, and they rapidly multiply and accumulate in the body. Symptoms of lymphoma can include swollen lymph nodes, fever, night sweats, weight loss, and fatigue. Hodgkin's disease is a type of cancer that affects the lymphatic system. It is characterized by the presence of a type of cell called a Reed-Sternberg cell. Symptoms of Hodgkin's disease may include swollen lymph nodes, fever, night sweats, and weight loss.

## 3. CFD Modeling of Lymphatic Flow

CFD modeling of lymphatic flow involves the use of numerical techniques to simulate the flow of lymphatic fluid through the body. This type of fluid dynamics simulation is often used to evaluate the effectiveness of new treatments or medical devices for treating diseases such as lymphedema or to better understand the mechanisms of lymphatic drainage. Lymphedema is an incurable form of tissue swelling that can occur if there is a disruption to lymphatic pumping. CFD models have been used to study the behavior of lymphatic fluid in complex geometries, to identify potential areas of obstruction or stagnation, and to develop improved methods for lymphatic drainage.

CFD modeling of lymphatic valves has been used to investigate the effects of various factors on the structure and function of these valves. The results of such studies can provide valuable insights into the mechanism of valve closure and its influence on the effectiveness of lymph drainage. In particular, CFD modeling can be used to study the effect of valve geometry, permeability of the valves, flow rate, and blood pressure on the closure mechanism of the lymphatic valves. Additionally, CFD models can be used to investigate the effect of factors such as tissue stiffness and wall thickness on the valve's closure dynamics. Such studies can help to better understand how these valves function *in vivo* and may lead to the development of new treatments to improve the efficiency of lymphatic drainage. CFD modeling can also be used to investigate the effects of different treatments, such as the use of drugs or devices, on the flow of lymphatic fluid. CFD simulations can provide insight into the dynamics of the lymphatic system and may be used to optimize the design of lymphatic valves and other treatments.

CFD modeling provides a way to study the lymphatic vessels and nodes without having to physically examine them. Creating three-dimensional fluid-structure interaction models of the secondary lymphatic valves and image-based models of lymph nodes increases the understanding of how the lymph flows through the system. This is beneficial due to the small size and fragility of the vessels and nodes, making it too difficult to physically study them. An idealized geometry of the lymphatic valve and its surrounding wall was analyzed using COMSOL Multiphysics, a commercial finite element analysis (FEM) software. Through the implementation of a material model that captures the highly non-linear stress-strain behavior of the wall, two-dimensional and three-dimensional models are being utilized to explore the closure behavior of the valve (Cooper, 2016). COMSOL Multiphysics was used to create image-based models of murine lymph nodes. Selective plane illumination microscopy and synchrotron micro-computed tomography were employed to take the necessary images, which are then employed to determine the geometry and permeability of the lymph node tissue to the flow of lymph. A linear relation is established between the gray scale values in the images and the permeability value, assuming the lighter the grayscale, the denser the material and thus having lower permeability. The model demonstrates that the flow pathways of the fluid are impacted by the internal structure of the node (Fig. 4).

The study of novel lipid-based drug particles in lymphatic networks has nowadays received attention due to its improved release control and better chemical stability over colloidal drugs (Ahn and Park, 2016). The lymphatic network's ability to absorb large molecules and avoid first-pass liver metabolism renders it a desirable route for the delivery of therapeutic substances. In the majority of cases, the drug is injected into the subcutaneous tissue, and then it must traverse the tissue space before being drained off by the lymphatics. However, it is difficult to monitor the transportation of drug molecules via this path, and thus, computational models of lymphatic drainage can be a great help in evaluating the effectiveness of a proposed delivery system. A

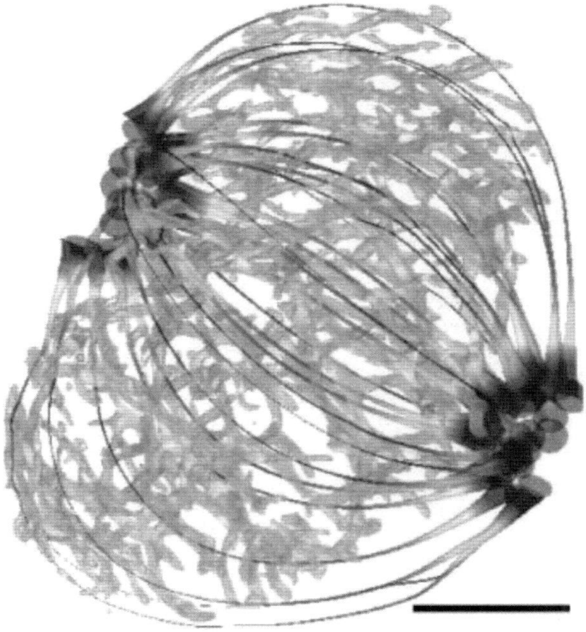

**Figure 4.** Stream tubes showing the pathways of fluid flow through lymph nodes [Cooper, 2016].

three-dimensional computational model of the peripheral lymphatic network and its nearby interstitium based on anatomical data, has been developed by Jayathungage et al. (2022) to quantify how drug particle size, physiological flow rates, and injection details influence uptake and transit times.

Introducing a multi-layer microfluidic platform that allows users to culture engineered blood and lymphatic micro vessels while providing independent control of blood, lymphatic, and interstitial fluid pressures is a current trend (Selahi and Jain, 2022). Lee et al. (2022) measured fluid velocity using optical microscopy methods in correlation with varying rates of interstitial fluid flow in response to applied input pressures of blood, lymphatic, and interstitial pressure. These results agree with CFD models. The use of microfluidic and computational platforms will enable us to analyze important fluid mechanical characteristics and cellular processes that are linked to diseases such as lymphedema and solid cancer that are caused by imbalances in fluids.

## 4. Splanchnic Fluid Flow

The splanchnic circulation is responsible for the distribution of blood from the abdominal viscera to the liver and then back to the heart. It also facilitates the movement of nutrients and hormones throughout the body. Splanchnic fluid flow is regulated by the autonomic nervous system and hormones and is important for maintaining homeostasis. Splanchnic motility is the movement of the viscera and

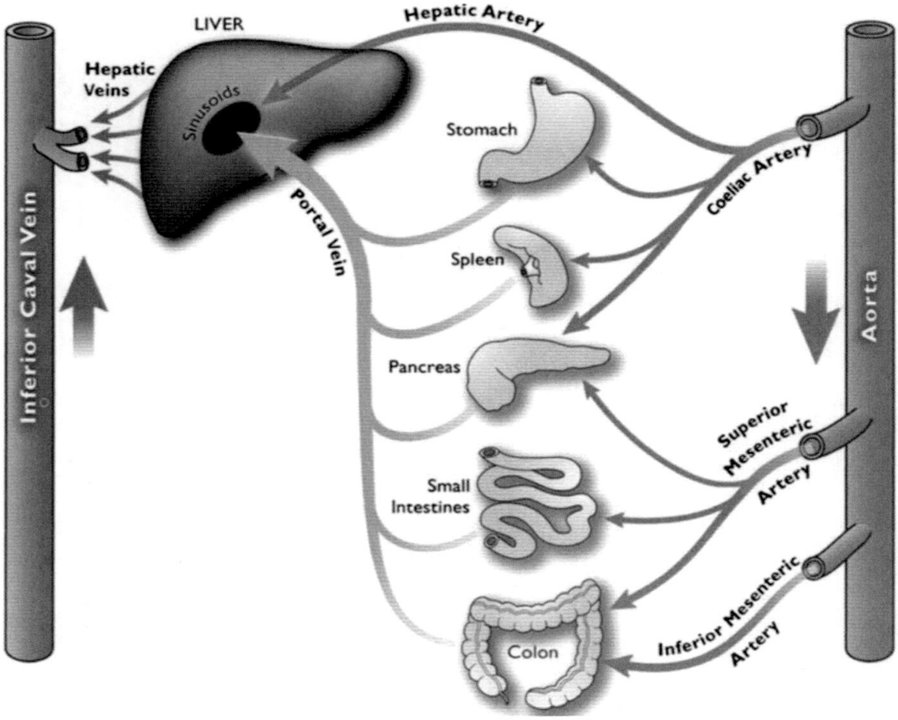


**Figure 5.** Schematic representation of the splanchnic circulation [Harper and Chandler, 2016].

is controlled by the enteric nervous system, hormones, and the autonomic nervous system. It is important for digestion and absorption of nutrients. The splanchnic circulation is composed of six parallel circulations, namely—gastric, small intestinal, colonic, pancreatic, hepatic, and splenic (Fig. 5). The celiac artery, Superior Mesenteric Artery (SMA), and Inferior Mesenteric Artery (IMA) are the three major arteries that supply these organs, and they join together to form a network of smaller arteries (Harper and Chandler, 2016).

The complexity of the circulatory system of certain splanchnic organs is due to the presence of intramural circulation. It is just as important to redistribute the total blood flow between the intramural blood vessels as it is to have sufficient total blood flow. Various extrinsic and intrinsic factors can affect the splanchnic circulation. External influences such as the cardiovascular system, autonomic nervous system, and neurohumoral agents can affect the hemodynamic conditions. Internal mechanisms, such as the properties of the vasculature, metabolites, intrinsic nerves, paracrine substances, and hormones, are also at play. All of these regulatory mechanisms work together to restrict any drastic changes to tissue perfusion (Parks and Jacobson, 1985).

## *4.1 Anatomy and Physiology of Splanchnic Blood Flow (SBF)*

The splanchnic circulation is the major source of blood flow to the liver and gastrointestinal tract, which helps to regulate the body's metabolic and endocrine functions. The hepatic artery, originating from the celiac artery, supplies oxygenated blood to the liver (Fig. 5). The hepatic veins are the main veins that drain oxygen-depleted blood from the liver. The portal vein is a large vein that carries venous blood from the intestine, spleen, and pancreas to the liver. This vein also helps to regulate the flow of hormones and nutrients to and from the organs. The splenic artery, arising from the celiac artery, supplies oxygenated blood to the spleen, and the splenic veins drain oxygen-depleted blood from the spleen. The superior mesenteric artery is a branch of the abdominal aorta that supplies oxygenated blood to the small intestine, and the superior mesenteric vein drains oxygen-depleted blood from the small intestine. The anatomy of the splanchnic circulation is complex, but the vessels involved are crucial for the proper functioning of the digestive and metabolic systems. Proper regulation of the splanchnic circulation is essential for maintaining homeostasis in the body.

The blood goes through many small, sinusoidal channels in the liver before ultimately exiting via the hepatic veins and entering the vena cava, which leads to the heart. The diagram reveals an unusual pattern of blood flow in the gastrointestinal system. The oxygen-rich blood runs through multiple parallel pathways to the stomach, spleen, small intestine, large intestine, pancreas, and liver. It then moves through the portal vein, into the liver and out of it through the hepatic veins, and finally into the inferior vena cava. Around 30% of the cardiac output is sent to the liver via the celiac, superior mesenteric and inferior mesenteric arteries. The portal vein carries 1300 mL/minutes of blood to the liver and 500 mL/minutes of oxygenated blood is pumped to the organ by the hepatic artery while fasting.

The splanchnic blood flow is regulated by the autonomic nervous system, hormones, and local factors. The sympathetic nervous system plays an important role in the regulation of splanchnic blood flow. Under resting conditions, the sympathetic nervous system acts to constrict the splanchnic blood vessels, thus reducing the blood flow to the digestive organs. This is known as the sympathetic tone. In response to a meal, the sympathetic tone is reduced and parasympathetic nervous system activity is increased, leading to an increase in splanchnic blood flow. Hormonal factors, including insulin, cortisol, and glucagon, also play an important role in regulating splanchnic blood flow. Insulin causes vasodilation in the splanchnic vessels, increasing blood flow and facilitating the absorption of glucose from the gut. Cortisol and glucagon, on the other hand, cause vasoconstriction and reduce blood flow to the splanchnic organs. Local factors, such as the presence of metabolites, also play a role in splanchnic blood flow. For example, the presence of metabolites in the gut can cause vasodilation, leading to an increase in blood flow.

The amount of blood that flows through the splanchnic area is usually 30 milliliters per minute for every 100 grams of tissue, which makes up approximately 15–30% of the total blood output from the heart and plays a key role in the cardiovascular system. It is a major reservoir and regulator of blood volume, and body activities like exercise can reduce the amount of blood flow in the splanchnic circulation. The amount of blood flow to the splanchnic circulation may be reduced to < 10 ml per min per 100 g of tissue in times of low cardiac output (during fasting), or it can reach a maximum of 250 ml per min per 100 g of tissue post-meal (at peak hyperaemia). The splanchnic circulation is thus highly adaptable. The complexity of how physiological regulation of SBF works is considerable, yet the academic discussion mainly concentrates on three circulatory aspects: intrinsic, extrinsic and humoral.

## *4.2 Anatomy and Physiology of Gastrointestinal (GI) Flows*

The gastrointestinal (GI) tract comprises of glands and hollow organs extending from the mouth to the anus. These organs help to chop food into small pieces, lubricate it, start digestion of carbohydrates and fats in the mouth, and then act as a conduit for the stomach in the oesophagus. The stomach serves as a temporary storage for food and also produces acid and protease to initiate the digestive process. It has a capacity of approximately 1 L. The small intestine is the location for the absorption of nutrients and minerals; it is roughly 7 m long in adults. In an adult human, the large intestine has a length of approximately 1.5 m and its main purpose is to reabsorb electrolytes, such as chloride, sodium, and magnesium, and absorb a large quantity of liquids – 90% of the absorbed water (Kumar et al., 2019). This intestine also stores the feces before they are expelled from the body. Additionally, the salivary glands, liver, and pancreas, which all produce digestive enzymes, are located in this intestine.

The physiology of gastrointestinal flow involves the movement of food and fluids through the digestive tract. This process is regulated by a complex network of hormones, neurotransmitters, and neural pathways. Ingested food is broken down into nutrients by mechanical and chemical means in the stomach before being passed through the intestines. Intestinal contractions help to move the food and fluids through the digestive tract and into the small intestine. In the small intestine, further breakdown of the food occurs and absorption of nutrients into the bloodstream takes place. From here, the food and fluids continue their journey through the remaining parts of the gastrointestinal tract until they are eventually excreted. The process of digestion is regulated by hormones such as gastrin, secretin, and cholecystokinin. These hormones stimulate the release of digestive juices and enzymes, which break down the food into smaller particles. They also help to regulate the rate of gastrointestinal flow. Neurotransmitters like serotonin, acetylcholine, and dopamine also play important roles in regulating gastrointestinal flow. The autonomic nervous system, which consists of the sympathetic and parasympathetic nervous systems, helps to regulate digestive activity by modulating the tone of the gastrointestinal muscles.

The smooth muscle membrane in the gastrointestinal tract produces electrical potential at between 5 and 15 mV, leading to the creation of 'slow waves' that control the rhythmical contraction of the gastrointestinal muscles. Typically, the stomach has a contraction rate of about 3 cycles per minute, and the duodenum has a frequency of 12 times per minute. The ileum usually contracts 8–9 times each minute (Helmcke, 2020).

The gastrointestinal (GI) tract can be thought of as a continuous tube that extends from the mouth to the anus. This tube undergoes multiple processes in order to break down food and absorb the nutrients that the body requires. Mastication (chewing) is accomplished by a jaw muscle force, with a pressure of 55 pounds on the incisors and 200 pounds on the molars. This action helps digestion by breaking the food down into smaller pieces, thus increasing its surface area followed by swallowing (deglutition). The stomach has three main duties when it comes to motor functions: it can store large amounts of food prior to processing; it mixes the food with its secretions to create a semi-liquid substance known as chyme; and it slowly empties the chyme into the small intestine so it can be more effectively digested and absorbed.

The small intestine's movements are divided into Mixing Contractions and Segmentation Contractions. The latter is caused by electrical impulses that make the longitudinal and circular muscles contract, resulting in a section of the intestinal wall (1–2 cm) being contracted, which causes the lumen to be narrowed. The maximum frequency of this type of contraction is 12 times per minute. Peristaltic waves cause propulsive movements in the small intestine that move chyme toward the anus. This motion is at a speed of 0.5–2.0 cm/second and is more rapid in the initial section and slower in the end section of the small intestine. The large intestine has motions that are generally slow, owing to its primary responsibility for the absorption of water and electrolytes. These movements can be categorized in the same way as those of the small intestine, namely, mixing and propelling. Various glands produce secretions to enable digestion and protection of the gastrointestinal (GI) tract. Single-cell mucous glands, pits, tubular glands, salivary glands, the pancreas, and the liver are responsible for producing digestive enzymes, mucus for lubrication, and other essential secretions for proper digestion and absorption. The use of enzymes in chemical digestion facilitates the breaking down of carbohydrates, fats, and proteins into tiny molecules, alongside water, electrolytes, and other substances that can be absorbed. Once these molecules have been reduced to a size suitable for absorption, they are transported from the intestinal wall to the blood and lymphatic capillaries.

The intestines cycle daily through about 9000 mL of liquid, which includes 2000 mL of ingested fluid and 7000 mL of secretions from the gastrointestinal tract and related organs. The majority of this fluid, 98%, is taken back into the body and only 0.2 L is released through stools (Kaw and Mahan, 2015).

## *4.3 Fluid Mechanics of Gastrointestinal Diseases*

This sub-section discusses some of the common gastrointestinal disorders that are seen globally, including acid reflux (heartburn), vomiting, diarrhea (both acute and chronic), and peptic ulcer disease. There are more than fifteen of such GI diseases.

Acid reflux (commonly known as heartburn) is a condition in which stomach acid flows back up into the esophagus and causes a burning sensation in the chest or throat. Symptoms can include a sour taste in the mouth, a sore throat, chest pain, and difficulty swallowing. Esophageal pH monitoring is the optimal way to diagnose acid reflux, which is when stomach acid travels backward in the esophagus, causing it to have a pH of less than 4.

Nausea and vomiting are uncomfortable symptoms that can occur due to a variety of conditions, including motion sickness, viral infections, food poisoning, certain medications, tumors, overindulging, or head injuries and surgery. If a person has severe vomiting, this may lead to fluctuations in their body's electrolytes and dehydration.

Peptic ulcer disease is caused by the acidic juices of the stomach damaging the lining of the stomach or small intestine. Symptoms of this condition include GI bleeding (usually the most noticeable sign), stomach perforation that can lead to severe bleeding and infection of abdominal organs, and obstruction of the gastric outlet that leads to vomiting without bile.

The World Health Organization (WHO) defines diarrhea as having loose or watery stools at least three times in a single day. This is a frequently encountered health problem throughout the world, particularly in children. If it is present for a period of fewer than two weeks, it is known as acute diarrhea, while chronic diarrhea is an issue that lasts for more than a month. Usually, the cause of acute cases is a virus. The rapid transit of stools through the large intestine is a symptom of diarrhea, which is when the intestine doesn't have enough time to absorb the necessary water, ions, and electrolytes. Treating diarrhea involves managing water and electrolyte levels in the body, as it can lead to dehydration and electrolyte imbalances, similar to vomiting.

## 4.4 CFD Modeling of Splanchnic and Gastric Fluid Flow

Computational Fluid Dynamics is a powerful tool for simulating complex fluid flow phenomenon. It can be used to predict the distribution of splanchnic fluids in a variety of organs and tissues. CFD can be used to evaluate the effects of physiological and environmental factors on splanchnic fluid flow dynamics and to assess the impact of interventions on the flow rate, shear stress, and pressure gradients in these systems. CFD can also be employed to study the nature of turbulence and its effect on splanchnic fluid properties. Additionally, CFD can be used to design new medical devices and instrumentation for splanchnic fluid analysis and optimization.

Understanding the dynamics of gastric contents during digestion is also becoming an area of increasing research due to their role in food disintegration and the subsequent bioavailability of nutrients and bioactive compounds. There are new opportunities to enhance our understanding of how gastrointestinal motility affects the function of the digestive tract in health and disease by numerically analyzing the dynamics of food digesta systems under various physiological conditions. However, this approach is challenging due to several issues related to numerical and modeling aspects, such as the intricate geometry, motility, and physics of the digestive process. Furthermore, a good *in-vivo* comprehension of the interaction between the properties of the digesta and the motor response of the GI wall is also essential for this approach.

Gastro-oesophageal reflux disease affects a considerable proportion of the population and is believed to be mainly caused by dysfunction of the gastro-esophageal junction (GOJ). McMahon et al. (2007) utilized CFD techniques to examine the flow patterns that occur at the GOJ to gain insights into the GOJ's role as a valve mechanism that prevents the backward flow of gastric contents into the oesophagus during a reflux event. They constructed a simplified two-dimensional CAD model of the GOJ geometry and imposed a negative pressure difference of 10 cm $H_2O$ column between the upper and lower boundaries. The study predicted the upward and steady flow of a water-like fluid across the GOJ under these conditions. The fluid flow's maximum velocity was almost 2 m/s at the narrowest part of the GOJ, which was 1.46 mm in diameter, and it formed a jet that curved towards the left side of the oesophagus.

In the postprandial state, the stomach's motility is distinguished by the occurrence of a sequence of periodic and rhythmic Antral Contraction Waves (ACWs). Within the antrum region, the gastric flow's fundamental dynamics involved the development of two primary flow patterns. These included a robust, jet-like motion with a retropulsive effect located at the most occlusive antral contraction wave (ACW) and sluggish recirculations (eddies) that occurred between consecutive waves. The retropulsive jet and eddies structures intensified with increasing distal propagation and occlusiveness of the Antral Contraction Wave (ACW), achieving velocities of up to 7.4 mm/s and 2 mm/s, respectively (Pal et al., 2004). Ferrura and Singh (2010) developed a three-dimensional CFD model of stomach region to investigate the shape and motility pattern of the stomach wall during digestion and the same was used to study the flow characteristics of gastric contents at different viscosities. With the increase of viscosity of the gastric contents, the main mechanism of driving digestion—the retropulsive jet-like motion and eddy structures were found reducing considerably and pressure was increased in the stomach region (refer to Fig. 6). This suggests that a highly viscous meals are poorly mixed while moving down to the gastric passages. In another CFD investigation, Ferrura and Singh (2011) showed that viscous effects of high viscous Newtonian fluid stopped the food particles immediately after being released into the slow gastric flow predicted at the location of the cardia. this result also supports the theory that gastric dilution may actually play a critical role in determining the rheological properties of gastric contents associated with highly viscous meals (Marciani et al., 2001).

In a recent study by Li and Jin (2021), the impact of gastric motility on the movement of gastric contents, including the mixing of gastric juices and food, was examined. To gain insights into the flow dynamics involved in the mixing process

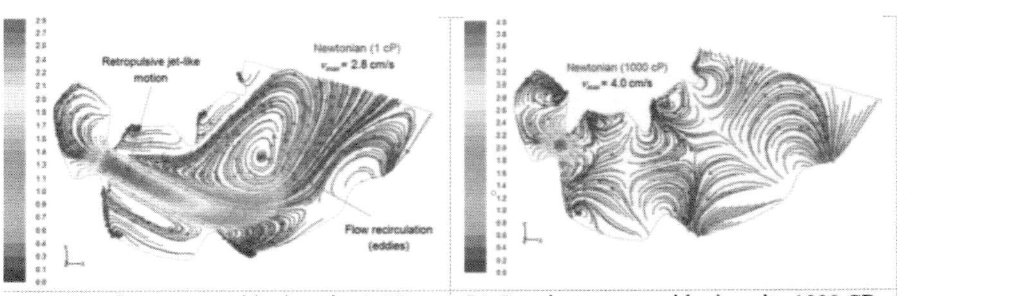


**(a)** Gastric contents with viscosity 1 CP **(b)** Gastric contents with viscosity 1000 CP

**Figure 6.** Effect of viscosity on the formation of the retropulsive-jet-like motion and eddy structures.

of liquid foods with gastric juice, a CFD model was developed by the authors. The numerical findings revealed that the Terminal Antral Contraction (TAC) plays a significant role in increasing the kinetic energy within the stomach. The TAC generates retropulsive jets and leads to a decrease in the pH of the liquid gastric contents. The simulation results highlighted that the density disparity between the food and stomach juice has substantial implications for the spatial distribution of gastric pH dynamics.

The application of CFD techniques presents new possibilities for enhancing our comprehension and modeling of the underlying mechanisms involved in human digestion. The CFD models have shown their capability in providing an improved understanding of the fluid dynamics that influence gastrointestinal (GI) function across various physiological conditions. Recent advancements in spatiotemporal mapping techniques and medical imaging technologies have proven valuable, as they offer unique means to visualize and characterize the motor and secretion functions of the GI wall when subjected to different diets. A multidisciplinary approach integrating the mathematics, science, engineering, and medicine are essential to register further progress in the computational research in splanchnic and gastric fluid flow.

## 5. Synovial Fluid Flow

Synovial fluid is a viscous fluid secreted by the membranes that line the joint cavities of synovial joints. It functions to reduce friction at the articulating surfaces of the joint, and to provide nutrients to the cartilage that lines the joint surfaces. It also helps to cushion and nourish the joint surfaces, and to transport away metabolic waste products. Synovial fluid flows in a laminar fashion in the joint capsule, and is propelled by the movement of the joint.

### *5.1 Anatomy and Physiology of a Synovial Joint*

A synovial joint is a type of joint that allows for movement between two bones (Fig. 7). It is the most common type of joint in the human body and is found in the knees, elbows, hips, and shoulders. Anatomically, a synovial joint consists of the following components:

*Articular Cartilage*: A thin layer of cartilage that covers the ends of the bones and helps to reduce friction between them. Articular cartilage is composed of a network of collagen fibers and proteoglycans, which are molecules that help to bind water and provide cushioning. This cushioning helps to absorb shock and reduce wear and tear on the joint.

*Synovial Membrane*: The synovial membrane is a thin layer of tissue that lines the joint capsule of a synovial joint. It is composed of two layers of cells, the inner layer of cells is called the synoviocytes and the outer layer of cells is called the fibroblasts. The synovial membrane secretes a lubricating fluid called synovial fluid which helps to reduce friction between the bones of the joint. The synovial membrane also contains blood vessels, nerves, and lymphatic vessels which provide nutrition and

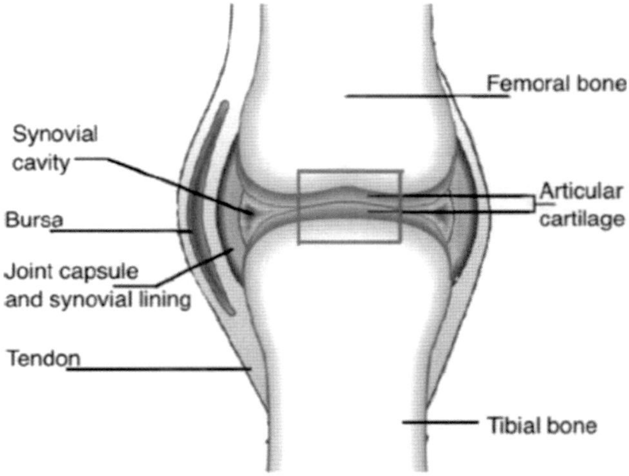


**Figure 7.** Schematic diagram of a synovial joint showing synovial cavity [Moghadam et al., 2015].

sensation to the joint. The synovial membrane also helps to protect the joint from infection and injury.

*Synovial fluid*: Synovial fluid lines the joint capsule and helps to reduce friction between the bones and other tissues in the joint (Fig. 8). It also helps to nourish the cartilage and provide nutrients to the joint. Synovial fluid helps to reduce the amount of wear and tear on the joint surfaces, allowing for smooth and pain-free movement.

*Joint Capsule*: The joint capsule is a fibrous membrane that surrounds a synovial joint and holds the bones together. It is composed of an outer layer of dense connective tissue and an inner layer of synovial membrane. The joint capsule is filled with synovial fluid.

*Ligaments*: Ligaments are strong bands of connective tissue that connect bones to other bones in a synovial joint. They provide stability and support to the joint and limit the range of motion. Ligaments are composed of collagen fibers and are usually found in pairs, with one ligament on either side of the joint. They are usually very tough and can withstand a lot of force. Ligaments also help to absorb shock and reduce friction between the bones. Without ligaments, the bones in a synovial joint would not be able to move properly and would be more prone to injury.

*Tendons*: Tendons are strong, fibrous connective tissues that attach muscles to bones. In a synovial joint, tendons are responsible for connecting the muscles to the bones and allowing for movement. Tendons are made up of collagen fibers and are surrounded by a sheath of connective tissue. The sheath helps to protect the tendon from friction and wear and tear. Tendons also help to stabilize the joint by providing tension and support. Without tendons, the joint would not be able to move properly. Tendons can become inflamed and cause pain if they are overused or injured.

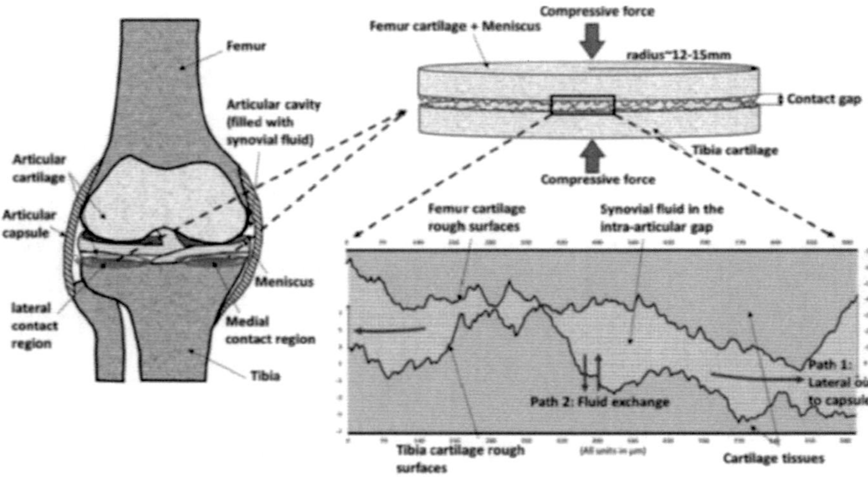


**Figure 8.** Schematic diagram of a synovial joint showing synovial fluid [Liao et al., 2019].

*Bursae*: Bursae are small fluid-filled sacs that are located around synovial joints. They are filled with synovial fluid and act as cushions between bones, tendons, and muscles. Bursae help reduce friction and provide a smooth surface for the joint to move on. They also help to absorb shock and reduce the amount of force that is placed on the joint. Bursae can become inflamed and cause pain and swelling in the joint. This is known as bursitis.

The physiology of a synovial joint involves the movement of the bones in relation to each other. This movement is made possible by the muscles and ligaments that attach to the bones. The muscles contract and relax to move the bones, while the ligaments provide stability and prevent excessive movement. The synovial fluid also helps to reduce friction between the bones and allows for smooth movement. Despite its roughness, the porous cartilage surface of articulating joints has an initial friction coefficient of approximately 0.01, which is better than any man-made bearing currently available (Kumar and Ateshian, 2015). The low friction and wear of cartilage is caused by a combination of hydrodynamic lubrication, boundary lubrication and biphasic lubrication through interstitial fluid pressurisation, known as mixed-mode lubrication.

## 5.2 Properties of Synovial Fluid

Synovial fluid, which is a non-Newtonian fluid (pseudo plastic), is likened to egg white. 'Syn' means like and 'ovial' means egg white. It is a viscous liquid, clear and straw colored, and is found in bone joints. The synovial fluid is a type of ultrafiltrate or plasma dialysate that has glucose, hyaluronic acid, and proteins in it. Synovial fluid is a pseudo-plastic, viscous liquid that serves a biomechanical purpose. It is used to reduce friction and wear on the cartilage of a synovial joint, as well as to absorb movement shocks. It is usually clear and straw-colored. The volume of synovial fluid in a healthy human knee joint is typically between 1 and 4 ml (Kumar et al., 2015). As the shear rate rises from 0.1 to 450 per second, the viscosity of the normal synovial fluid of a knee joint drops from over 10,000 to 100 mPa seconds. The range of shear modulus (G) is between 20 and 45 dyn/cm$^2$ when the shear rate is between 0 and 100 per second, and it rises moderately for higher shear rates. The ratio of the relaxation time of synovial fluid (the high shear viscosity divided by the low shear module) is quite high, ranging from 3 to 10 seconds (Kumar and Sharma, 2015). The increase in shear rate causes the hyaluronic acid molecules to become aligned, resulting in shear thinning. The deterioration of synovial fluid in rheumatoid arthritis causes it to no longer exhibit non-Newtonian behavior and its lubricating capabilities are diminished.

## *5.3 CFD Modeling of Synovial Fluid Flows*

CFD models can be used to simulate the flow of synovial fluid in the joint cavity. CFD models can be used to calculate the pressure and velocity of the fluid and the forces it exerts on the surrounding surfaces. CFD models can also be used to investigate the effects of changes in the dimensions of the joint cavity and the parameters of the fluid on the flow characteristics. This can be used to better understand the effects of joint pathology and to design better treatments for joint diseases.

Synovial fluid within the temporomandibular joint (TMJ) plays a vital role as a lubricant and shock absorber. It facilitates smooth movements of the jaw by reducing friction and providing cushioning to the articular cartilage and other tissues in the TMJ. During jaw movement, there is a regular pattern observed in the flow of synovial fluid within the TMJ. In a study conducted by Xu et al. (2012) on patients with small disk perforation, the upper compartment of the TMJ was examined using a finite-element fluid modeling approach during jaw opening-closing at various positions from 0 to 3 cm. The findings indicated that the fluid flow in the upper compartment generally exhibited an anticlockwise circulation, along with localized vortexes, when the jaw was opening up to 2 cm. However, at a jaw opening-closing distance of 3 cm, an abnormal flow field was observed, accompanied by changes in fluid pressure associated with the perforation. This abnormal flow field and pressure change pose a risk of further perforation expansion or rupture and hinder the self-repair process of the perforated disc.

In another investigation conducted by Xu et al. (2013), the upper compartment of the TMJ joint of a healthy individual was reconstructed in three dimensions. Finite-volume fluid dynamic modeling was utilized to investigate the fluid flow pattern and pressure distribution during jaw movements. The CFD simulation revealed that the synovial fluid within the upper compartment exhibited a regular flow pattern when the jaw was opening. As the jaw opened by 1 cm, there was an increase in fluid pressure and expansion of the posterior space (Fig. 9B). Along with the overall anticlockwise circulation of synovial fluid, localized vortices were observed at the anterior and posterior regions of the compartment, particularly prominent at the posterior portion. With a jaw opening of 2 cm, the posterior vortex became more distinct, while the anterior vortex disappeared (Fig. 9C). When the jaw reached a 3 cm opening, the anterior space vanished, and the posterior space took on a

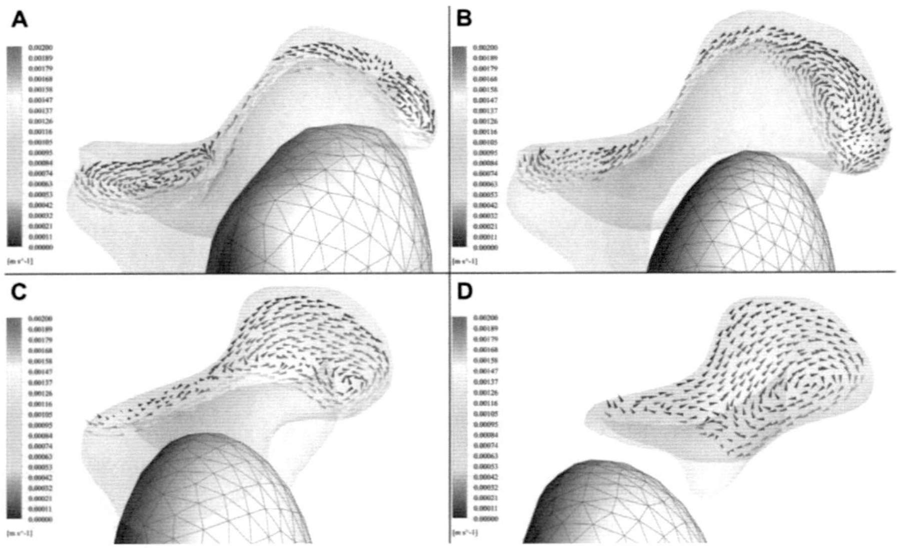


**Figure 9.** Flow velocity contours in the upper compartment of a normal TMJ. TMJ (A) in a closed-jaw position, and at jaw openings of (B) 1 cm, (C) 2 cm, and (D) 3 cm [Xu et al., 2013].

wedge-shaped appearance, tapering towards the anterior. The synovium accumulated in the posterior space, generating a typical anticlockwise circulation with a central vortex (Fig. 9D) (Fig. 2a–d illustrate these findings).

Moghadam et al. (2015) created a poroelastic model of the knee cartilage to investigate the temperature rise within the cartilage caused by dissipation and the influence of synovial fluid flow on heat transfer. The simulation results indicate that the local temperature in the knee cartilage experiences a noticeable increase as a result of its viscous nature. However, the flow of synovial fluid does not have a significant impact in preventing this phenomenon. This can be attributed to the low permeability of the cartilage and the moderate exchange of fluid at the cartilage surface during deformation.

Recently, Liao et al. (2019) conducted a validated numerical investigation to examine the fluid flow within cartilage contact gaps and quantify the permeability of lateral synovial fluid flow through these gaps. The study took into account the surface roughness of the cartilage. Two models were developed to analyze the flow at both macro- and micro-scales, utilizing Darcy's law at the macro-scale and the Navier-Stokes equation at the micro-scale. The numerical findings indicate that the height of the contact gap and the fluid pressure gradient are two crucial parameters that significantly influence the permeability of the gap. As the contact gap narrows, the permeability decreases, and notably, the permeability is highly sensitive to the fluid pressure gradient. Additionally, as the contact gap closes, the permeability gradually approaches that of the surrounding cartilage tissue, signifying the functional closure of the contact gap. In another research, Liao et al. (2020) focused on investigating the pathological alterations in fluid flow within cartilage contact gaps caused by changes in cartilage surface roughness and synovial fluid characteristics in osteoarthritic

(OA) conditions. The findings indicate that a rougher surface of OA cartilage tends to decrease the permeability of the contact gap by approximately 30 to 60%. Moreover, as the size of the contact gap decreases, the reduction in permeability becomes more pronounced, potentially leading to early fluid ultrafiltration into the tissue. Consequently, these changes in fluid flow behavior within contact gaps under OA conditions could significantly impact the functionality of articular joints.

## 6. Mucus Flow

Mucus flow is the production and release of mucus, a thick, slippery substance that is secreted by the mucous membranes of the body. Mucus is typically produced in the sinuses, nose, throat, lungs, and gastrointestinal tract. It is important for lubrication and protection against bacteria and other foreign particles. In the respiratory system, mucus helps to keep the airway clean and moist, preventing it from becoming dry and irritated. It also helps to trap and remove dust and other airborne particles, as well as pathogens. In the gastrointestinal tract, mucus helps to lubricate the lining and protect it from digestive acids and enzymes.

Mucus in the respiratory tract is the first line of defence, which ensures that the air received by the bronchi is clean. Coughing, voluntarily or involuntarily, helps to clear out the mucus and hence keeps the passage clean otherwise the passage may get blocked and the bronchi would not get any air. Due to coughing, a shear force acts on the surface of the mucus which propels the mucus towards the mouth and ultimately the person can spit it. Respiratory mucus is a heterogeneous fluid, mostly made up of water and macromolecule substances like mucin. Due to the presence of these macromolecules, the mucus represents the 'shear thinning phenomenon' which states that the mucus viscosity decreases as the shear rate increases (van der Schans, 2007; Mauroy et al., 2011; Volsko, 2013). Typically, airflow shear rate in the respiratory tract ranges from $10^{-1}$ to $10^{6}$ Hz. The mucus viscosity may increase to $10^{4} - 10^{6}$ times that of water at lower shear rates. The mucus viscosity is dramatically decreased at a shear rate near to the physiological maximum of $10^{3} - 10^{4}$ Hz, which is close to the water viscosity (Lai et al., 2009). The Goblet cells are responsible for the secretion of mucus in the respiratory passage. In healthy patients, the mucus thickness is around 30 μm, but in conditions of infection and bronchitis, the goblet cells tend to function inaccurately and the mucus thickness can reach 300 μm or more (Foster, 2002). Increased thickness of the mucus layer causes the congestion of respiratory tract and makes breathing difficult.

Mucus is a substance with viscoelastic properties, characterized by a complex behavior that deviates from Newtonian fluid dynamics. The apparent viscosity of mucus plays a critical role in the process of clearing it from the body. In individuals without health issues, mucus exhibits low viscosity and elasticity, enabling easy movement through ciliary action (Fahy and Dickey, 2010). However, various diseases can disrupt these properties, leading to significant challenges in clearing mucus through coughing. Paradoxically, excessively low elasticity in mucus can also impede effective cough clearance (King, 2006), emphasizing the importance of maintaining a dynamic balance between viscosity and elasticity. The composition of mucus consists of two distinct layers: the serous layer, also known as sol, which

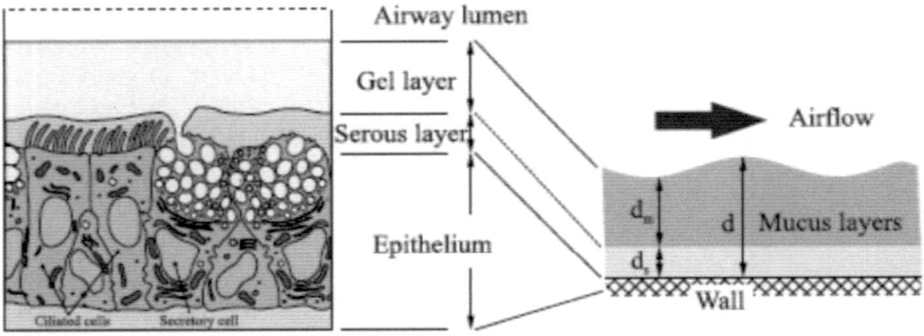


**Figure 10.** Epithelial and mucus layers in a human respiratory tract [Paz et al., 2019].

has low viscosity, and the mucus layer, referred to as gel, which exhibits high viscoelasticity (refer to Fig. 10). When a cough occurs, the gel layer glides over the sol layer, and the effectiveness of the cough is influenced by the thickness of the sol layer.

## *6.1 CFD Modeling of Mucus Flow*

CFD modeling of mucus flow in a human airway can be done to study the rheological behavior of mucus in airways and to analyze the flow patterns of the mucus. This can help in understanding the mechanisms of mucus flow, which is important for designing effective treatments for airway diseases such as asthma and cystic fibrosis. The model can be used to simulate the flow of mucus in both healthy and diseased airways, as well as to evaluate the effectiveness of different treatments. The model can also be used to predict the mucus flow patterns in different airway geometries, such as curved or branched airways. The model can be used to study the effects of physical parameters, such as airway size, mucus viscosity and surface tension, on the flow of mucus.

The clearance of mucus in the human airways is basically gas-liquid interaction phenomenon, which in CFD is widely dealt with 2 of the models namely, Eulerian Wall Film (EWF) model and the Volume of Fluid (VOF) model. Various studies have already been done to validate the movement and clearance of mucus by these models. Ren et al. (2022) worked on generation-0 to generation-5 geometry reconstructed using CT scan and used EWF Model for simulation the mucus flow. Effect of the coughing parameters on the mucus clearance was studied and it was concluded that the increase of Cough Expired Volume (CEV) is the best way for the clearance and not the Cough Peak Flow Rate (CPFR), which is good only in low shear rates. Using the EWF and VOF model, Paz et al. (2017, 2019) examined the effects of mucus viscosity, mucus thickness, cough flow rate on epiglottis and glottis on upper airway geometry. Cough efficiency was used to study the mucus clearance process. Cough efficiency refers to the ratio of mass of mucus before and after coughing.

Assistive techniques for coughing are frequently employed in patients with neuromuscular diseases, where respiratory muscle dysfunction is present (Kim

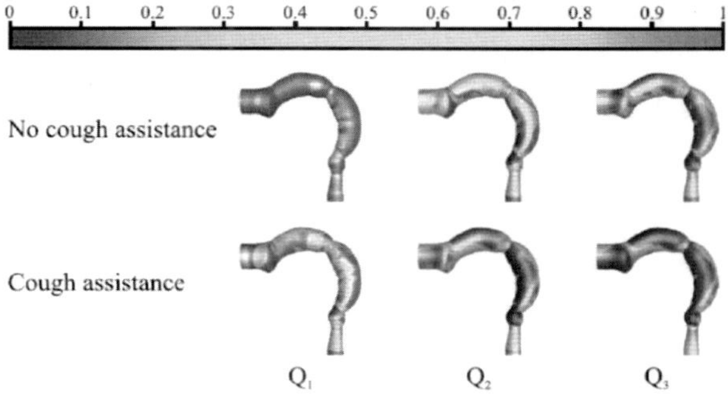


**Figure 11.** Dimensionless mucus thickness without and with cough assistance [Paz et al., 2017].

et al., 2016). These techniques involve the use of cough assistance devices, typically connected to the patient via an oronasal mask, which periodically apply inspiratory and expiratory pressure. These devices enhance cough flow and aid in the clearance process, although quantifying their effectiveness remains a challenge despite their clinical validation. The cough assistance devices generally operate within pressure ranges of approximately 40 cm $H_2O$ for inspiratory pressure and –40 cm $H_2O$ for expiratory pressure. Paz et al. (2017) expanded on this by applying pressures ranging from –4000 to +4000 Pa, with a 500 Pa increment, to replicate the effects of cough assistance devices. The improvement in Cough Efficiency (CE) associated with these devices was visually compared using contour plots of mucus thickness after coughing, as depicted in Fig. 11. This figure juxtaposed the first, second, and third quartiles of cough values in healthy individuals (Britton et al., 2014) with and without cough assistance, employing ±4000 Pa pressure. The analysis revealed that the primary mechanism contributing to cough clearance is a viscous shear force, while factors such as glottis closure time and epiglottis position did not exert a significant influence on CE. The employment of cough assistance devices led to an enhancement in CE, with the rate of improvement exhibiting logarithmic growth in correlation with the applied pressure. It was observed that a successful mucus clearance process can be achieved through coughing, even with a fixed glottis. However, substantial enhancement in CE results from the glottis closure.

## 7. Sweat Flow

Sweat is an important component of the human body's cooling system, and its production and evaporation help regulate body temperature. It is important to study how sweat is produced, transported across the skin, and then evaporated. The study of sweat in bio-fluid flow can provide insights into how the body maintains its temperature, how it responds to changes in the environment, and how it adapts to physical activity.

Sweat is composed of water, electrolytes, and small amounts of urea, lactate, and other waste products. Sweat is odorless, but when it comes into contact with bacteria on the skin, it can produce an unpleasant smell. Sweat is produced by the sweat glands in the skin and is released through the pores. Sweat helps to regulate body

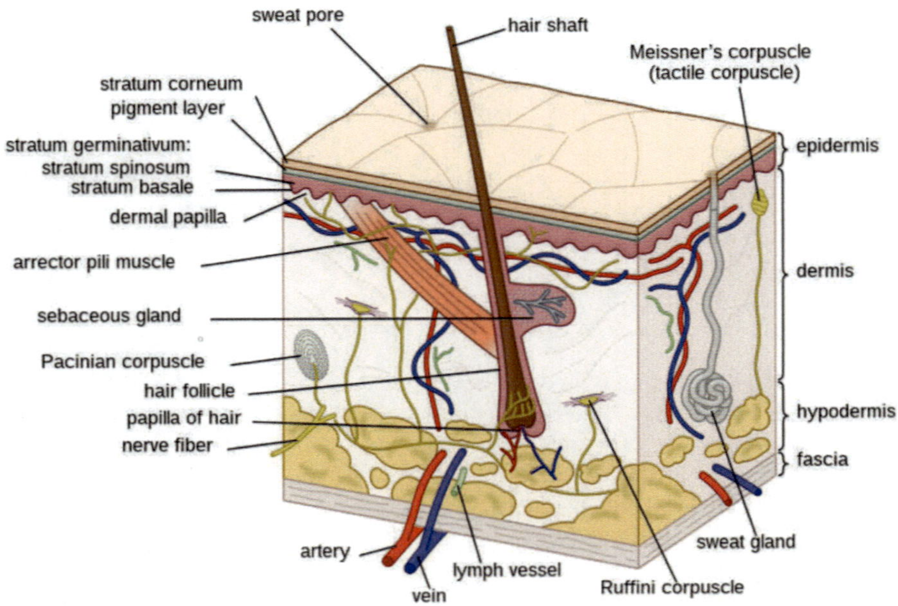


**Figure 12.** A cross-section of the human skin, with the sweat gland [Miloslava and Oldřich, 2009].

temperature by cooling the body when it evaporates. Sweat can be a sign of physical exertion, stress, or illness.

Sweat is produced by sweat glands, which are small tubular structures located in the skin (Fig. 12). Sweat glands are made up of two parts: an outer layer of cells called the epidermis, and an inner layer of cells called the dermis. The epidermis contains sweat glands that produce sweat, while the dermis contains sweat ducts that carry the sweat to the surface of the skin. The eccrine sweat gland, which is managed by the sympathetic nervous system, is responsible for controlling body temperature. When the internal temperature increases, the eccrine glands release water onto the skin, which is then evaporated to cool the body down. If eccrine glands are functioning across a large portion of the body (such as in horses, bears, and humans), they are a major factor in controlling body temperature.

## *7.1 Structure of Sweat Glands*

Sweat glands are small tubular structures located in the dermis layer of the skin. They are composed of two main parts: the secretory coil and the duct. The secretory coil is the portion of the gland that produces sweat, while the duct is the portion of the gland that transports the sweat to the surface of the skin.

The secretory coil is composed of two layers of cells: the myoepithelial cells and the secretory cells. The myoepithelial cells are responsible for secreting sweat, while the secretory cells are responsible for absorbing water and electrolytes from the blood. The sweat produced by the myoepithelial cells is then transported through the duct to the surface of the skin.

The sweat glands are regulated by the autonomic nervous system, which is responsible for controlling the body's involuntary functions. When the body is exposed to heat or stress, the autonomic nervous system signals the sweat glands to produce more sweat. This helps to cool the body down and regulate its temperature.

## *7.2 CFD Modeling of Sweat Flows*

CFD can be used to model sweat flow in order to better understand the mechanisms of sweat evaporation and cooling. CFD can provide detailed information about the velocity and temperature distribution of sweat, as well as the local evaporation rate. This information can be used to analyze the effectiveness of different cooling strategies and to design improved sweat-wicking fabrics. Additionally, CFD can be used to study the interactions between sweat and the environment, such as wind and convective heat transfer from the skin to the environment.

Canuto and Cimolin (2011) developed a thermodynamic model to describe the evolution of sweat in a porous medium when it comes into contact with a part of the human body. This model accurately represents the heat transfer phenomenon associated with evaporation. It consists of a set of evolution equations for three scalar variables: temperature, absolute humidity, and sweat. These variables are interconnected through the evaporation term. In Fig. 13, there is an illustration of an airflow passing through a porous tissue situated on a solid region that represents a general part of the body. The body releases heat and sweat from its contact interface, and the flow field is responsible for removing them through conduction and convection. The key phenomenon that the model focuses on is evaporation, which transforms sweat into water vapor. This vapor can be carried away from the fabric by the airflow.

Dixit and Gade (2015) employed the Simplified Thermoregulatory Bio-heat (STB) equation, which incorporates thermoregulatory activities such as vasomotion, sweating, and shivering, to provide a comprehensive and transient evaluation of thermal comfort. To achieve this, the STB equation was solved alongside the Navier-Stokes (N-S) equation using the CFD tool. The study reveals that the coupling of the STB model with CFD simulation can effectively address the requirements of assessing dynamic thermal comfort in scenarios where a human body is exposed to a thermal environment. Shih et al. (2020) developed a Human-Clothing-environment (HCE) model to simulate the transient thermophysiological response of the Human

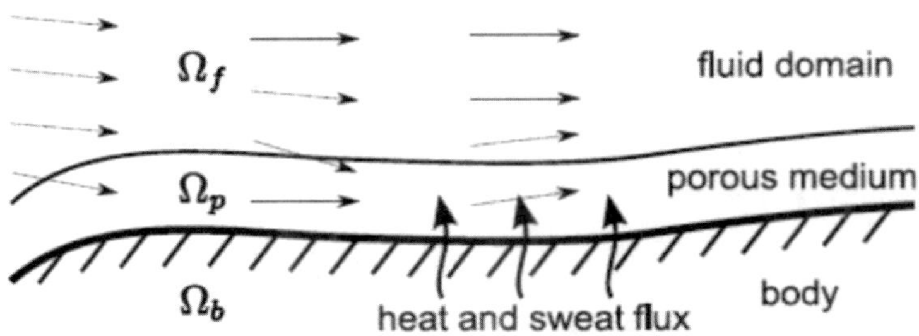


**Figure 13.** Thermodynamic sweat model [Canuto and Cimolin, 2011].

body during resting and exercising. The model considers heat and moisture transport within clothing. To capture the thermophysiological response of the human body, the bio-heat equation was utilized to predict the temperature distribution within different body tissues. The fabric, consisting of multiple pores filled with wet air, was treated as a porous medium. The developed model considered human movement, heat transfer, fiber absorption, and sweat evaporation within the porous fabric. Through CFD simulations, the transient behavior of heat and moisture transfer in the coupled system of the human body, fabric, and environment was observed. The numerical results demonstrated a close relationship between skin temperature and sweating rate with environmental temperature. Specifically, a 5°C increase in environmental temperature led to an approximate 1°C increase in skin temperature and a $4.4 \times 10^{-6}$ kg/m$^2$s increase in sweating rate at the end of the sweating period. During activities involving sweat production, the skin temperature gradually decreased due to sweat evaporation, and the change in skin temperature was largely unaffected by variations in environmental temperatures.

## 8. Urine Flow

Urine flow is the movement of urine through the urinary tract. It is a complex process that is regulated by hormones, nerves, and the muscles in the urinary system. The bladder and urethra are the two main organs involved in urine flow. The bladder stores urine until it is ready to be expelled, while the urethra is the tube that carries urine from the bladder to the outside of the body. The process of urine flow begins when the bladder muscles contract and the internal urethral sphincter relaxes, allowing urine to pass from the bladder into the urethra. As the bladder empties, the external urethral sphincter contracts to prevent urine from leaking out of the body. When the bladder is completely empty, the internal and external urethral sphincters close and urine flow stops.

Urine is a liquid waste product of the body that is composed of water, electrolytes, and waste products filtered from the blood. It is typically composed of 95 water, 2.5 urea, 1.5 chloride, 0.7 sodium, 0.6 potassium, 0.4 creatinine, and 0.2% other substances. Urine also contains small amounts of proteins, hormones, and other organic molecules. The density and dynamic viscosity of urine are found as 1017 kg/m$^3$ and 0.8293 mPa at 37°C, respectively (Kren et al., 2001; Inman et al., 2013).

### *8.1 Anatomy and Physiology of a Urinary System*

The anatomy of urinary system consists of the following organs.

*Kidneys*: The kidneys are two bean-shaped organs located on either side of the spine in the lower back. They are responsible for filtering waste and toxins from the blood, producing urine, and regulating electrolyte balance.

*Ureters*: The ureters are two thin tubes that connect the kidneys to the bladder. They transport urine from the kidneys to the bladder.

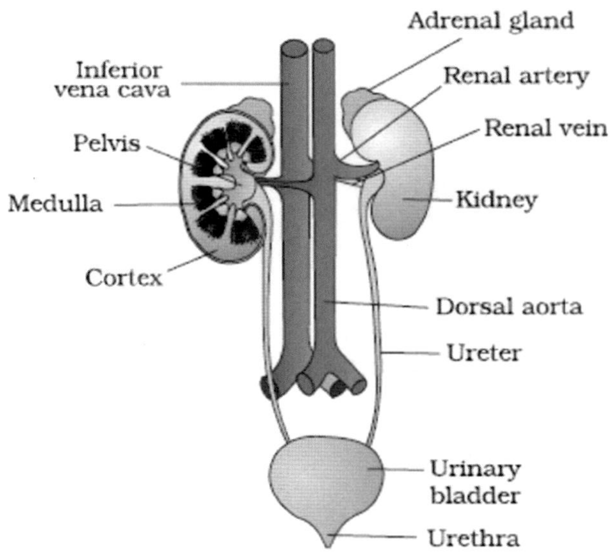


**Figure 14.** Human urinary system.

*Bladder*: The bladder is a muscular organ located in the lower abdomen. It stores urine until it is ready to be expelled from the body.

*Urethra*: The urethra is a tube that connects the bladder to the outside of the body. It is responsible for carrying urine out of the body.

The urinary system is responsible for the production, storage, and elimination of urine. It is composed of the kidneys, ureters, urinary bladder, and urethra as shown in Fig. 14. The kidneys are the main organs of the urinary system. They are responsible for filtering waste products from the blood and producing urine. Urine is composed of water, electrolytes, and waste products such as urea and creatinine. The kidneys also produce hormones that regulate blood pressure, red blood cell production, and calcium metabolism. The ureters are two tubes that connect the kidneys to the bladder. They transport urine from the kidneys to the bladder. The bladder is a muscular organ that stores urine until it is ready to be eliminated. The walls of the bladder are composed of smooth muscle that contracts and relaxes to control the release of urine. The urethra is a tube that connects the bladder to the outside of the body. It carries urine from the bladder to the outside of the body during urination. The urinary system works together to filter waste products from the blood, produce urine, and eliminate it from the body.

## *8.2 CFD Modeling of Urine Flows*

In CFD modeling, urine is considered as a steady, incompressible Newtonian fluid. CFD modeling of urine flow in the ureter can be used to study the flow behavior of urine through the ureter and to investigate the causes and effects of any blockages or obstructions. CFD models can be used to analyze the impact of different parameters such as the ureter's diameter, the angle of the ureter, the concentration of salts and proteins in the urine, and the flow velocity. The models can also be used to analyze the effect of medical interventions such as stenting or balloon angioplasty. The results from the CFD models can be used to evaluate the effectiveness of various treatments and to determine the best course of action.

Niu and Chang (2007) performed an initial numerical simulation of the lower urinary system using CFD. The simulation aimed to study the distribution of urine flow velocity, wall pressure, and shear stress along the urethra. The CFD model was based on MRI scans and uroflowmetry data from a healthy female, assuming rigid walls. The numerical results revealed the occurrence of vigorous secondary flows near the end of the urethra as the inflow rate increased. Additionally, oscillatory patterns were observed in the pressure and shear stress distributions near the beginning section of the urethra at peak flow rates.

Nowadays, a double 'J' stent is commonly used for patients with ureteral stenosis or renal stones who have undergone lithotripsy. This stent is equipped with multiple side holes along its shaft, which serve as alternative pathways for urine flow. Kim et al. (2015) conducted a study to assess the flow rate and pattern around these side holes in a curved model of a stented ureter, using human anatomy as a basis. A straight model was also utilized for comparison. The study found that the total flow rate was higher in stents with a larger number of side holes. In the curved ureter, the inflow and outflow through the side holes were more active compared to the straight ureter. This indicates that even in a non-stenosed or non-occluded ureter, there is still flow through the side holes. Additionally, when the diameter of the ureter changed, the in-stent flow rate remained consistent, while the extraluminal flow rate was higher in ureters with a larger diameter. Kim et al. (2022) conducted another study to examine the impact of vesicoureteral reflux (VUR) on the upper and lower urinary tracts in the presence or absence of ureteral stenosis and a Double J Stent (DJS). They created a model of the entire urinary tract with a DJS implant and investigated various factors related to the urinary system, including flow rates, flow patterns, distribution of Wall Shear Stress (WSS), the force exerted by reflux urination, and the helicity of the bladder. The helicity within the bladder varied based on the presence of stenosis and the opening and closing of the urethra. By inserting a ball into the stent, the reflux within the stent was reduced by 40%. The primary factor contributing to VUR was the opening and closing of the vesicoureteral junction by the detrusor muscle. The largest urine reflux (11.7 mL/s) into the kidney occurred when the detrusor muscle was relaxed. The study suggests that ureteral stenosis affects VUR and decreases urine reflux, while the insertion of a ball in the stent reduces urine reflux through the stent's lumen.

Posterior Urethral Valve (PUV) is the primary cause of obstruction in the Lower Urinary Tract (LUT) among children. Although endoscopic valve ablation is recognized as the most effective initial treatment, some patients still experience long-term impairment in renal function. Current non invasive techniques like uroflowmetry and ultrasound fail to provide crucial hydraulic details associated with the critical structure during urination. Chen et al. (2022) addressed this issue by creating three simplified three-dimensional models of the LUT were created,

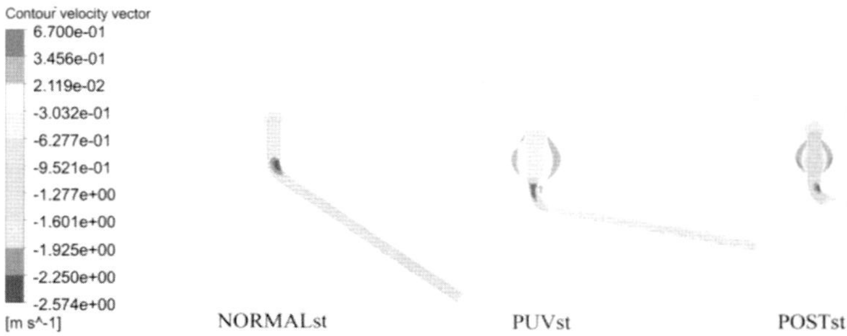


**Figure 15.** Velocity distribution in different lower urinary tract models [Chen et al., 2022].

namely NORMALst, PUVst, and POSTst. These models were based on geometrical parameters and assumptions. At an inlet pressure and outlet pressure of 5000 Pa (including contributions from intra-abdominal pressure and detrusor pressure) and 0 Pa respectively, as determined from previous urodynamic results, the urine flow rates were measured as follows: NORMALst – 18.08 ml/s, PUVst – 11.14 ml/s, and POSTst – 12.16 ml/s. Analysis of the NORMALst model revealed a high-velocity region in the dorsal urethra of the PU-4. In the PUVst model, this high-velocity region shifted to the ventral part of the PU-4, accompanied by significant reverse flow in the dorsal side and marginal area of PU-1. The POSTst model exhibited a similar high-velocity region in the dorsal part of the PU-4, with a more pronounced reverse flow in the marginal region of PU-1 (refer to Fig. 15). The findings from CFD suggest that the dilated posterior urethra hampers urine delivery efficiency and increases hydraulic energy loss both before and after surgery.

## References

Ahn, H. and Park, J. (2016). Liposomal delivery systems for intestinal lymphatic drug transport. Biomater Res. 20(1): 36.

Britton, D., Benditt, J.O., Merati, A.L., Miller, R.M., Stepp, C.E., Boitano, L. et al. (2014). Associations between laryngeal and cough dysfunction in motor neuron disease with bulbar involvement. Dysphagia 29(6): 637–646.

Burton-Opitz, R. and Nemser, R. (1917). The viscosity of lymph. American Journal of Physiology. Legacy Content 45(1): 25–29.

Canuto, C. and Cimolin, F. (2011). A sweating model for the internal ventilation of a motorcycle helmet. Comput. Fluids 43(1): 29–37. https://doi.org/10.1016/j.compfluid.2010.07.004.

Chen, H.S., Liu, X., Zhang, Z.C. et al. (2022). Computational fluid dynamics modeling approaches to assess lower urinary tract hydraulic dynamics in posterior urethral valve before and after endoscopic valve ablation: a pilot study. World J. Urol 40: 505–511. https://doi.org/10.1007/s00345-021-03875-3.

Cooper, L. (2016). Investigations of lymphatic fluid flow. University of Southampton, Faculty of Engineering and the Environment, Doctoral Thesis, 238pp.

Dixit, A. and Gade, U. (2015). A case study on human bio-heat transfer and thermal comfort within CFD. Building and Environment 94(1): 122–130.

Dixon, J.B., Greiner, S.T., Gashev, A.A., Cote, G.L., Moore, Jr., J.E. and Zawieja, D.C. (2006). Lymph flow, shear stress, and lymphocyte velocity in rat mesenteric prenodal lymphatics. Microcirculation 13.

Fahy, J.V. and Dickey, B,F. (2010). Airway mucus function and dysfunction. N Engl. J. Med. 363(23): 2233–2247.

Ferrua, M.J. and Paul Singh, R. (2011). Understanding the fluid dynamics of gastric digestion using computational modeling. Procedia Food Science 1: 1465–1472. https://doi.org/10.1016/j.profoo.2011.09.217.

Ferrua, M.J. and Singh, R.P. (2010). Modeling the fluid dynamics in a human stomach to gain insight of food digestion. J. Food Sci. 75(7): R151–62. https://doi.org/10.1111/j.1750-3841.2010.01748.x.

Ferrua, M.J. and Singh, R.P. (2013). Computational modeling of gastrointestinal fluid dynamics. *In*: Cheng, L., Pullan, A. and Farrugia, G. (eds.). New Advances in Gastrointestinal Motility Research. Lecture Notes in Computational Vision and Biomechanics, vol 10. Springer, Dordrecht. https://doi.org/10.1007/978-94-007-6561-0_13.

Foster, W.M. (2002). Mucociliary transport and cough in humans. Pulmonary pharmacology & Therapeutics 15(3): 277–282.

Harper, D. and Chandler, B. (2016). Splanchnic circulation. BJA Education 16(2): 66–71. https://doi.org/10.1093/bjaceaccp/mkv017.

Helmcke, T. (2020). Anatomy and physiology of the gastrointestinal tract. *In*: StatPearls [Internet]. Treasure Island (FL): StatPearls Publishing.

Inman, B.A., Etienne, W., Rubin, R., Owusu, R.A., Oliveira, T.R., Rodriques, D.B. et al. (2013). The impact of temperature and urinary constituents on urine viscosity and its relevance to bladder hyperthermia treatment. Int. J. Hyperthermia 29(3): 206–210. https://doi.org/10.3109/02656736.2013.775355.

Jayathungage, T.D., Don, Suresh, V., Cater, J.E. and Clarke, R.J. (2022). Modelling uptake and transport of therapeutic agents through the lymphatic system. Computer Methods in Biomechanics and Biomedical Engineering 25(8): 861–874, DOI: 10.1080/10255842.2021.1984434.

Kaw, R. and Mahan, L.K. (2015). Food and Nutrition (2nd ed.). St. Louis, MO: Elsevier.

Kim, K.W., Choi, Y.H., Lee, S.B., Baba, Y., Kim, H.H., Suh, S.H. et al. (2015). Numerical analysis of the effect of side holes of a double J stent on flow rate and pattern. Bio-Medical Materials and Engineering 26(1): S319–S327. https://doi.org/10.3233/BME-151319.

Kim, K.W., Park, S.H., Im, G., Lee, S.B., Baba, Y., Lee, C. et al. (2022). CFD study on vesicoureteral reflux in the urinary tract with double J stent. Computers in Biology and Medicine 145: 105456. https://doi.org/10.1016/j.compbiomed.2022.105456.

Kim, S.M., Choi, W.A., Won, Y.H. and Kang, S-W. (2016). A comparison of cough assistance techniques in patients with respiratory muscle weakness. Yonsei Med J. 57(6): 1488–1493.

King, M. (2006). Physiology of mucus clearance. Paediatr Respir Rev. 7(1): 212–214.

Knipe, H., Worsley, C., Yu, Y. and Hacking, C. (28 Dec 2013). Thoracic duct. Reference article, Radiopaedia.org (Accessed on 18 Feb 2023). https://doi.org/10.53347/rID-26567.

Kren, J., Horak, M., Zat'Ura, F. and Rosenberg, M. (2001). Mathematical model of the male urinary tract. Biomed Pap Med Fac Univ Palacky Olomouc Czech Repub 145(2): 91–96. https://doi.org/10. 5507/bp.2001.021.

Kumar, A. and Ateshian, G.A. (2015). Lubrication of articular cartilage. pp. 545–564. *In*: Biomechanics of the Musculoskeletal System. John Wiley & Sons, Ltd.

Kumar, A. and Sharma, S. (2015). Rheological properties of synovial fluid: A review. International Journal of Biological Macromolecules 78: 545–553. doi:10.1016/j.ijbiomac.2015.04.053.

Kumar, S. and Saini, S. (2018). Lymphatic flow: A review. International Journal of Applied Engineering Research 13(2): 845–851.

Kumar, V., Abbas, A.K. and Aster, J.C. (2019). Robbins and Cotran Pathologic Basis of Disease. 10th ed. Philadelphia: Elsevier. eISBN: 9780323609937.

Kumar, V., Abbas, A.K., Aster, J.C. and Fausto, N. (2015). Robbins and Cotran Pathologic Basis of Disease (9th ed.). Philadelphia, PA: Elsevier Saunders.

Lai, S.K., Wang, Y.Y., Wirtz, D. and Hanes, J. (2009). Micro-and macrorheology of mucus. Advanced Drug Delivery Reviews 61(2): 86–100.

Lee, G., Huang, S.A., Aw, W.Y., Rathod, M.L., Cho, C., Ligler, F.S. et al. (2022). Multilayer microfluidic platform for the study of luminal, transmural, and interstitial flow. Biofabrication 14: 025007. https://doi.org/10.1088/1758-5090/ac48e5.

Li, C. and Jin, Y. (2021). A CFD model for investigating the dynamics of liquid gastric contents in human-stomach induced by gastric motility. Journal of Food Engineering 296: 110461. https://doi.org/10.1016/j.jfoodeng.2020.110461.

Liao, J., Smith, D.W., Miramini, S., Thibbotuwawa, N., Gardiner, B.S. and Zhang, L. (2019). The investigation of fluid flow in cartilage contact gap. Journal of the Mechanical Behavior of Biomedical Materials 95: 153–164. https://doi.org/10.1016/j.jmbbm.2019.04.008.

Liao, J.J., Miramini, S., Liu, X. and Zhang, L. (2020). Computational study on synovial fluid flow behaviour in cartilage contact gap under osteoarthritic condition. Computers in Biology and Medicine 123: 103915. https://doi.org/10.1016/j.compbiomed.2020.103915.

Liao, J.J., Smith, D.W., Miramini, S., Thibbotuwawa, N., Gardiner, B.S. and Zhang, L. (2019). The investigation of fluid flow in cartilage contact gap. J. of the Mechanical Behavior of Biomedical Materials 95: 153–164. https://doi.org/10.1016/j.jmbbm.2019.04.008.

Lubopitko Encyc. Corp. (2016). Lymphatic circulation. In Lubopitko Encyclopedia. http://encyclopedia.lubopitko-bg.com/Lymphatic_Circulation.html.

Marciani, L., Gowland, P.A. and Spiller, R.C. (2001). Effect of meal viscosity and nutrients on satiety, intragastric dilution and emptying assessed by MRI. American Journal of Physiology Gastrointestinal and Liver Physiology 280: G1227–1233.

Mauroy, B., Fausser, C., Pelca, D., Merckx, J. and Flaud, P. (2011). Toward the modeling of mucus draining from the human lung: role of the geometry of the airway tree. Physical Biology 8(5): 056006.

Mazzoni, M.C., Skalak, T.C. and Schmid-Schönbein, G.W. (1987). Structure of lymphatic valves in the spinotrapezius muscle of the rat. Blood Vessels 24: 304–312.

McMahon, B.P., Odie, K.D., Moloney, K.W. and Gregersen, H. (2007). Computation of flow through the oesophagogastric junction. World J. Gastroentero 13(9): 1360–1364.

Miloslava, E. and Oldřich, E. (2009). Přehled anatomie. Praha: Galén/Karolinum. ISBN 978-80-7262-612-0.

Moghadam, M.N., Abdel-Sayed, P., Camine, V.M. and Pioletti, D.P. (2015). Impact of synovial fluid flow on temperature regulation in knee cartilage. Journal of Biomechanics 48(2): 370–374. https://doi.org/10.1016/j.jbiomech.2014.11.008.

Moore, J.E. Jr. and Bertram, C.D. (2018). Lymphatic system flows. Annu. Rev. Fluid Mech. 50: 459–82. https://doi.org/10.1146/annurev-fluid-122316-045259.

Niu, Y.Y. and Chang, D.Y. (2007). CFD Simulation of shear stress and secondary flows in urethra. Biomedical Engineering: Applications, Basis and Communications 19(2): 117–127.

Pal, A., Indireshkumar, K., Schwizer, W., Abrahamsson, B., Fried, M. and Brasseur, J.G. (2004). Gastric flow and mixing studied using computer simulation. Proc R Soc Lond B 271: 2587–2594.

Parks, D.A. and Jacobson, E.D. (1985). Physiology of the Splanchnic circulation. Arch Intern Med. 145(7): 1278–1281. https://doi.org/10.1001/archinte.1985.00360070158027.

Paz, C., Suárez, E. and Vence, J. (2017). CFD transient simulation of the cough clearance process using an Eulerian wall film model. Computer Methods in Biomechanics and Biomedical Engineering 20(2): 142–152. http://dx.doi.org/10.1080/10255842.2017.1360872.

Paz, C., Suárez, E., Vence, J. and Cabarcos, A. (2019). Analysis of the volume of fluid (VOF) method for the simulation of the mucus clearance process with CFD. Computer Methods in Biomechanics and Biomedical Engineering 22(5): 547–566. https://doi.org/10.1080/10255842.2019.1569637.

Ren, S., Cai, M., Shi, Y., Luo, Z. and Wang, T. (2022). Influence of cough airflow characteristics on respiratory mucus clearance. Physics of Fluids 34(4): 041911.

Selahi, A. and Jain, A. (2022). Engineered models of the lymphatic vascular system: Past, present, and future. Microcirculation 00: e12793. https://doi.org/10.1111/micc.12793.

Shih, S.H., Shih, Y.C., Wang, L.C., Shiah, S.W. and Hu, S.C. (2020). The model development and case study for transient thermal response of the human body interacting with the coupled fabric-environment system. Case Studies in Thermal Engineering 18: 100592. https://doi.org/10.1016/j.csite.2020.100592.

van der Schans, C.P. (2007). Bronchial mucus transport. Respiratory Care 52(9): 1150–1158.

Volsko, T.A. (2013). Airway clearance therapy: finding the evidence. Respiratory Care 58(10): 1669–1678.

Xu, Y., Zhan, J., Zheng, Y., Han, Y., Zhang, Z., Xi, Y. et al. (2012). Synovial fluid dynamics with small disc perforation in temporomandibular joint. J. of Oral Rehabilitation 39: 719–726. https://doi.org/10.1111/j.1365-2842.2012.02307.x.

Xu, Y., Zhan, J.M., Zheng, Y.H., Han, Y., Zhang, Z.G. and Xi, C. (2013). Computational synovial dynamics of a normal temporomandibular joint during jaw opening. J. Formos Med. Assoc. 112(6): 346–51. https://doi.org/10.1016/j.jfma.2012.02.015.

# Chapter 10

# Nanoparticle-Based Drug Delivery

*Keshav S. Moharir*

## 1. Introduction

Drug delivery systems are means of carrying pharmacologically active or initially inactive prodrugs to the actual site of action (Vlachopoulos et al., 2022). Several factors affect the therapeutic efficacy of the active moiety, ranging from the physico-chemical properties of the drug and the route of administration to the disease conditions of the patients. The conventional or classical drug delivery methods were not spatiotemporal and frequently resulted in either sub-therapeutic or excess plasma drug concentration. The optimum therapeutic efficacy was compromised from conventional routes and dosage forms owing to changes in pH, enzymatic degradation, first pass effect and several biological barriers. To overcome these limitations of conventional dosage forms, novel drug delivery systems were introduced through the gradual evolutionary process, the most important being nanomedicine powered by nanotechnology-based approaches (Alshammari et al., 2023). Nanotechnology bridges the gap between biological and physical sciences with applications of nano-level structures and nanophases with special reference to drug delivery systems. This has further created a profound influence on the research areas of microfluidics, theranostics, biosensors, tissue engineering, drug delivery, and organoid design.

Nanotechnology has played significant roles in various fields of health and medicine such as cardiology, immunology, oncology, endocrinology, ophthalmology, pulmonology, etc. Among drug delivery, it is being used for specialized systems for brain targeting, gene delivery, tumour targeting, and intracellular drug delivery (Moharir et al., 2022). This is achieved through optimized Nanoparticles (NPs), which

Department of Pharmaceutics, Gurunanak College of Pharmacy, Nari Road, Nagpur, Maharashtra, India.
Email: keshavmoharir@gmail.com

DOI: 10.1201/9781003287636-10

exhibit several benefits over other conventional dosage forms. NPs can overcome inherent biological barriers such as mucosal layer, intracellular trafficking, and drug wastages at unintended sites and can maximize drug delivery to targeted cells and intracellular cell organelles. Apart from these, NPs improve the solubility and stability of moieties, reduce toxicity and improve therapeutic efficacy by enhancing the circulation time of cargo. Although with several advantages, certain grey areas remain that can be admitted to factors influencing normal NP functioning (Bonifacio et al., 2014). These may be drug-related properties, formulation characteristics, local tissue-specific peculiarities and the nature of the excipients used. However, to overcome these problems, NPs have been further refined with the development of alternative drug delivery pathways, the use of complex drug release strategies, stealth mechanisms to overcome scavenging mechanisms of reticuloendothelial system (RES) and removal by immune cells. One step ahead, the NPs are now being evaluated for personalized medicines with different RNAs, genetic materials, CRISPR Cas9 moieties, peptides, and proteins (Hamimed et al., 2022). Being in submicron size from 10 nm to 100 nm, the NPs have entirely different material properties that make them unique nanocarriers for the desired cargo. Different types of NPs can be tuned to carry drug molecules specifically targeted toward desired organs, or they can be carved into 'smart precision nanomedicines' for the disease under consideration. The NPs used for drug delivery systems utilize one or more polymers to give desired characteristic properties. These polymers in NPs may be natural, semi-synthetic or synthetic. They can be manufactured with different methods to achieve the desired size, shape, and near-to-ideal carrier prerequisites. The overall utility of NPs in drug delivery approaches has been positive, which further requires resolving biosafety-related issues.

This chapter summarizes the basics of NPs as drug delivery systems covering their types, mechanisms of drug release, modalities of formulation systems, applications, and challenges. Further, it also highlights biosafety issues and possible toxicity associated with NPs.

## 2. Benefits of Nanoparticle-based Drug Delivery Systems

The approval of more than a hundred nanoparticle-based formulations and a hundred more in clinical trials at the commercial level is a testimonial to the advantages of nanomedicine over conventional dosage forms and drug delivery systems (Anselmo and Mitragotri, 2019). Nanoparticles as carriers provide huge advantages that can be summarized as -

- Enhanced oral bioavailability
- Decreased drug toxicity
- Enhanced solubility
- Increased stability of drug and formulation
- Increased drug targeting possibility
- Increment in surface area
- Increased patient compliance

- Decreased drug resistance
- Enhanced rate of dissolution
- Reduction in dose required
- Controlled drug delivery
- Protection from surrounding detrimental environmental conditions
- Combination of therapeutic and diagnostic agents simultaneously
- Ability to skip the immune system

The advantages of NP-based formulations over conventional formulations can be attributed to their improved properties. These value additions include increased surface area, high aspect ratio, improved cargo carrying capacity and stability, surface functionalization, smart bio-responsive or chemical-responsive nature, enhanced electrical, thermal, and optical properties, etc. (Khan et al., 2019a). In addition to their materials, NPs vary in sizes, shape, and dimensions. Technically, an NP can be zero-dimensional, such as nanodots, with only one parameter forming one dimension like graphene, two-dimensional having length and breadth, for instance, carbon nanotubes, or three-dimensional, namely height, length, and breadth, such as gold NPs. These NPs when further modified as complex systems provide additional benefits—change the multiple entry and drug release pathways, maximize the therapeutic efficacy against specific macromolecules, cell cycle specificity for specific cell cycle and overcome the mechanisms of drug resistance.

Figure 1 gives an idea about wide scope of NPs in drug delivery systems and beyond. NPs also enjoy applications in therapeutics and diagnostics – known as theranostics, advanced in imaging for disease progression and early diagnosis, and tissue engineering.

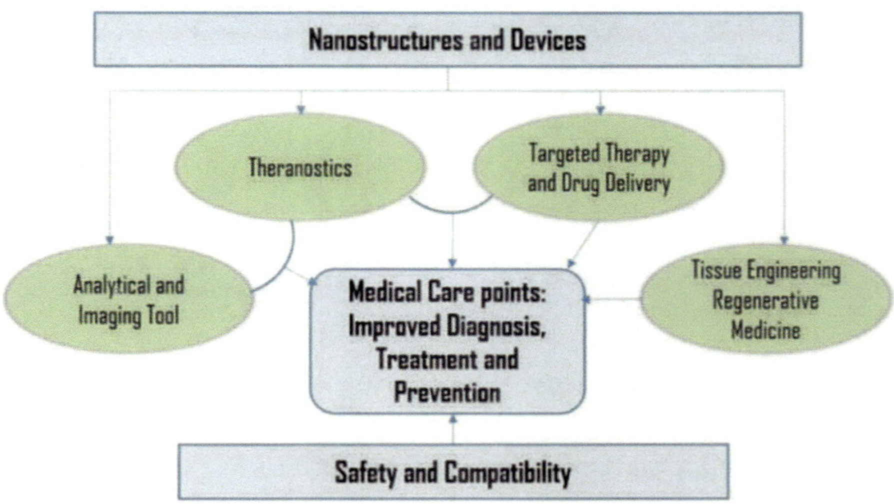


**Figure 1.** A broader view of nanoparticles in drug delivery and allied applications. Adapted from (Patra et al., 2018).

## 3. Types of Nanoparticles

There are several aspects with which NPs can be classified. One aspect covers their inherent chemical nature such as organic NPs (solid lipid, micellar, dendrimers) and inorganic NPs (gold, silver, silica, quantum dots, magnetic iron oxide) while another aspect gives thought to size and shape (spherical, rods, crystals, flowers, cylinders, nanotubes, porous, mesoporous, non-porous). Yet another approach to the classification of NPs talks about how they are manufactured – top-down or bottom-up approach. In addition, the type of polymeric materials employed such as natural or synthetic, degradable, or biodegradable, simple block polymer or multi-branched also influence their selection in the final formulation. New to the list of classifications, NPs can be classified as one-dimensional, two-dimensional, and three-dimensional nanoparticles. The drug properties, desired drug release pattern, target organ or tissue, dose, and manufacturing method govern the selection of main NP-forming components. This section further gives details of NP classification.

The therapeutic advantages of NPs are dependant on the absorption, distribution, metabolism, and excretion of drug-loaded NPs and further their plasma drug concentration. In addition, the NPs need to stay for a longer time to maximize the pharmacological actions of drug moiety. Accordingly, Fig. 2 describes the overall pharmacokinetics and pharmacodynamics of drug-loaded NPs influencing their efficacy.

### a. Solid lipid nanoparticles

Solid Lipid Nanoparticles (SLN) were introduced to overcome the drawbacks of conventional colloidal drug delivery systems such as micelle carriers, liposomes, polymeric micro, or nano particulate. The interest behind reasons development of lipid-based systems include:

a) improvement oral drug bioavailability and maintenance of plasma drug concentration,

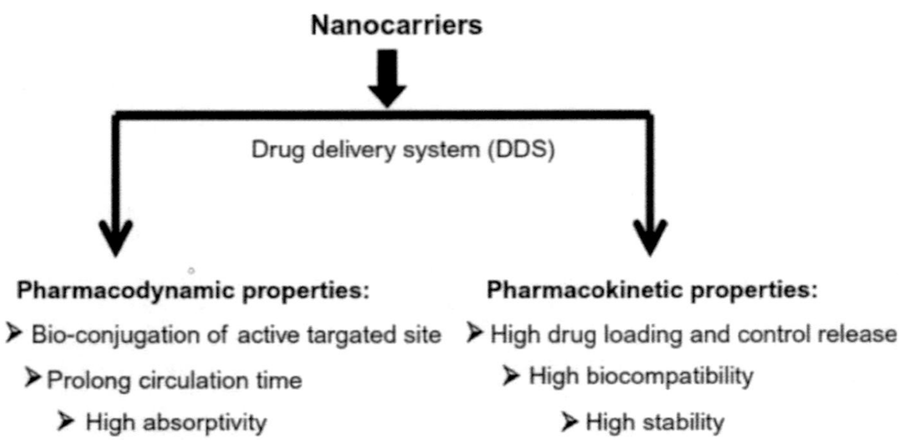


**Figure 2.** Features showing the pharmacokinetic-pharmacodynamic axis of nanoparticles used in drug delivery systems.

b) better characterization and compatibility of lipid excipients and
c) ease of scale up production

SLN consist of phospholipids monolayer as hydrophobic solid core, which is dispersed in surfactant or aqueous solution. The hydrophobic lipid chain aligns with the fat matrix, which can contain drug molecules inside them (Kaur et al., 2008). The key advantages of SLN nanocarrier include:

a) improved site-specific drug delivery
b) solventless system can be prepared
c) improved control over drug release kinetics
d) easy of manufacturing, sterilization, and scale up

In addition, the SLN have improved physicochemical properties such as surface charge, particle size distributions, zetapotential, morphology, surface area of particles, improved dissolution rate, drug loading capacity, and mobility of active ingredients, etc. that makes SLN much sought nanocarrier for drug delivery purpose. The literature shows several applications of SLN. For instance, incorporation of poorly water-soluble drugs such as cyclosporin-A and paclitaxel in SLN improved their oral bioavailability (Satapathy et al., 2021). This higher bioavailability is mainly attributed to increased surface area, improved rate of dissolution, and higher concentration of drug in GIT fluids. Rich lipid content allows SLN to stay in the intestinal wall thus increasing the residence time in GIT, as in case of docetaxel (Mosallaei et al., 2013).

**b. Nanostructured-lipid carriers**

SLN has only solid lipid as lipid content, hence allowing low drug loading. This low capacity of drug payload is due to the internal crystal lattice structure and is followed by drug expulsion. Thus, there was a need for the development of a better lipid-based drug delivery system with high drug loading capacity. The second-generation lipid nanoparticles were developed and introduced as Nanostructured-Lipid Carriers (NLC) to enhance drug loading. NLCs are binary systems with lipids existing in binary form, i.e., solid as well as liquid, leveraging more area for drug incorporation. The *in-vitro* drug release behaviour from NLC results in increased half-life and prevents enzymatic degradation (Tamjidi et al., 2013). The basic composition of NLC is lipid, aqueous part, and surfactant. Depending on liquid and solid mixture composition, NLC can be classified into:

a) imperfect NLC – large solid lipid than liquid lipid, high drug loading due to internal arrangement
b) multiple fat-in-water or oil-in-solid type – high oil or fat in the solid lipid mixture, improved solubility, and stability
c) The amorphous type, which is composed by mixing some particular lipids (hydroxyoctacosanyl hydroxystearate or isopropyl myristate, etc.) with solid lipids. It prevents drug expulsion using β-modification during storage.

The NLC-based drug delivery systems show several great potentials from different routes of administration. Hydrophobic drugs incorporated into NLC show improved bioavailability when given by the *parenteral* route. Owing to the presence of mucosa and rich blood supply, drugs in NLC, when given through the nasal route, have better absorption, mainly those that degrade in GIT acid and enzymes. By *oral* route, lipophilic drugs are easily entrapped into the NLC core to resolve solubility problems. In addition, it is now established that NLC increases bioavailability and prolongs plasma drug concentrations by longer circulation time (Date et al., 2011). When inhaled as a particle as a spray, a clinically important drug concentration is achieved through the pulmonary route for the treatment of lung-related diseases. Major NLC advantage is reflected in the ability of drug-loaded NLC to cross BBB to reach *brain* vasculature and endothelial cells owing to fast cellular uptake. Baicalein NLC with PEG, polysorbate-80, or another ligand such as transferrin, OX26 antibody, and low-density lipoprotein was proved as a superior approach for drug delivery to the brain (Tsai et al., 2012). NLC was proven to be an excellent candidate for drug delivery to the blood-brain capillaries for Parkinson's and Alzheimer's disease (Chen et al., 2016).

**c. Liposomes**

Liposomes are lipid-based mono or multi-lamellar vesicles, explored most extensively and at a much higher stage of development among all NPs for targeted and novel drug delivery. These are synthesized by the hydration of phospholipids. Structurally, liposomal phospholipid vesicles envelop the inner aqueous cores. The polar head of phospholipids are oriented in the core region and externally, non-polar tails are present (Sutton et al., 2007).

Depending upon number of bilayers and their size, the main three types of liposomes include:

a) *Multilamellar vesicles* - consist of multiple lipid bilayers interspaced by aqueous regions. These are heterogeneous in size, ranging from a few hundreds to thousands of nm in diameter.

b) *Small unilamellar vesicles* - consist of a single lipid bilayer surrounding the inner aqueous space, size range less than 100 nm.

c) *Large unilamellar vesicles* - consist of a single lipid bilayer surrounding the inner aqueous space with diameters usually larger than 100 nm.

In their structure, hydrophilic as well as hydrophobic drugs can be encapsulated in the aqueous core and in the lipid bilayer simultaneously, making liposomes ideal candidates for drug delivery. The main methods of liposome preparation include microfluidics, thin film hydration, solvent injection, ether injection, reverse-phase evaporation, and detergent depletion method. Each of them has its own advantages and disadvantages. Liposomes enhance the movement of lipophilic drugs (for instance, paclitaxel) that dissolve in the outer lipid membrane of the NPs to reach the targeted cells (Guzman et al., 1996). Such NPs ensure drug depots with prolonged release pharmacokinetics and long stay at the target locations.

### d. Metal-organic frameworks (MOFs)

These are novel nanohybrid drug carriers with porous structures of organic framework, formed by the strong covalent bonds of the metal ion and multidentate linkers or bridging ligands. MOFs show remarkable properties such as flexibility, low density, high surface area (up to 7000 $m^2g^{-1}$), tunable pore size, and ability to diversity in metal bridges or linkers. The organic linkers in MOFs possess large storage pores and several adsorption sites. On receiving external stimuli, pore sizes can be modified. The encapsulation efficiency of MOFs can be increased by surface modification owing to their high porosity (Pettinari et al., 2017).

stimuli-sensitive MOFs have high drug loading efficacy and release capacity, low cytotoxicity, and tunable release kinetics of guest drug molecules. MOFs show promising utility in drug delivery due to

a) Targeted drug delivery at specific sites
b) transport of drugs efficiently across the barriers
c) ability to deliver drugs at intracellular sites
d) multipurpose use – theranostics in single formulation (Yusuf et al., 2022)

Although useful and effective with the noteworthy performance of MOFs in the laboratory, converting them into economical and safe biomedicine is still caught by regulatory hurdles.

Figure 3 collectively describes the pictorial representation of different types of NPs that are commonly used for drug loading. According to the type, advantages and limitations are also mentioned.

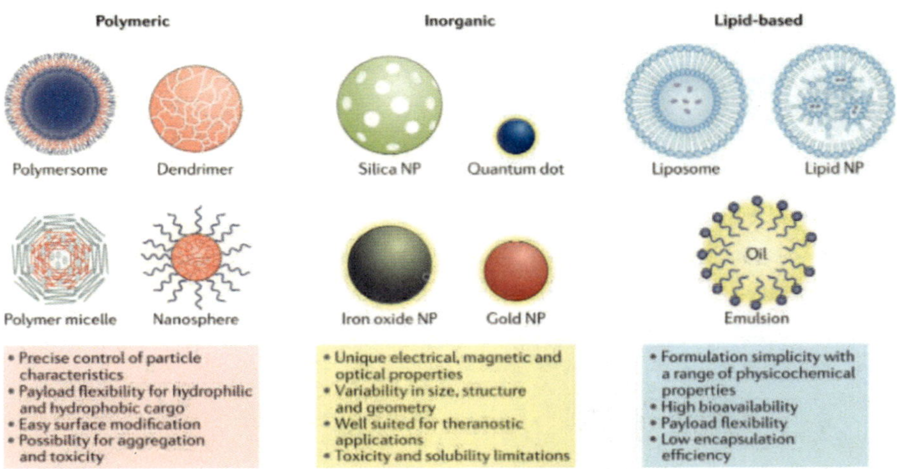


**Figure 3.** Different types of nanoparticles used in drug delivery applications with advantages and limitations. Adapted from (Mitchell et al., 2021).

**e. Micellar nanoparticles**

Micelles and allied micellar drug delivery systems are spherical self-aggregates of repeating monomer units with hydrophilic tail and hydrophobic head group. The polar head group has a strong affinity for water molecule with the surfactant molecule projecting its tail towards the aqueous interface. This arrangement has lower energy than a free monomer, making it a stable preparation. At Critical Micelle Concentration (CMC), micelles can improve drug bioavailability and solubility, as demonstrated by several drug carriers. The length of an amphiphilic moiety, manufacturing parameters and solvent type, micelles show a variety of morphologies such as rod-like, spherical, vesicular, and lamellar shape, and play a noteworthy role in the improvement of drug pharmacokinetics (Sultana and Ismail, 2016)

**f. Noisome**

Niosomes are potent micellar NPs with size ranging from 1- to 1000 nm, wherein drug molecules are entrapped within the vesicles. It embeds bilayers by the fabrication of biodegradable or biocompatible non-ionic surfactants such as polyglycerol, alkyl, and cholesterol dispersed in aqueous media. Unlike liposomes, niosomes have better stability and compatibility since lipid is replaced by non-ionic surfactants. Niosomes can improve drug-entrapped circulation, drug release pattern, biodistribution, and stability of drugs during shelf life and drug delivery from formulation (Sharma et al., 2009). The advantages of niosomes include (Samed et al., 2018):

a) transdermal delivery of drugs due to higher rate of penetration
b) excellent bioavailability via ophthalmic drug delivery
c) peptides can be delivered safely with lowest toxicity and improved stability

**g. Nanoemulsions**

Nanoemulsion is another type of NP based on surfactant as a nanocarrier. It is suitable as a carrier for hydrophobic drug delivery. This biphasic system is kinetically stable at sub-micron particle size (Jaiswal et al., 2015). Nanoemulsion has garnered interest in drug delivery systems due to following reasons –

a) It seldom shows issues of creaming or sedimentation
b) Provide large surface area and hence more drugs can penetrate when given as transdermal formulation
c) Low surfactant quantity required hence low toxicity risk
d) Fine droplet size can prevent coalescence

As an example, Amphotericin B is a poorly water-soluble drug and needs improved dissolution characteristics. In the form of marketed nanoemulsion AmB—Intralipid®, encapsulation improves drug solubility and, hence, better bioavailability compared to the original pure drug (Nasr et al., 2012).

Polymeric micelles are composed of amphiphilic block co-polymers assembled to form supramolecular core-shell structures called 'polymeric micelles.' These are formed as aggregates inside the solution, which are spheroidal in shape, with the hydrophobic cores protected from water by a crown of hydrophilic groups. Usually, the size of polymeric micelles is below 100 nm and finds major use in the systematic delivery of hydrophobic drugs (Xu et al., 2013). Due to the hydrophilic surface, the system is protected from uptake by RES. They have another advantage of trapping contrast agents or drugs inside the hydrophobic cores. The cargo can also be covalently linked to the components of the micelle structure. Their utility in drug delivery systems is proven due to high stability at physiological conditions, versatile and high loading capacity, high circulation time and accumulation of drug at the target site, surface functionalization or modification for conjugation of ligands and prolongation of drug release (Liu et al., 2020).

**h. Carbon based NPs**

Carbon-containing NPs represent two classes - Fullerenes and carbon nanotubes. Fullerene includes NPs with an empty globular-spherical frame. Carbon nanotubes are enlarged, tubular, 1–2 nm in size. The carbon nanotubes (CNT) appear in three forms - as rolled sheet alone, dual, or multiple walls in the form of Single-walled nanotubes (SWNTs), double-walled nanotubes (DWNTs), and multi-walled nanotubes (MWNTs) (Elliott et al., 2013).

CNT biocompatibility can be improved by chemically treating their surfaces with different agents. For example, by covalent attachment of PAMAM dendrimers, amphiphilic di-block co-polymers, PEG coating on the surface or dispersion within a hyaluronic acid matrix (Arsawang et al., 2011). Drug release from CNTs can be tuned through electrical or chemical properties. In one research, the drug release from CNTs was controlled by covering open ends of CNTs were with polypyrrole (PPy) sheets. The drug loading on CNTs is achieved by either of the three methods – chemi-sorption on the CNT surface by hydrophobic and electrostatic interactions, and hydrogen bonds, attachment to functionalized CNTs, and encapsulation of the drug inside CNTs. However, shape, size and geometry of CNTs can be toxic due to high aspect ratio and presence of graphite (Smart et al., 2006).

**i. Metal NPs**

Several metallic NPs have emerged as good drug delivery carriers in recent years. Numerous metals have been explored as NPs for development of efficient drug delivery systems, though gold and silver NPs are important for biomedical purpose. Surface modification and functionalization of metal NPs with various ligands has been tried for improved drug delivery. Some of these ligands include peptides, sugars, protein, and DNA.

Gold nanoparticles have characteristic physical and chemical properties for drug delivery and diagnosis. The gold core is inert with low toxicity. Monodispersed gold NPs synthesize is easy, with size range from 1 to 150 nm. In addition, gold properties can be utilized to initiate the drug release at a remote location.

**j. Magnetic NPs**

Magnetic NPs are actively investigated for Magnetic Resonance Imaging (MRI) contrast agents and targeted drug delivery. Colloidal iron oxide NPs, such as Super Paramagnetic Iron Oxides (SPIO) and ultrasmall superparamagnetic iron oxides (USPIO), are the most commonly investigated magnetic NPs for drug delivery applications. These NPs are typically composed of nano-sized magnetite ($Fe_3O_4$) or maghemite ($\gamma$-$Fe_2O_3$) covered with protective polymeric coating. Further, these NPs have crystal structures with close-packed cubic lattices created by oxygen ions and iron ions intercalated between them.

**k. Dendrimer NPs**

Dendrimers are well-defined, monodisperse, homogeneous structures consisting of unimolecular architecture with three distinct regions: (1) a core consisting of molecule having minimum two identical chemical functions; (2) branches from the core, with repeating units and branch junction, the repetition being in geometrical progression resulting in radially concentric layers known as "generations"; and (3) numerous terminal functional groups on the dendrimer surface (Pan et al., 2005). The initial dendrimer was synthesized by divergent method using poly(amidoamine) (pAMAM) polymer. Malik et al. (1999) generated pAMAM with 3.5 generations and entrapped cisplatin (20–25%w/w) capable of controlled drug release *in-vitro*. Improved solubilization of cisplatin demonstrated significant *in-vivo* antitumour activity in melanoma mice, while free cisplatin showed minimal activity. Encapsulation of drug in dendrimers strongly depends on the drug and the structural features. In addition, drug release is passively controlled by hydrogen bonding, electrostatic interactions, hydrophobicity, and steric hindrance, which are further governed at the interface of the dendrimer surface and aqueous environment (Zhang et al., 2010).

**l. Silica-based NPs**

Silica NPs have been used for bioimaging and drug delivery due to their high surface area-to-volume ratio, easy synthesis, easy surface functionalization, uniform morphology, controlled size, and proven biocompatibility. The silica NPs, can be presented in two forms, NPs with solid core/shell silica and mesoporous silica. Mesoporous silica NPs has high surface area 600–1000 $m^2/g$ and pore volume in the range 0.6–1.0 mL/g (Kestens et al., 2016a). This allows higher drug loading and controllable drug release. Mesoporous NPs show dual surfaces with drug loading capacities, both interior and exterior which can be further used for its dual drug targeting modalities. For successful drug delivery, precision over diameter, targeting geometry and loading capacity is necessary (De Temmerman et al., 2012).

## *3.1 Polymeric NPs*

Polymers used in polymeric NPs can be synthetic, semisynthetic, or natural. Natural polymers containing NPs are generally biocompatible and non-toxic, but they can suffer from stability issues in the vicinity of various biological membranes. In recent eras, newer polymers have been explored for the testing of drug delivery NPs. Such synthetic polymers include poly(D,L-lactide), polycyanoacrylate, poly(lactide-co-glycolide), and PLA or poly(lactid acid). Chitosan is the most explored natural polymer, others being gelatin, cellulose, albumin, pectin, gums, and alginate to prepare NPs. However, drawbacks associated with natural polymers include batch-to-batch variations, susceptibility to degradation and in some cases potential antigenicity (Chen et al., 2020).

Polymeric NPs give an alternative to other types of NPs due to advantageous properties like nonimmunogenicity, biocompatibility, biodegradability, and nontoxicity. These are colloidal carriers in the size range of 10 nm–1 μm, comprising of synthetic or natural polymers. Polymeric NPs are a wider class, combining different NPs such as matrix systems (nanospheres) and vesicular systems (nanocapsules). In nanocapsules, the drug is confined as a core inside the cavity, surrounded by a polymeric membrane, while in nanospheres, the drug is dispersed homogeneously in the polymer matrix. Polymeric NPs are regarded as a matrix system in which the matrix in uniform in terms of polymer and drug. Besides this, polymeric NPs can be formulated as a combination of matrix and membrane-bound reservoir systems (Herrero-Vanrell et al., 2005). The choice of polymer selection and ease of modulation of drug release from polymeric NPs make them ideal drug delivery systems for cancer therapy, vaccines development, contraceptives and targeted antibiotics delivery (Elsabahy and Wooley, 2012).

## 4. Characterization of Nanoparticles

Characteristic properties of nanoparticles include the morphology, size, and surface charge using technically advanced techniques such as Atomic Force Microscopy (AFM), Scanning Electron Microscopy (SEM) and Transmission Electron Microscopy (TEM) (Lamprecht et al., 2014). *In vivo* distribution and stability of NPs is largely influenced by properties such as the size distribution, surface charge, and average particle diameter. Surface charge also affects physical stability and re-dispersibility of the nanoparticle dispersion along with their *in vivo* performance (Durán-Lobato et al., 2020; José Alonso, 2004). Thus, it is necessary to know how nanoparticle characteristics influence the overall performance of NP-based drug delivery systems.

### *4.1 Particle Size*

Characterizations of NPs are mainly evaluated by the particle size, particle size distribution and NP morphology. The electron microscopy ascertains the size, shape, and contours of the NPs. Various tools are employed to understand the pharmacokinetics and biodistribution of NPs in drug delivery. It is now an established fact that the particle size of NPs has a cognizable effect on the rate of drug release (Khan et al., 2019b). Particle size and size distribution affect the *in-vivo* drug distribution, toxicity, biological fate, and ability to target targeting specific tissue or organ. These properties also influence drug release, drug loading and stability of nanoparticles. NPs with 100 nm size showed a 2.5-fold greater uptake than 1 μm microparticles and a 6-fold greater uptake than 10 μm microparticles by Caco-2 cells. A larger surface area-to-volume ratio is associated with smaller particle size; hence, more drug loading occurs associated with better control over drug release. However, larger particles have lower surface area, thus less drug loading accompanied by slower drug release. Thus, control of particle size provides a means of tuning drug release rates. At the same time, smaller particles have greater risk of aggregation in longer shelf life, dispersion, and transport. Dynamic Light Scattering (DLS) and photon-correlation spectroscopy are the most commonly used techniques to measure particle size and size distribution.

## *4.2 Surface Properties of Nanoparticles*

Nanoparticle surface hydrophobicity determines the extent of blood components that bind this surface. Therefore, hydrophobicity influences the *in-vivo* fate of NPs. But once in the blood, surface non-modified conventional NPs are rapidly opsonized and immediately cleared by the immune and lymphatic systems. To increase the circulation time and reduce quick cellular uptake, it is necessary to coat or decorate the surface of NPs with hydrophilic agents. This is commonly achieved by polyethylene glycol (PEG), polyethylene oxide, tween 80 or other hydrophilic copolymers. Surface coating PEG molecules has been shown to avoid opsonization and phagocytosis of NPs by immune cells (Olivier, 2005).

## *4.3 Zeta Potential of NPs*

The zeta potential ($\zeta$) of NPs describes the surface charge and associated stability of nanoparticles. Depending upon the composition and dispersion medium used, the electrical potential of nanoparticles changes and so does its stability. Drug-loaded NPs with zeta potential above $\pm$ 30 mV indicate stability in suspension. At this zeta potential, the surface charge prevents NP aggregation. Further details reveal the proper encapsulation of charged core material at the centre (Couvreur et al., 2002).

## *4.4 Surface Hydrophobicity*

Techniques such as contact angle measurements, hydrophobic interaction chromatography, adsorption of probes, and biphasic partitioning, etc. are commonly used to determine surface hydrophobicity of NPs. Recently, sophisticated analytical techniques and tools for the analysis of surface properties have been introduced, such as X-ray photon correlation spectroscopy. This not only determines surface hydrophobicity but also helps to identify chemical groups on the surface of NPs. Surface-modified NPs have changed characteristics with respect to solubility, affinity, and ability to skip RES (McSweeney et al., 2018).

## *4.5 Drug Loading*

Successful drug-loaded NPs have practically high drug-loading capacity, reducing the matrix material quantity in the formulation. Conventionally, there are two methods of drug loading onto NPs. The incorporation method wherein is incorporated during nanoparticle formulation, and the adsorption/absorption/layering method in which the drug is loaded after nanoparticle formation; is often achieved by digesting the NPs with the rich drug solution. Entrapment efficiency and drug loading depends on the drug solubility in the excipient matrix material (polymer itself or liquid dispersion medium), which further depends on the matrix composition, drug-polymer interactions, molecular weight, polymer-drug ratio, and operational conditions of the manufacturing methods. An important point to ponder here is that the greatest loading efficiency for macromolecules, proteins or drugs encapsulated in NPs is observed when loaded at or near their isoelectric point. In the case of smaller molecules, improved drug loading occurs with significant ionic interactions between the drug and the polymer matrix materials (Picker and Hoag, n.d.).

### *4.6 Drug Release*

Several factors are connected with the rate of drug release from NPs. The important and significant among them are (a) drug solubility (b) partitioning of drug from matrix or surface (c) diffusion of drug from the matrix of NP (d) rate of NP matrix erosion, dissolution, or degradation (e) maintenance of sink condition (f) surrounding biochemical situation like pH, ionic concentration, and disease conditions, etc. The kinetics of drug release are also governed to some extent by shape, size, and crystalline/amorphous nature of the drug and polymer. Spherical NPs show predictable drug release than tube or irregularly shaped NPs. Drug release from water soluble polymers is found to be faster for BCS class I and II drugs (Mahajan et al., 2022). Initial, rapid 'burst release' is observed in case of weakly bound drugs. At the same time, the longer the circulation time of NPs, the higher the drug release. However, the scenario changes when NPs are membrane-coated instead of matrix. The drug release, in this case, is mainly partitioning or a combination of partitioning and diffusion. Sometimes, the excipients in the NPs affect drug release. For instance, the presence of poly ethylene oxide-polypropylene oxide co-polymer (PEO-PPO) is known to enhance the rate of drug release by reducing drug interaction with chitosan, the main polymer in the NPs (Calvo et al., 1997).

### *4.7 Surface Morphology and Structure*

At the nanoscale level, the morphology of NPs is found to affect the quality of formulations in terms of stability, drug release and drug loading (Dreaden et al., 2012; Ridolfo et al., 2021). Commonly used techniques to study surface structures of NPs include Scanning Electron Microscopy (SEM), TEM, Atomic Force Microscopy (AFM), X-Ray Diffraction (XRD), Differential Scanning Calorimetry (DSC) and BET. Electron microscopy reveals the two-dimensional structures with contours. Atomic force microscopy measures localized forces, mainly adhesion strength, magnetic forces, grain structure, surface roughness and important mechanical properties. This helps to understand the interaction between the drug and NP matrix. X-ray diffraction is important for knowing crystalline or amorphous structure of NPs. From amorphous NPs, drug release is faster as compared to highly crystalline structures. Differential Scanning Calorimetry (DSC) is a thermal technique that measures the energy emitted or absorbed by a sample as a function of a change in temperature. DSC is used to determine the phase diagram, enthalpy, entropy, and specific heat of the NPs, thus revealing overall stability. BET procedure lets us understand the porosity of the NPs. More porous nature is known to accommodate high drug concentration, like mesoporous silica (Kestens et al., 2016b).

# 5. Applications of Nanoparticles in Drug Delivery

With several advantages of NPs over conventional drug delivery systems, NPs have been used successfully in pre-clinical experimentations. Not only this, several have shown clinical efficacy and as a result, FDA, EMA, and MHRA have approved several NP based nanomedicines. The drug delivery from NPs can take two mechanisms:

a) *Passive targeting* - passive targeting involves drug delivery to the site by natural action; a common example is an accumulation of chemotherapeutic drugs in solid tumours. This leads to enhanced vascular permeability in the vicinity of tumour tissues in comparison with healthy tissues. In passive targeting, ligand-receptor activity can be selective; hence targeting at the intended site is possible. NPs can encounter multiple obstacles moving to their target. These obstacles can be mucosal barriers, nonspecific uptake of the NPs and unintended delivery of the drug (due to uncontrolled release).
b) *Active targeting* - Active targeting needs some changes in the NP formulation, such as surface functionalization, ligand addition, prodrug preparation, etc., of drug carriers that selectively recognize receptors on the surface of the host cells. This can be achieved by specific targeting of target tissues with NPs by several means such as antibody attachment on the NP surface or changing surface charge by chemical treatment.

This part of the chapter briefly gives glimpses of NP applications in drug delivery.

## *5.1 Oral Delivery of Peptides and Proteins*

Polymer-based NPs can encapsulate the bioactive molecules and protect from hydrolytic and enzymatic degradation; for example, polymeric NPs loaded with insulin showed blood glucose reduction in diabetic rats up to 14 days after oral administration. This shows that NPs protect insulin from bacterial gut flora, proteolytic enzymes such as trypsin, pepsin, and chymotrypsin, and viscous mucous (Brandtzaeg et al., 1997). Ligand such as lectins have been shown to improve oral absorption of peptides. Chitosan poly (acrylate) NPs improve oral absorption of peptides by enhancing paracellular drug transport across the membrane (Schipper et al., 1999).

## *5.2 Gene Therapy*

Genetic material delivery as drug candidate has received considerable attention and NPs have served this mostly in the form of vaccines. Martens et al. developed retinal gene delivery to overcome the barriers. Nucleic acid was incorporated in hyaluronic acid polyplex for intravitreal injection. PEG-coated complex indicated uptake by endocytosis with improved bioavailability (Martens et al., 2015). In another study, hyaluronic acid-coated chitosan NPs complex of hyaluronic acid-chitosan complex (HA-CS NPs) reported reduced immunogenic response with HA as compared to alginate, starch, and gelatin (Almalik et al., 2017).

## *5.3 Reduction of Drug-related Toxicities*

Drugs undergoing hepatic metabolism at high rates produce toxic metabolites. So, bypassing first-pass metabolism in the liver and absorption from the lymphatic pathway can reduce the production of toxic metabolite and reduce hazardous effects. Elmowafy et al. reported a formulation of carbamazepine incorporating beeswax NPs. Hepatic toxicity after 2 months of administration was reduced significantly owing to bypassing of liver metabolism due to the longer time of NP circulation in the blood and controlled release of carbamazepine (Elmowafy et al., 2018).

## *5.4 Cancer Therapy*

Current chemotherapy is non-specific and mainly aimed at destroying all of the rapidly dividing cancerous cells. The disadvantage of this therapy is that the body's other non-cancerous cells are also killed, such as hair follicles and intestinal epithelium. With smart NPs, targeted drug delivery to the cancerous cells avoids activity on the normal cell sites (Shen et al., 2016).

NPs in cancer therapeutics using dendrimers has shown to improve the therapeutic index of cytotoxic drugs with the use of biocompatible excipients, and the surface modification with PEGylation, glycosylation, acetylation, and with various ligands (Cheng et al., 2011). Liang et al. (2017) reported dual drug delivery for Non-Small-Cell Lung Cancer (NSCLC) treatment. Gemcitabine and paclitaxel were incorporated in NLCs carrying *N*-Acetyl-d-Glucosamine (NAG) as linker of two drugs. NPs exhibited higher drug uptake by A549 cell lines and more delivery towards tumour cells as compared to normal cells.

## *5.5 Diagnostic Testing*

The use of NPs for diagnostic purposes is being explored in imaging and tracer studies. These mainly involve MRI scans and photodynamic early diagnosis of cancer (Wolfbeis, 2015). NPs in the form of quantum dots, which can be customized to code for specific colours and guide in disease condition progress, have been developed (Emerich and Thanos, 2006). These diagnostic NPs have good photostability and tunable emission spectrum. Thus, bioimaging helps in differential diagnosis, early detection as well as disease progression. In recent development of two decades, theranostic agents are by combining both drug and imaging agent in a smart NP formulation. This aids in monitoring the therapeutic response and assessment of drug targeting and localization at the target site (Bhojani et al., 2010).

### 5.6 HIV and AIDS Treatment

Human Immunodeficiency Virus (HIV) infection if not addressed timely, leads to Acquired Immune Deficiency Syndrome (AIDS). This lately destroys immune system and increases susceptibility to other infections. The conventional treatment involves antiretroviral therapy with number of daily pills. Polymeric NP-based treatment can reduce this burden effectively. Nanotechnology has played a significant role in the delivery of antiretroviral drugs and the improvement of compliance (Jayant and Nair, 2016). Several studies report that NPs with antiretroviral drugs could effectively target macrophages and monocytes *in vitro*. Destache et al. (2009) reported Poly(Lactic-co-Glycolic acid) (PLGA) NPs that encapsulated three antiretroviral drugs, ritonavir, efavirenz, and lopinavir. The NP system provided sustained drug release for 4 weeks (28 days), while unbound drugs were eliminated in 48 h (Mallipeddi and Rohan, 2010).

### 5.7 Transdermal Drug Delivery

Transdermal Drug Delivery system (TDD) offers suitable route for drugs having low oral bioavailability. TDD avoids detrimental GIT conditions as well as first pass metabolism. TDD avoids drug overdosing and fluctuation in plasma drug concentrations on oral dosing. NPs in the form of NLCs have shown promising results when given by transdermal route. Mendes et al. (2019) developed donepezil NLCs and embedded them into Carbopol 940 gel. With the use of series of permeation enhancers, transdermal permeability was improved by over 10% as compared to oral route (Lin and Duh, 2016).

### 5.8 Pulmonary Drug Delivery

Pulmonary route has biological barriers such as mucus, ciliated cells and alveolar macrophages which reduce and hurdle pulmonary drug delivery. Ciliated cells clear the drug targeted at upper respiratory tract, while the drugs intended for lower respiratory tract, are phagocytosed and digested by alveolar macrophages. Owing to their particle size, lipid nanocarrires can reach lower respiratory tract achieving high drug accumulation, diffusion, and absorption (Jaques and Kim, 2000). Patlolla et al reported pharmacokinetics and toxicity of celecoxib loaded NLCs for targeting respiratory tract. Drug release from NPs was in a controlled manner while particle size range was in aerodynamic diameter suitable for lung delivery. The most of nebulized NLCs were located in alveolar region of mice. Pharmacokinetics study presented constant plasma drug levels for 6 h (Patlolla et al., 2010).

Table 1 gives a representative list of drug loaded NPs approved by FDA and EMA. The table is not comprehensive, however gives an idea about areas in which

**Table 1.** Different types of FDA and/or EMA approved NPs used in drug delivery systems and their applications.

| Type of nanocarrier | Type of NP | Marketed/Brand name | Active component | Manufacturing company | Applications | Reference |
|---|---|---|---|---|---|---|
| Polymer-based nanomedicine | Nanoemulsion | Diprivan® | Propofol | AstraZeneca LP | anaesthesia maintenance For patient under critical care | (Terblanche and Coetzee, 2008) |
| | Polymer-protein conjugate | Adagen® | Adenosine deaminase | Enzon Pharmaceuticals Inc | immunodeficiency disorder with adenosine deaminase | (Booth and Gaspar, 2009) |
| | Polymer-based | Oncaspar® | L-asparaginase | Enzon Pharmaceuticals Inc | Acute lymphoblastic leukemia | (Ettinger, 1995) |
| | Polymer-protein conjugate | Neulasta® | Filgrastim | Amgen, Inc | Neutropenia | (Duncan, 2005) |
| | Micelles | Genexol-PM® | Paclitaxel | Lupin Ltd | Breast cancer | (Oerlemans et al., 2010) |
| | Polymer steroid mixture | Zilretta® | Triamcinolone acetonide | Flexion Therapeutics | Osteoarthritis of knees | (Paik et al., 2019) |
| Inorganic nanoparticles | Iron nanoparticles | Venofer® | Iron sucrose | Lutipold Pharmaceuticals, Inc | Iron deficiency associated with chronic kidney disease | (Bhandari et al., 2015) |
| | Iron nanoparticles | Dexferrum® | Iron dextran | American Regent | Iron deficiency associated with chronic kidney disease | (Hood et al., 2000) |
| | Hafnium oxide NPs | Hensify® | Hafnium oxide | Nanobiotix | squamous cell carcinoma - Locally advanced | (Bonvalot et al., 2019) |
| Nanocrystals | Nanocrystal | Ritalin LA® | Methylphenidate hydrochloride | Novartis | Attention deficit hyperactivity disorder (ADHD) in children | (Van den Driessche et al., 2017) |
| | Nanocrystal | Tricor® | Fenofibric aid | Abott Laboratories | Hyperlipidemia | (Saurav et al., 2012) |
| | Nanocrystal | Invega® | Paliperidone palmitate | Janssen Pharmaceuticals | Schizophrenia | (Nagino et al., 2011) |
| | Nanocrystal | Megace® ES | Megestrol acetate | Par Pharmaceuticals | Anorexia and cachexia | (Tuca et al., 2013) |

*Table 1 contd. ...*

*...Table 1 contd.*

| Type of nanocarrier | Type of NP | Marketed/Brand name | Active component | Manufacturing company | Applications | Reference |
|---|---|---|---|---|---|---|
| Lipid based | Lipid complex | Abelcet® | Amphotericin B | Liposome Co | Mucormycosis, aspergillosis and invasive fungal infections | (Rust and Jameson, 1998) |
| | PEGylated liposome | Caelyx® | Doxorubicin hydrochloride | Janssen Pharmaceutica NV | AIDS-related Kaposi's sarcoma, Metastatic breast cancer and ovarian cancer | (Ranson et al., 2001) |
| | Unilamellar liposome | AmBiosome® | Amphotericin B | NeXstar Pharmaceuticals | Fungal infections – candidiasis, Aspergillosis, and cryptococcosis resistant to amphotericin B | (Boswell et al., 1998) |
| | Liposome | Myocet® | Doxorubicin hydrochloride and an anthracycline cytotoxic agent | Teva Pharmaceutical Industries Ltd | Metastatic breast cancer | (Brucker et al., 2016) |
| | Liposome | Lipodox® | Doxorubicin hydrochloride | Sun Pharmaceutical Industries Ltd. (SPIL) | ovarian cancer, Kaposi's sarcoma, and multiple myeloma | (Chou et al., 2015) |

NPs based formulation research is heading and the diseases which need better formulations as compared to conventional drug delivery systems.

## 6. Challenges in the Development of Nanoparticles for Drug Delivery

Successful development of drug-loaded NP development is a time-consuming process that requires meticulous and curated optimization at every step in the manufacturing followed by validation of analytical testing procedures. This is necessary because time-tested, regulatory agency-approved evaluation procedures are not yet established that address the risk/benefit ratio of NPs. The need of the hour is product-based, comprehensive preclinical data and clinical test protocols that are accepted worldwide. Although some research is being carried out in terms of organoids, spheroids, and organ-on-chip for ready evaluation to mimic human volunteer organs, this is still in the stage of infancy. Nanoparticles as drug carriers need to address some issues such as biosafety, toxicity, immune response, and pharmacology with reproducible data and clinical outcomes (Havel et al., 2016). The toxicity issue is at the forefront since sub-micron particle size can cross BBB, placental barrier, and move freely across the cells. This challenge is further escalated by the type of NPs because organic, lipid-based, inorganic, metallic, and magnetic exhibit different material properties and hence diverse pharmacokinetic and pharmacodynamic profiles. Another technological challenge is technology transfer and scale-up of nanoparticle production. NP production has a challenge in terms of scaling up production batch size from the laboratory to industry, mainly because the behaviour of NPs during production in small quantities is predictable while at large scale the collective properties differ to some extent (Đorđević et al., 2022).

Despite all odds, the use of NP-based drug delivery approaches is going to be increased, the latest example being COVID-19 vaccines utilizing nanoparticle-based vaccine delivery platforms. Since most of the FDA ratified nano-based agents are cancer-linked, NPs show a bright future in cancer theranostics. In total, the use of NP-based drug delivery systems for a range of ailments and the associated market is bound to grow in the coming years (Metselaar and Lammers, 2020).

## 7. Future Prospects and Market Outlook

Nanoparticles serve the purpose of not just the conveyance of medicines but also facilitating diagnostic and therapeutic monitoring of drugs throughout the entire cancer therapy. Image-guided drug delivery combines scanning such as MRI with drug-loaded NPs to track their targeting behaviour and pharmacokinetics. Remarkable progress in the field of NP-based formulations has been notable for the past two decades, with scores of product approvals by EMA and FDA. NP-drug formulations provide relief in a wide range of treatments and diagnostic avenues for complex diseases, such as cancer, ocular diseases, and lung diseases. The commercially marketed NP-based formulations include lipid-containing nanomedicines, polymer-based formulations, inorganic nanoparticles, nanocrystals, and nanoparticles containing genetic materials such as mRNA, nucleotides, viral particles, DNA, and proteins. These are revolutionary formulations and have a significantly positive impact on the healthcare system. However, since the formulations containing therapeutic agents are much more complex, it is necessary to address issues regarding safety, quality, and efficacy in alignment with the guidelines laid down by FDA, ICH, and EMA. In addition, extensive studies on the detailed characterization of NPs, their cost-benefit analyses, and preclinical plus clinical testing are needed on an urgent basis. Hence, based on the reports of previous studies, coupled with future rigorous studies and stringent regulatory guidelines to make available safe and effective treatment will yield nano-based drug delivery systems an exclusive solution for unanswered clinical needs. With the rosy picture of NP-based drug delivery systems, certain points elude its market success, namely, translation from the preclinical to clinical stage and cost-effective feasibility for investors.

## 8. Conclusion

NPs make an attractive destination for deliverables of drug molecules and, nowadays, biosimilars. The applications of NPs containing therapeutic moieties go beyond disease treatment or modification, for instance, biosensing, diagnostics and bio imaging. Nanomaterials have also employed successfully in a wide range of applications both *in-vitro* and *in-vivo*. This chapter summarized a brief description of NPs, their types, characterization, applications in drug delivery and challenges. The discussion on characterization using techniques such as DSC, XRD, AFM, TEM, and, SEM have shown that NPs have sizes that vary from a small nanometre to 100 nm. While the applications established need more in-depth analysis for safety and toxicity, NP-based drug delivery systems need to solve several obvious questions, including significant and long-term health impacts of NPs and potentially scalable, replicable manufacturing processes in future.

## References

Almalik, A., Benabdelkamel, H., Masood, A., Alanazi, I.O., Alradwan, I., Majrashi, M.A. et al. (2017). Hyaluronic acid coated chitosan nanoparticles reduced the immunogenicity of the formed protein corona. Scientific Reports 7(1): 10542. https://doi.org/10.1038/s41598-017-10836-7.

Alshammari, M.K., Alshehri, M.M., Alshehri, A.M., Alshlali, O.M., Mahzari, A.M., Almalki, H.H. et al. (2023). Camptothecin loaded nano-delivery systems in the cancer therapeutic domains: A critical examination of the literature. Journal of Drug Delivery Science and Technology 79: 104034. https://doi.org/https://doi.org/10.1016/j.jddst.2022.104034.

Anselmo, A.C. and Mitragotri, S. (2019). Nanoparticles in the clinic: An update. Bioengineering & Translational Medicine 4(3): 1–16. https://doi.org/10.1002/btm2.10143.

Arsawang, U., Saengsawang, O., Rungrotmongkol, T., Sornmee, P., Wittayanarakul, K., Remsungnen, T. et al. (2011). How do carbon nanotubes serve as carriers for gemcitabine transport in a drug delivery system? Journal of Molecular Graphics and Modelling 29(5): 591–596. https://doi.org/https://doi.org/10.1016/j.jmgm.2010.11.002.

Bhandari, S., Kalra, P.A., Kothari, J., Ambühl, P.M., Christensen, J.H., Essaian, A.M. et al. (2015). A randomized, open-label trial of iron isomaltoside 1000 (Monofer®) compared with iron sucrose (Venofer®) as maintenance therapy in haemodialysis patients. Nephrology Dialysis Transplantation 30(9): 1577–1589. https://doi.org/10.1093/ndt/gfv096.

Bhojani, M.S., Van Dort, M., Rehemtulla, A. and Ross, B.D. (2010). Targeted imaging and therapy of brain cancer using theranostic nanoparticles. Molecular Pharmaceutics 7(6): 1921–1929.

Bonifacio, B.V., da Silva, P.B., Ramos, M.A. dos S., Negri, K.M.S., Bauab, T.M. et al. (2014). Nanotechnology-based drug delivery systems and herbal medicines: a review. International Journal of Nanomedicine 1–15.

Bonvalot, S., Rutkowski, P.L., Thariat, J., Carrère, S., Ducassou, A., Sunyach, M.-P. et al. (2019). NBTXR3, a first-in-class radioenhancer hafnium oxide nanoparticle, plus radiotherapy versus radiotherapy alone in patients with locally advanced soft-tissue sarcoma (Act.In.Sarc): a multicentre, phase 2–3, randomised, controlled trial. The Lancet Oncology 20(8): 1148–1159. https://doi.org/10.1016/S1470-2045(19)30326-2.

Booth, C. and Gaspar, H.B. (2009). Pegademase bovine (PEG-ADA) for the treatment of infants and children with severe combined immunodeficiency (SCID). Biologics: Targets and Therapy 349–358.

Boswell, G.W., Buell, D. and Bekersky, I. (1998). AmBisome (Liposomal Amphotericin B): A comparative review. The Journal of Clinical Pharmacology 38(7): 583–592. https://doi.org/https://doi.org/10.1002/j.1552-4604.1998.tb04464.x.

Brandtzaeg, P., Berstad, A.E., Farstad, I.N., Haraldsen, G., Helgeland, L., Jahnsen, F.L. et al. (1997). Mucosal immunity—a major adaptive defence mechanism. Behring Institute Mitteilungen 98: 1–23. http://europepmc.org/abstract/MED/9382729.

Brucker, J., Mayer, C., Gebauer, G., Mallmann, P., Belau, A.K., Schneeweiss, A. et al. (2016). Non-pegylated liposomal doxorubicin for patients with recurrent ovarian cancer: A multicentric phase II trial. Oncology Letters 12(2): 1211–1215. https://doi.org/10.3892/ol.2016.4740.

Calvo, P., Remuñan-López, C., Vila-Jato, J.L. and Alonso, M.J. (1997). Chitosan and chitosan/ethylene oxide-propylene oxide block copolymer nanoparticles as novel carriers for proteins and vaccines. Pharmaceutical Research 14: 1431–1436.

Chen, Y., Pan, L., Jiang, M., Li, D. and Jin, L. (2016). Nanostructured lipid carriers enhance the bioavailability and brain cancer inhibitory efficacy of curcumin both *in vitro* and *in vivo*. Drug Delivery 23(4): 1383–1392. https://doi.org/10.3109/10717544.2015.1049719.

Chen, Y., Wei, C., Lyu, Y., Chen, H., Jiang, G. and Gao, X. (2020). Biomimetic drug-delivery systems for the management of brain diseases. Biomaterials Science 8(4): 1073–1088. https://doi.org/10.1039/C9BM01395D.

Cheng, Y., Zhao, L., Li, Y. and Xu, T. (2011). Design of biocompatible dendrimers for cancer diagnosis and therapy: current status and future perspectives. Chemical Society Reviews 40(5): 2673–2703.

Chou, H., Lin, H. and Liu, J.M. (2015). A tale of the two PEGylated liposomal doxorubicins. pp. 1719–1720. *In*: OncoTargets and Therapy (Vol. 8). Dove Medical Press Ltd. https://doi.org/10.2147/OTT.S79089.

Couvreur, P., Barratt, G., Fattal, E. and Vauthier, C. (2002). Nanocapsule technology: a review. Critical Reviews™ in Therapeutic Drug Carrier Systems 19(2).

Date, A.A., Vador, N., Jagtap, A. and Nagarsenker, M.S. (2011). Lipid nanocarriers (GeluPearl) containing amphiphilic lipid Gelucire 50/13 as a novel stabilizer: fabrication, characterization and evaluation for oral drug delivery*. Nanotechnology 22(27): 275102. https://doi.org/10.1088/0957-4484/22/27/275102.

De Temmerman, P.-J., Van Doren, E., Verleysen, E., Van der Stede, Y., Francisco, M.A.D. and Mast, J. (2012). Quantitative characterization of agglomerates and aggregates of pyrogenic and precipitated amorphous silica nanomaterials by transmission electron microscopy. Journal of Nanobiotechnology 10(1): 1–11.

Destache, C.J., Belgum, T., Christensen, K., Shibata, A., Sharma, A. and Dash, A. (2009). Combination antiretroviral drugs in PLGA nanoparticle for HIV-1. BMC Infect Dis. 2009 Dec 9; 9: 198. doi: 10.1186/1471-2334-9-198.

Đorđević, S., Gonzalez, M.M., Conejos-Sánchez, I., Carreira, B., Pozzi, S., Acúrcio, R.C. et al. (2022). Current hurdles to the translation of nanomedicines from bench to the clinic. Drug Delivery and Translational Research 12(3): 500–525. https://doi.org/10.1007/s13346-021-01024-2.

Dreaden, E.C., Alkilany, A.M., Huang, X., Murphy, C.J. and El-Sayed, M.A. (2012). The golden age: gold nanoparticles for biomedicine. Chemical Society Reviews 41(7): 2740–2779. https://doi.org/10.1039/C1CS15237H.

Duncan, R. (2005). Nanomedicine gets clinical. Materials Today 8(8, Supplement): 16–17. https://doi.org/https://doi.org/10.1016/S1369-7021(05)71032-4.

Durán-Lobato, M., Niu, Z. and Alonso, M.J. (2020). Oral delivery of biologics for precision medicine. Advanced Materials 32(13): 1901935. https://doi.org/10.1002/adma.201901935.

Elliott, J.A., Shibuta, Y., Amara, H., Bichara, C. and Neyts, E.C. (2013). Atomistic modelling of CVD synthesis of carbon nanotubes and graphene. Nanoscale 5(15): 6662–6676. https://doi.org/10.1039/C3NR01925J.

Elmowafy, M., Shalaby, K., Badran, M.M., Ali, H.M., Abdel-Bakky, M.S. and Ibrahim, H.M. (2018). Multifunctional carbamazepine loaded nanostructured lipid carrier (NLC) formulation. International Journal of Pharmaceutics 550(1–2): 359–371.

Elsabahy, M. and Wooley, K.L. (2012). Design of polymeric nanoparticles for biomedical delivery applications. Chemical Society Reviews 41(7): 2545–2561. https://doi.org/10.1039/C2CS15327K.

Emerich, D.F. and Thanos, C.G. (2006). The pinpoint promise of nanoparticle-based drug delivery and molecular diagnosis. Biomolecular Engineering 23(4): 171–184.

Ettinger, A.R. (1995). Pegaspargase (Oncaspar). Journal of Pediatric Oncology Nursing 12(1): 46–48. https://doi.org/10.1177/104345429501200110.

Guzman, L.A., Labhasetwar, V., Song, C., Jang, Y., Lincoff, A.M., Levy, R. et al. (1996). Local intraluminal infusion of biodegradable polymeric nanoparticles: a novel approach for prolonged drug delivery after balloon angioplasty. Circulation 94(6): 1441–1448.

Hamimed, S., Jabberi, M. and Chatti, A. (2022). Nanotechnology in drug and gene delivery. Naunyn-Schmiedeberg's Archives of Pharmacology 395(7): 769–787. https://doi.org/10.1007/s00210-022-02245-z.

Havel, H., Finch, G., Strode, P., Wolfgang, M., Zale, S., Bobe, I. et al. (2016). Nanomedicines: From bench to bedside and beyond. The AAPS Journal 18(6): 1373–1378. https://doi.org/10.1208/s12248-016-9961-7.

Herrero-Vanrell, R., Rincon, A.C., Alonso, M., Reboto, V., Molina-Martinez, I.T. and Rodriguez-Cabello, J.C. (2005). Self-assembled particles of an elastin-like polymer as vehicles for controlled drug release. Journal of Controlled Release 102(1): 113–122.

Hood, S.A., O'Brien, M. and Higgins, R. (2000). The safety of intravenous iron dextran (Dexferrum [R]) during hemodialysis in patients with end stage renal disease. Nephrology Nursing Journal 27(1): 41–43.

Jaiswal, M., Dudhe, R. and Sharma, P.K. (2015). Nanoemulsion: an advanced mode of drug delivery system. 3 Biotech 5(2): 123–127. https://doi.org/10.1007/s13205-014-0214-0.

Jaques, P.A. and Kim, C.S. (2000). Measurement of total lung deposition of inhaled ultrafine particles in healthy men and women. Inhalation Toxicology 12(8): 715–731.

Jayant, R. and Nair, M. (2016). Nanotechnology for the treatment of NeuroAIDS. J. Nanomed Res. 3(1): 00047.

José Alonso, M. (2004). Nanomedicines for overcoming biological barriers. Biomedicine & Pharmacotherapy 58(3): 168–172. https://doi.org/https://doi.org/10.1016/j.biopha.2004.01.007.

Kaur, I.P., Bhandari, R., Bhandari, S. and Kakkar, V. (2008). Potential of solid lipid nanoparticles in brain targeting. Journal of Controlled Release 127(2): 97–109. https://doi.org/https://doi.org/10.1016/j.jconrel.2007.12.018.

Kestens, V., Roebben, G., Herrmann, J., Jämting, Å., Coleman, V., Minelli, C. et al. (2016a). Challenges in the size analysis of a silica nanoparticle mixture as candidate certified reference material. Journal of Nanoparticle Research 18(6): 171. https://doi.org/10.1007/s11051-016-3474-2.

Kestens, V., Roebben, G., Herrmann, J., Jämting, Å., Coleman, V., Minelli, C. et al. (2016b). Challenges in the size analysis of a silica nanoparticle mixture as candidate certified reference material. Journal of Nanoparticle Research 18(6): 171. https://doi.org/10.1007/s11051-016-3474-2.

Khan, I., Saeed, K. and Khan, I. (2019a). Nanoparticles: Properties, applications and toxicities. Arabian Journal of Chemistry 12(7): 908–931. https://doi.org/https://doi.org/10.1016/j.arabjc.2017.05.011.

Khan, I., Saeed, K. and Khan, I. (2019b). Nanoparticles: Properties, applications and toxicities. Arabian Journal of Chemistry 12(7): 908–931. https://doi.org/https://doi.org/10.1016/j.arabjc.2017.05.011.

Lamprecht, C., Hinterdorfer, P. and Ebner, A. (2014). Applications of biosensing atomic force microscopy in monitoring drug and nanoparticle delivery. Expert Opinion on Drug Delivery 11(8): 1237–1253. https://doi.org/10.1517/17425247.2014.917078.

Liang, Y., Tian, B., Zhang, J., Li, K., Wang, L., Han, J. et al. (2017). Tumor-targeted polymeric nanostructured lipid carriers with precise ratiometric control over dual-drug loading for combination therapy in non-small-cell lung cancer. Int J Nanomedicine. 2017 Mar 2; 12: 1699–1715. doi: 10.2147/IJN.S121262.

Lin, W.J. and Duh, Y.S. (2016). Nanostructured lipid carriers for transdermal delivery of acid labile lansoprazole. European Journal of Pharmaceutics and Biopharmaceutics 108: 297–303. https://doi.org/https://doi.org/10.1016/j.ejpb.2016.07.015.

Liu, X., Li, C., Lv, J., Huang, F., An, Y., Shi, L. et al. (2020). Glucose and $H_2O_2$ dual-responsive polymeric micelles for the self-regulated release of insulin. ACS Applied Bio Materials 3(3): 1598–1606.

Mahajan, K., Thakur, N., Goswami, M., simran, kundal and Arora, I. (2022). Polymer based nanoparticles for BCS class II drugs—A mini review. Materials Today: Proceedings. https://doi.org/https://doi.org/10.1016/j.matpr.2022.12.007.

Malik, N., Evagorou, E.G. and Duncan, R. (1999). Dendrimer-platinate: a novel approach to cancer chemotherapy. Anticancer Drugs 1999 Sep; 10(8): 767–76. PMID: 10573209.

Mallipeddi, R. and Rohan, L.C. (2010). Progress in antiretroviral drug delivery using nanotechnology. International Journal of Nanomedicine 5: 533–547.

Martens, T.F., Remaut, K., Deschout, H., Engbersen, J.F.J., Hennink, W.E., van Steenbergen, M.J. et al. (2015). Coating nanocarriers with hyaluronic acid facilitates intravitreal drug delivery for retinal gene therapy. Journal of Controlled Release 202: 83–92. https://doi.org/https://doi.org/10.1016/j.jconrel.2015.01.030.

McSweeney, M.D., Wessler, T., Price, L.S.L., Ciociola, E.C., Herity, L.B., Piscitelli, J.A. et al. (2018). A minimal physiologically based pharmacokinetic model that predicts anti-PEG IgG-mediated clearance of PEGylated drugs in human and mouse. Journal of Controlled Release 284: 171–178.

Mendes, I.T., Ruela, A.L.M., Carvalho, F.C., Freitas, J.T.J., Bonfilio, R. and Pereira, G.R. (2019). Development and characterization of nanostructured lipid carrier-based gels for the transdermal delivery of donepezil. Colloids and Surfaces B: Biointerfaces 177: 274–281. https://doi.org/10.1016/j.colsurfb.2019.02.007.

Metselaar, J.M. and Lammers, T. (2020). Challenges in nanomedicine clinical translation. Drug Delivery and Translational Research 10(3): 721–725. https://doi.org/10.1007/s13346-020-00740-5.

Mitchell, M.J., Billingsley, M.M., Haley, R.M., Wechsler, M.E., Peppas, N.A. and Langer, R. (2021). Engineering precision nanoparticles for drug delivery. Nature Reviews Drug Discovery 20(2): 101–124. https://doi.org/10.1038/s41573-020-0090-8.

Moharir, K.S., Kale, V. and Kurakula, M. (2022). Chapter 11 - Nanotherapeutic systems for drug delivery to brain tumors. pp. 217–238. *In*: Mahajan, N.M., Saini, A., Raut, N.A. and Dhoble, S.J. (eds.). Photophysics and Nanophysics in Therapeutics. Elsevier. https://doi.org/https://doi.org/10.1016/B978-0-323-89839-3.00002-6.

Mosallaei, N., Jaafari, M.R., Hanafi-Bojd, M.Y., Golmohammadzadeh, S. and Malaekeh-Nikouei, B. (2013). Docetaxel-loaded solid lipid nanoparticles: preparation, characterization, *in vitro*, and *in vivo* evaluations. Journal of Pharmaceutical Sciences 102(6): 1994–2004. https://doi.org/10.1002/jps.23522.

Nagino, K., KOO, T. and Harada, Y. (2011). Pharmacological properties of paliperidone ER (INVEGA) and results of its clinical studies. 日本薬理学雑誌 137(6): 245–254.

Nasr, M., Nawaz, S. and Elhissi, A. (2012). Amphotericin B lipid nanoemulsion aerosols for targeting peripheral respiratory airways via nebulization. International Journal of Pharmaceutics 436(1): 611–616. https://doi.org/https://doi.org/10.1016/j.ijpharm.2012.07.028.

Oerlemans, C., Bult, W., Bos, M., Storm, G., Nijsen, J.F.W. and Hennink, W.E. (2010). Polymeric micelles in anticancer therapy: targeting, imaging and triggered release. Pharmaceutical Research 27(12): 2569–2589. https://doi.org/10.1007/s11095-010-0233-4.

Olivier, J.-C. (2005). Drug transport to brain with targeted nanoparticles. NeuroRx 2(1): 108–119.

Paik, J., Duggan, S.T. and Keam, S.J. (2019). Triamcinolone acetonide extended-release: a review in osteoarthritis pain of the knee. Drugs 79(4): 455–462. https://doi.org/10.1007/s40265-019-01083-3.

Pan, B., Gao, F. and Gu, H. (2005). Dendrimer modified magnetite nanoparticles for protein immobilization. Journal of Colloid and Interface Science 284(1): 1–6. https://doi.org/https://doi.org/10.1016/j.jcis.2004.09.073.

Patlolla, R.R., Chougule, M., Patel, A.R., Jackson, T., Tata, P.N.V. and Singh, M. (2010). Formulation, characterization and pulmonary deposition of nebulized celecoxib encapsulated nanostructured lipid carriers. Journal of Controlled Release 144(2): 233–241.

Patra, J.K., Das, G., Fraceto, L.F., Campos, E.V.R., Rodriguez-Torres, M. del P., Acosta-Torres, L.S. et al. (2018). Nano based drug delivery systems: recent developments and future prospects. Journal of Nanobiotechnology 16(1): 71. https://doi.org/10.1186/s12951-018-0392-8.

Pettinari, C., Marchetti, F., Mosca, N., Tosi, G. and Drozdov, A. (2017). Application of metal – organic frameworks. Polymer International 66(6): 731–744. https://doi.org/https://doi.org/10.1002/pi.5315.

Picker, K.M. and Hoag, S.W. (n.d.). Characterization of the Thermal Properties of Microcrystalline Cellulose by Modulated Temperature Differential Scanning Calorimetry.

Ranson, M.R., Cheeseman, S., White, S. and Margison, J. (2001). Caelyx (stealth liposomal doxorubicin) in the treatment of advanced breast cancer. Critical Reviews in Oncology/Hematology 37(2): 115–120. https://doi.org/https://doi.org/10.1016/S1040-8428(00)00107-4.

Ridolfo, R., Tavakoli, S., Junnuthula, V., Williams, D.S., Urtti, A. and van Hest, J.C.M. (2021). Exploring the impact of morphology on the properties of biodegradable nanoparticles and their diffusion in complex biological medium. Biomacromolecules 22(1): 126–133. https://doi.org/10.1021/acs.biomac.0c00726.

Rust, D.M. and Jameson, G. (1998). The novel lipid delivery system of amphotericin B: drug profile and relevance to clinical practice. Oncology Nursing Forum 25(1): 35–48.

Samed, N., Sharma, V. and Sundaramurthy, A. (2018). Hydrogen bonded niosomes for encapsulation and release of hydrophilic and hydrophobic anti-diabetic drugs: an efficient system for oral anti-diabetic formulation. Applied Surface Science 449: 567–573.

Satapathy, M.K., Yen, T.-L., Jan, J.-S., Tang, R.-D., Wang, J.-Y., Taliyan, R. et al. (2021). Solid Lipid Nanoparticles (SLNs): An advanced drug delivery system targeting brain through BBB. Pharmaceutics 13(8). https://doi.org/10.3390/pharmaceutics13081183.

Saurav, A., Kaushik, M. and Mohiuddin, S.M. (2012). Fenofibric acid for hyperlipidemia. Expert Opinion on Pharmacotherapy 13(5): 717–722. https://doi.org/10.1517/14656566.2012.658774.

Schipper, N.G.M., Vårum, K.M., Stenberg, P., Ocklind, G., Lennernäs, H. and Artursson, P. (1999). Chitosans as absorption enhancers of poorly absorbable drugs: 3: Influence of mucus on absorption enhancement. European Journal of Pharmaceutical Sciences 8(4): 335–343. https://doi.org/https://doi.org/10.1016/S0928-0987(99)00032-9.

Sharma, S.K., Chauhan, M. and Anilkumar, N. (2009). Span-60 niosomal oral suspension of fluconazole: formulation and *in vitro* evaluation. Asian Journal of Pharmaceutical Research and Health Care 1(2): 142–156.

Shen, B., Ma, Y., Yu, S. and Ji, C. (2016). Smart multifunctional magnetic nanoparticle-based drug delivery system for cancer thermo-chemotherapy and intracellular imaging. ACS Applied Materials & Interfaces 8(37): 24502–24508.

Smart, S.K., Cassady, A.I., Lu, G.Q. and Martin, D.J. (2006). The biocompatibility of carbon nanotubes. Carbon 44(6): 1034–1047. https://doi.org/https://doi.org/10.1016/j.carbon.2005.10.011.

Sultana, N. and Ismail, K. (2016). Specific ion effects of chloride vis-à-vis acetate, propionate and butyrate counterions on the cetylpyridinium headgroup at the micelle–solution and air–solution interfaces. Journal of Molecular Liquids 213: 145–152. https://doi.org/https://doi.org/10.1016/j.molliq.2015.11.016.

Sutton, D., Nasongkla, N., Blanco, E. and Gao, J. (2007). Functionalized micellar systems for cancer targeted drug delivery. Pharmaceutical Research 24: 1029–1046.

Tamjidi, F., Shahedi, M., Varshosaz, J. and Nasirpour, A. (2013). Nanostructured lipid carriers (NLC): A potential delivery system for bioactive food molecules. Innovative Food Science & Emerging Technologies 19: 29–43. https://doi.org/https://doi.org/10.1016/j.ifset.2013.03.002.

Terblanche, N. and Coetzee, J.F. (2008). A comparison of induction of anaesthesia using two different propofol preparations. Southern African Journal of Anaesthesia and Analgesia 14(6): 25–29. https://doi.org/10.1080/22201173.2008.10872573.

Tsai, M.-J., Wu, P.-C., Huang, Y.-B., Chang, J.-S., Lin, C.-L., Tsai, Y.-H. et al. (2012). Baicalein loaded in tocol nanostructured lipid carriers (tocol NLCs) for enhanced stability and brain targeting.

International Journal of Pharmaceutics 423(2): 461–470. https://doi.org/https://doi.org/10.1016/j.ijpharm.2011.12.009.

Tuca, A., Jimenez-Fonseca, P. and Gascón, P. (2013). Clinical evaluation and optimal management of cancer cachexia. Critical Reviews in Oncology/Hematology 88(3): 625–636. https://doi.org/https://doi.org/10.1016/j.critrevonc.2013.07.015.

Van den Driessche, C., Bastian, M., Peyre, H., Stordeur, C., Acquaviva, É., Bahadori, S. et al. (2017). Attentional lapses in attention-deficit/hyperactivity disorder: blank rather than wandering thoughts. Psychological Science 28(10): 1375–1386. https://doi.org/10.1177/0956797617708234.

Vlachopoulos, A., Karlioti, G., Balla, E., Daniilidis, V., Kalamas, T., Stefanidou, M. et al. (2022). Poly(Lactic Acid)-based microparticles for drug delivery applications: an overview of recent advances. Pharmaceutics 14(2). https://doi.org/10.3390/pharmaceutics14020359.

Wolfbeis, O.S. (2015). An overview of nanoparticles commonly used in fluorescent bioimaging. Chemical Society Reviews 44(14): 4743–4768. https://doi.org/10.1039/C4CS00392F.

Xu, W., Ling, P. and Zhang, T. (2013). Polymeric micelles, a promising drug delivery system to enhance bioavailability of poorly water-soluble drugs. Journal of Drug Delivery 2013.

Yusuf, V.F., Malek, N.I. and Kailasa, S.K. (2022). Review on metal–organic framework classification, synthetic approaches, and influencing factors: applications in energy, drug delivery, and wastewater treatment. ACS Omega 7(49): 44507–44531. https://doi.org/10.1021/acsomega.2c05310.

Zhang, B., Chen, Q., Tang, H., Xie, Q., Ma, M., Tan, L. et al. (2010). Characterization of and biomolecule immobilization on the biocompatible multi-walled carbon nanotubes generated by functionalization with polyamidoamine dendrimers. Colloids and Surfaces B: Biointerfaces 80(1): 18–25. https://doi.org/https://doi.org/10.1016/j.colsurfb.2010.05.023.

# Chapter 11

# Advancements in Cardiovascular Assistive Devices and Future Innovations for Post-COVID Care

*Yamini Singh,*[1] *Tej Pratap,*[2] *Ambak Kumar Rai*[3] and *Abhishek Kumar Tiwari*[1,*]

## 1. Introduction

Cardiovascular disease (CVD) is a global health concern. According to the World Health Organization (WHO) data, ischemic heart disease has emerged among the top ten global fatal diseases (Kaptoge et al., 2019). Acute coronary syndrome, coronary heart disease, and coronary artery disease are all examples of CVD. Congestive Heart Failure (CHF) may occur when the heart's capacity to pump blood gradually diminishes. These disorders result in more fatalities globally each year than all types of cancers. Early detection of left ventricular dysfunction has been one of the most contentious heart failure-related topics. Some individuals clearly express overt heart failure clinical symptoms, whereas others can still silently maintain the left

[1] Department of Applied Mechanics, Motilal Nehru National Institute of Technology Allahabad, Prayagraj-211004, Uttar Pradesh, India.

[2] Department of Mechanical Engineering, Motilal Nehru National Institute of Technology Allahabad, Prayagraj-211004, Uttar Pradesh, India.

[3] Department of Biotechnology, Motilal Nehru National Institute of Technology Allahabad, Prayagraj-211004, Uttar Pradesh, India.

* Corresponding author: aktiwari@mnnit.ac.in; abhiti1989@gmail.com

DOI: 10.1201/9781003287636-11

ventricular systolic function. Nonetheless, a person with myocarditis, valve disease, diabetes, high blood pressure, or obesity may always be at risk of getting severe heart failure. Elderly individuals are more prone to encounter heart failure than younger people. There are unquestionably a wide variety of other variables. causing heart failure. Nevertheless, the etiology of heart failure reformed over the time and coronary artery disease became a common cause of heart failure. Prevention and cure of heart failure do not have a single solution; however, early diagnosis and treatment have reduced heart failure-related mortality and also improved life expectancy. Out of these, CardiacAssist Devices (CADs), also known as mechanical circulatory support systems, have proven useful in the therapeutic therapy of heart problems.

Cardiac assist devices first emerged in the early 1960s, and thereafter, significant research and development efforts have been made for their advancement and expansion (Morlacchi and Nelson, 2011). There is now a wide range of cardiac assist devices which has unique benefits and assist clinicians and cardiologist significantly in the clinical management of patients. These devices also have certain drawbacks, which will be discussed later in this chapter. Nevertheless, all assistive devices are developed with the common aim of providing safe and consistent circulatory assistance when required. Clinical management difficulties increase when care is required for higher age groups and longer duration in young age groups. This motivates us to keep looking for the solution to the cardiovascular challenge to avoid clinical criticality and emotionally suffering cardiac patients. The development of CADs has some key factors, such as decreased device malfunctioning issues and removing the restriction imposed by percutaneous tethers. Moreover, CADs shall be appropriate for long-term or permanent usage. The lives of cardiac patients who depend on life-supporting equipment for years at a time can be significantly improved if present patient care methods are consistently replaced with coordinated, beneficial, and useful physician-patient interface systems (Han and Trumble, 2019).

The failure of CADs is caused by coagulation, bleeding, stroke, and infections. Certain clinical problems have been more frequent both during and after the COVID-19 epidemic. Novel coronavirus illness (COVID-19), whose severity produces cytokine storm, hyperinflammation, and a rise in the signs of serious cardiac damage, was caused by infection with the severe acute respiratory syndrome novel coronavirus. It was noticed that COVID-19 has accelerated cardiovascular issues, increasing morbidity and mortality in individuals with existing heart disease. Cardiovascular problems during COVID-19 have been connected to pulmonary embolism, Takotsubo cardiomyopathy, arrhythmia, myocardial injury, and myocarditis, among other complications. Patients with LVAD implants during COVID-19 have experienced more difficulties during acute infections than other cardiovascular patients. Hemodynamical and hematological alterations in blood during COVID-19 had an adverse effect on the functioning of LVAD implants. Therefore, a careful clinical management strategy is required during the care of such a population. As SARS-CoV-2 infection has caused a variety of cardiovascular and hematologic alterations, including low flow and suction events along with lowered pulsatility index, slow functioning of LVAD has also been noticed (Fang and O'Gara, 2015). A multidisciplinary strategy with knowledgeable doctors, nurses, and LVAD coordinators is necessary for the treatment of LVAD patients in COVID-19. Close monitoring of clinical status, device parameters and functioning, and laboratory findings are the foundation of the treatment. Indeed, a certain design consideration is required to lower the complexity involved during COVID-19 infection among patients so that adequate medical treatment can be planned for them. This review attempts to highlight the advancement in cardiovascular assist devices along with the complications that occurred in patients with such devices during or after recovery from COVID-19.

## 2. Cardiovascular Assist Devices

### *2.1 Ventricular Assist Devices (VAD)*

A Ventricular Assist Device has emerged as a trustworthy and secure treatment option for heart patients, which has been frequently used in the clinical management of end-stage heart failure. Nearly 22,000 implantations have been done to the present date. Newer generation devices with effective medical management and technology advancements improved patient outcomes and boosted destination therapy use (Brown and Angela, 2022). The most pressing issue with ventricular assist device therapy is the occurrence of adverse events, which persists despite this improvement. Therefore, continuous improvement in such devices is needed to reduce the damaging load on the devices. We are looking forward to ventricular assist device technology being adopted as a replacement for heart transplantation. In that case, this will only depend on our technical ability to lessen the risk associated with adverse pathological conditions. During the pandemic, there is a severe lack of information on cardiac problems related to COVID-19 patients. A sudden increase in heart patients has been observed in post-COVID-19. This is a serious challenge to clinicians and researchers.

There are two main forms of VADs, which are a left chamber assist device (LVAD) and a right chamber assist device (RVAD), and when both of these are used simultaneously, they are referred to as biventricular assist devices (BIVAD). LVAD facilitates the ventricle's blood flow to the artery through which the body receives oxygen-rich blood from the heart and is the most widely used VAD (Dandel and Hetzer, 2018) (Fig. 1). RVADs are temporarily utilized only for short-term right ventricle support after LVAD implantation or other heart procedures. A companion RVAD helps the right ventricle in pumping blood to the lungs. Moreover, VADs can be categorized based on the implanted locations. For example, the pump and power source of a connective tissue VAD is located outside the body through a tiny incision in the abdomen, and tubes connect the pump to the digestive system. This type of VAD could also be employed to provide immediate support before, during, or after surgery.

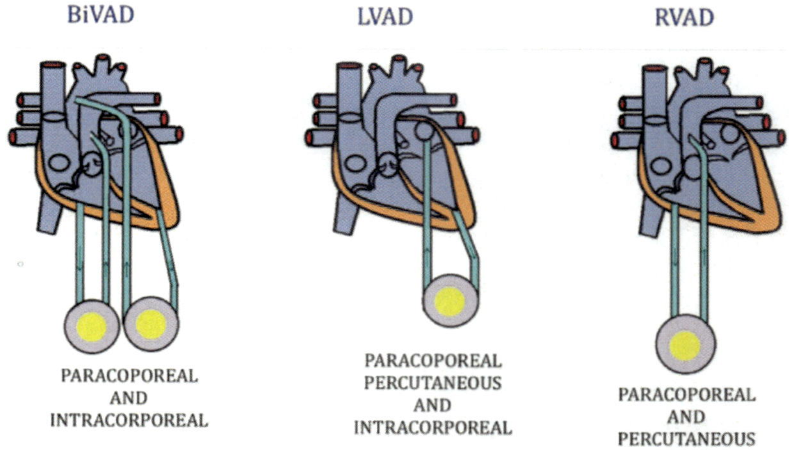


**Figure 1.** Classification of ventricular assistive devices.

## *2.2 Left Ventricular Assistive Devices (LVADs)*

Open-heart surgeries became feasible with the development of LVADs. This clinical surgery involves the pumping of blood to the rest of the body using a mechanically and battery-powered LVAD (Cid-Menendez and Dar, 2020). This device helps the left ventricle, i.e., the heart's pumping chamber. LVADs are generally utilized in:

- *Bridge-to-transplant therapy*: Patients undergoing this therapy had their lives saved while they awaited a heart transplant. Patients employ LVADs, which have, in some cases, been able to resuscitate failing hearts and avoid the urgent need for transplantation until the heart is ready. There may be certain patients who are not typically eligible for heart transplants, and therefore, they undergo destination therapy. In this situation, patients receive long-term care with LVAD, which extends and enhances the quality of their lives (Prinzing et al., 2016).

## *2.3 Implantable Cardioverter Defibrillator (ICD)*

ICD is a subcutaneous and battery-powered device that tracks the heart rate. Tiny wires link the heart to the ICD together. The device also shocks the heart with electric jolts to keep the heart back into a normal rhythm when the heart is beating too swiftly and erratically. ICDs have been effective in preventing sudden death in individuals who have been diagnosed with persistent ventricular tachycardia or fibrillation. Studies suggest that ICDs can prevent cardiac arrest in high-risk people who are at risk but have not yet suffered life-threatening ventricular arrhythmias.

## *2.4 Pacemaker*

A pacemaker works by electronically stimulating the heart muscle to maintain a regular pulse and rhythm. Syncope (fainting episodes), hypertrophic cardiomyopathy, and congestive heart failure are a few disorders that are frequently treated with pacemakers. Single-chamber pacemaker uses one lead in the heart's ventricles or the atria, whereas dual-chamber pacemakers place one lead in the right ventricle and one lead in the heart's atria. Biventricular pacemakers employ three leads at three different locations: one in the right atrium, the other in the right ventricle, and then one near the left ventricle for smooth functioning of the heart. Similar to a standard pacemaker, biventricular pacemakers use a third wire to provide electrical impulses to the heart while resynchronizing the beats of the left lower chambers of the heart, or ventricles. These pacemakers, often referred to as cardiac resynchronization devices, are implanted when the left chamber of the heart doesn't beat in unison, and the patient's symptoms aren't adequately relieved by medicine. In such a case, the resynchronization device coordinates the contraction of the left ventricle. The application of these devices also needs to be protected using a defibrillator. Biventricular defibrillator devices are also available. These hybrid devices help in maintaining a constant heartbeat constant in addition to speeding up or slowing down an irregular heartbeat.

## *2.5 Stents*

Drug-eluting and biodegradable stents are second and third-generation replacements for bare-metal stents. Advanced stents have been produced which has a unique set of geometries, architectures, backbone materials, and other properties. Figure 2 presents generation-wise advancements in stent development. DES (Drug-Eluding Stents) development is a recent innovation in interventional cardiology. Different drugs that target thrombosis, platelet activation, inflammation, and VSMC proliferation were tested.

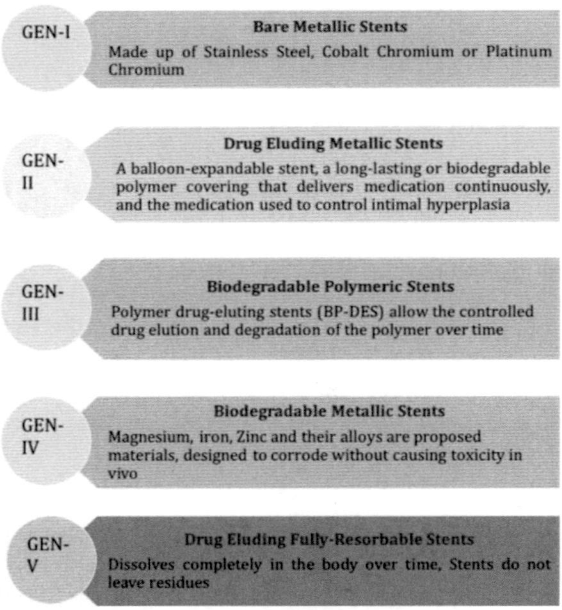


**Figure 2.** Generation-wise advancements in stent development.

The coating of BMS with materials like gold, which is considered to be inert; carbon, which resembles diamond; phosphorylcholine (PC), which mimics the cell membrane; and heparin, which prevents thrombosis, among many others, had no positive effects. Numerous hormone receptors, such as those for estrogen, glucocorticoids, and mineralocorticoids, showed very minor impacts when they were activated or antagonistic.

## 3. Challenges with Cardiovascular Assistive Devices

Ventricular assist devices encounter serious challenges such as blood clots, hemorrhage, infection, and device malfunction. Left Ventricular Assist Devices (LVADs) sometimes expose the right heart to failure risk (Slaughter et al., 2010). Recent VAD models have attempted to overcome a few of these challenges. A number of challenges, which are highlighted in the next sub-sections, still exist and need to be handled in the design of VADs.

### *3.1 Thrombosis*

VADs generally deal with blood. Contact of blood during flow with the device surface may cause thrombosis. Blood clots occur more frequently than the usual circumstances. Therefore, clotting shall be avoided in these devices (Nosé, 1998). Blood clots can impede blood vessels and block the blood flow to essential organs, including the brain. Pumping of blood with clots becomes difficult and puts extra effort into the functioning of the heart. Blood clots can result in deadly consequences (such as stroke) or other serious issues. Patients implanted with, therefore, sometimes require anticlotting medicine for the duration of the disease. Provisions are indeed required to avoid thrombosis in VADs.

### *3.2 Gastrointestinal Bleeding*

VAD implantation may cause bleeding also during and after surgery. Anticoagulant agents also raise the risk of gastrointestinal bleeding. Therefore, it is a challenge to maintain a balance between the risk of bleeding and anticlotting drugs. One should follow the doctor's instructions for taking prescriptions exactly.

### *3.3 Infections*

An internal pump connects to an external power source in implantable VADs. Transcutaneous VAD links the body's pump to the heart through the skin. A hole or multiple holes in the skin connect the parts inside the patient's body to the elements outside it in both forms of VADs. The chances of infection at the implantation site increase through these holes in the skin. There are tubes that permanently connect to the outside through the skin, and hence the probability of getting an unwanted infection is relatively high. The medical team must keep a careful eye if the patient exhibits any infection-related symptoms, such as discomfort around the VAD site, fluid drainage, or fever. In addition to all these conditions, as mentioned earlier, other circumstances can also severely affect the functioning of VADs in several ways.

For example, the pumping action may fail, the power supply may get disturbed, or the device component may stop working properly due to blood clot handling. Therefore, improved continuous flow VADs are required to mitigate the pathological cardiovascular conditions.

## 4. COVID-19 Outbreak and Cardiac Complexities

COVID-19 acute disease progression involves early infection, pulmonary disease, and severe hyperinflammation (Ahmad et al., 2022). Moreover, there is a possibility that the stages will significantly overlap. SARS-CoV-2 invades the lung parenchyma during the first infection and grows. At this stage, the immune system of the body mounts its first line of defense, which is manifested by mild constitutional symptoms, especially through the actions of monocytes and macrophages. Inflammatory processes cause vasodilation, increased endothelial permeability, and leukocyte recruitment, which are all associated with collateral tissue harm. Hypoxia, pulmonary harm, and cardiovascular stress are caused by these mechanisms. Systemic inflammation has also been noticed in a few individuals (despite declining virus levels) with an increase in host inflammatory response. Systemic inflammation increases the possibility of injury to distant organs. The findings of myocarditis in COVID-19 indicate that the heart is one of the targets of systemic inflammation. The lung is, although, the site of the original injury where the rash starts and spreads. However, the elevated inflammatory response can harm other organs, including the heart. The ACE-2 host cell receptor for viral surge protein enables infection by COVID-19 etiologic agents in the heart, vascular organs, and circulating cells (Wong et al., 2021).

More than 1.8 million people have been afflicted with COVID-19, which has taken the lives of more than 100,000 people (Reyes, 2020). It also overtaxed the healthcare systems due to the high frequency of catastrophic performances. Heart injury may be a potential pathogenic mechanism producing severe illness and death since a considerable portion of patients have high cardiac troponin levels. High troponin levels are clinically linked to higher mortality in COVID-19 patients. Respiratory symptoms are, although the first defined disorder, COVID-19 has also severely affected the other systems, such as the nervous and cardiovascular systems. Among additional thrombotic consequences, venous thrombosis, vasculitis, cardiomyopathy, and stroke have all also been widely noticed (Table 1). Markers that are linked to cardiovascular complexity during COVID-19 are summarized as follows:

### *4.1 Elevated Troponin Level*

Patients with COVID-19 have been found to have elevated troponin levels, which have been directly connected to tragic cardiovascular outcomes (Henry and Kereiakes, 2022). This condition can be related to a number of processes, including unmasked CAD, microangiopathy, viral myocarditis, and cytokine-driven cardiac injury (Fig. 3).

**Table 1.** Case studies on complications associated with cardiac assistive devices.

| Patient details | Device implanted | Complications | Biomarkers | Diagnostic tools used |
|---|---|---|---|---|
| 56-year-old female (Frick et al., 2020) | Left ventricular assist device (LVAD) [HeartMate II] | • Power spikes<br>• Symptomatic heart failure<br>• Pump thrombosis | • Double lactate<br>• Dehydrogenase (LDH) levels | • Initial Computed Tomography (CT) scan presented bilateral ground glass opacity<br>• Thromboelastogram |
| 61-year-old patient (Piperata et al., 2020) | LVAD (HeartMate III) | • Acute right heart failure<br>• Left pleural effusion | • Increased weight<br>• Peripheral edema<br>• Severe dyspnea | • Transthoracic echocardiogram<br>• Chest X-ray<br>• Blood and radiological tests |
| 72 years old man (Piperata et al., 2020) | LVAD | • Infection of the LVAD cable's exit place<br>• Post-ischemic dilated cardiomyopathy | ???? | • Computed tomography-scan<br>• Chest x-ray |
| 32-year-old man (Samuels et al., 2021) | Heartmate II™ LVAD (Abbott Laboratories, Abbott Park, IL, USA) | • Hemolysis,<br>• Pigment nephropathy<br>• Acute kidney damage as a result of pump thrombosis | • Decline in Hemoglobin (Hgb), Platelets (PLT), and haptoglobin<br>• Increase in Lactate Dehydrogenase (LDH)<br>• Elevation in Blood Urea Nitrogen (BUN)<br>• Elevation in in Creatinine (Cr), and D-Dimer | • Chest radiography |
| 43-year-old male patient (Samuels et al., 2021) | LVAD implantation (Heartmate-2 Abbott, Thoratec, Pleasanton, CA) | • Subpleural and peripheral haziness<br>• Light ground-glass opacities on both sides<br>• Bilateral pretibial edema<br>• Dyspnea | • Elevated C-reactive Protein (CRP) measures)<br>• Slight renal insufficiency<br>• Level of interleukin- spiked | • COVID-19 Polymerase Chain Reaction (PCR) swab test<br>• Transthoracic echocardiographic (TTE) test<br>• Thoracic computed tomography (CT) |
| 51-year-old male patient (Samuels et al., 2021) | LVAD implantation (HeartWare®, Medtronic, MN, USA) | • Decreased breath sounds on both sides and a mechanical device sound on the left | • $S_pO_2$ was at 89%<br>• Oxygen support arterial $O_2$ saturation at 95.4%<br>• Elevated Levels of lactate dehydrogenase<br>• Elevated Levels of D-dimer<br>• Elevated Levels of CRP | • X-ray revealed bilateral opacities suggestive of infection.<br>• Bilateral diffuse ground-glass opacities on CT scan. |

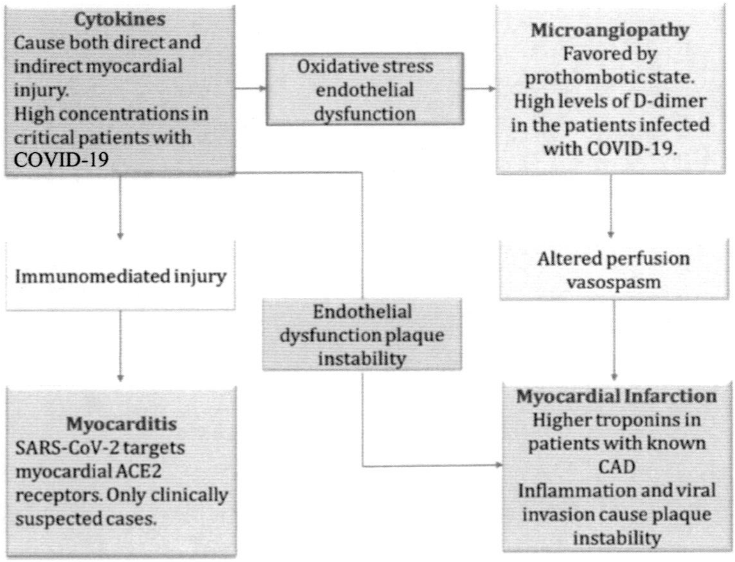


**Figure 3.** Causes of elevated troponin level (adapted from Hosseinipour et al., 2017).

- Cytokines are small proteins that play a crucial role in controlling the activity of other immune systems and blood cells; in COVID-19, cytokines are present in higher amounts, which results in oxidative stress on cardiac tissue.
- Such a state causes prothrombotic conditions and high levels of D-dimer.
- Changing vessel diameter due to higher troponins in patients with CAD cause myocardial infarction.
- Plaque instability and inflammation are the resultant of COVID-19 viral infection because SAR-CoV-2 targets the ACE2 receptors of myocardial tissue.
- Such a cascade of reactions is the estimated cause of the malfunctioning of cardiac assistive devices in COVID-19 patients.

### *4.2 Pump Thrombosis*

Due to hypercoagulability and a prothrombotic state, pulmonary embolism and thrombosis caused negative outcomes and increased mortality in severely sick COVID-19 patients. These conditions might have led to coagulopathy and disseminated intravascular coagulation. Consequently, COVID-19 patients have significant pulmonary and endothelial inflammation and enhanced thrombosis (Connors and Levy, 2020; Kichloo et al., 2020). Hypercoagulation syndrome causes a catastrophic decrease in the ventilation-to-perfusion ratio. This has clinical relevance as sepsis and hypoxia release von Willebrand factor can independently cause platelet aggregation in ICU patients with COVID-19. This factor had been high in the blood of critical care unit patients.

### *4.3 Idiopathic Cardiomyopathy*

There is evidence that patients with novel coronavirus infection had elevated blood troponin levels, which is substantially related to COVID-19 prognosis (Imazio et al., 2020). The direct binding of the coronavirus to its ACE-2 is also abundantly expressed in the heart and thus may cause myocardial damage in COVID-19 patients. Epicardial or intramyocardial small coronary artery thrombotic blockage from increased coagulability, myocarditis from generalized inflammation, epicardial or intramyocardial ischemia from circulatory and respiratory failure, etc., are all examples of myocardial injuries. Prolonged immunological activation following COVID-19 viral infection is found to increase patients' risk of developing dilated cardiomyopathy. Several cardiac symptoms, including myocardial damage, myocardial infarction, arrhythmias, heart failure, cardiomyopathy, shocks, thromboembolism, and cardiac arrest, have been seen in COVID-19 patients both during and after post-COVID-2019. Patients with pre-existing cardiac diseases such as coronary heart disease, high blood pressure, diabetes mellitus, and irregular coagulation are more prone to have these cardiovascular adverse effects.

## 5. Complications in Cardiac Devices Post-Covid-19

Arrhythmias, myocardial infarction, and myocarditis are the main cardiovascular complications in COVID-19 patients (Ranard et al., 2020). Tables 1 and 2 present the case studies of COVID-19 with assistive devices along with complications.

### *5.1 Failure of the Right Ventricle*

One of the adverse effects in COVID-19 patients has been acute Right Ventricular (RV) dysfunction. RV dysfunction has occurred in 20% of COVID-19 patients who had cardiovascular complications or severe illness (Saeed and Rajani, 2021). Acute RV failure can occur alone or in conjunction with biventricular failure during coronavirus disease (Haddad et al., 2008). Biventricular failure typically occurs due to myocarditis or generalized ischemia due to sepsis, whereas pulmonary embolism, pulmonary illness, pneumonia, or ARDS may result in single RV failure. LVAD implants experience higher long and short-term morbidity and mortality as a result of RV dysfunction, as they lack RV support other than in biventricular devices.

### *5.2 Hematologic Alterations*

COVID-19 patients experience abnormal blood coagulation profiles, which can have a variety of causes. Hematologic signs of COVID-19 also include abnormal blood coagulation profile, severe Disseminated Intravascular Coagulation (DIC), and venous and arterial thromboembolic occurrences (Kaur et al., 2021). Patients with COVID-19 who are severely unwell have been shown to have mild thrombocytopenia, an increased prothrombin time (PTT), and an enhanced activated partial thromboplastin time (Iba et al., 2020). A high prevalence of thrombotic events observed in COVID-19 patients can be because of a number of factors, such as the activation of thrombotic factors, cytokine storm, and oxidative stresses. A much

**Table 2.** Patient history with LVAD and the outcome during COVID-19 (Adapted from Zakrzewski et al., 2021).

| Characteristic | Patient 1 | Patient 2 | Patient 3 | Patient 4 | Patient 5 | Patient 6 |
|---|---|---|---|---|---|---|
| **Demographics** | | | | | | |
| Age, years | 64 | 61 | 61 | 56 | 47 | 59 |
| BMI | 31.6 | 21.9 | 19.4 | 37.2 | 31 | 30 |
| Sex | Male | Male | Male | Male | Female | Male |
| Blood group | A | B | O | O | A | O |
| Race/Ethnicity | White | Black | Hispanic | Black | Black | Black |
| **Medical characteristics** | | | | | | |
| Smoking history | ✓ | ✓ | ✓ | × | ✓ | × |
| Diabetes mellitus | ✓ | × | × | ✓ | ✓ | ✓ |
| Lung disease | × | ✓ | × | ✓ | × | ✓ |
| Hypertension | ✓ | ✓ | ✓ | ✓ | ✓ | ✓ |
| Chronic kidney disease stage ≥3 | ✓ | × | ✓ | × | ✓ | ✓ |
| Prior stroke | × | × | × | ✓ | × | × |
| **INTERMACS profile** | | | | | | |
| LVAD device | HM3 | HM3 | HM3 | HM3 | HM3 | HM3 |
| Time from +SARS-CoV-2 PCR to LVAD, days | 20 | 194 | 61 | 7 | 18 | 132 |
| LVAD intention (BTT or DT) | DT | DT | DT | DT | DT | DT |
| Supplemental oxygen | × | ---- | × | ✓ | × | × |
| **Post-LVAD outcomes** | | | | | | |
| Duration on LVAD support, days (IQR) | 33 | 140 | 45 | 194 | 243 | 172 |
| Days from LVAD surgery to extubation | 30 | 14 | 3 | 7 | 14 | 38 |
| Days from LVAD surgery to discharge | 30 | 73 | 16 | 50 | 55 | 71 |
| Pump thrombosis/ malfunction | | | | | | |
| Renal Recovery | | × | × | ✓ | ✓ | ✓ |
| CRRT | ✓ | ---- | ---- | ✓ | ✓ | ✓ |
| RVAD support | × | × | × | ✓ | ✓ | × |
| Stroke | ✓ | ✓ | × | × | ✓ | × |
| Tracheostomy | ✓ | × | × | × | × | ✓ |
| Alive at the end of follow-up | Not Survived | Survived | Survived | Survived | Survived | Survived |

higher frequency of pulmonary embolism in COVID-19 may be due to endothelial dysfunction and enhanced inflammatory process in the lung. It is worth mentioning that COVID-19 patients have hardly experienced bleeding, and the same has also not been reported in literature. COVID-19 patients with LVAD implants undergo antithrombotic therapy (Zakrzewski et al., 2021). Patients with LVADs who have an acute COVID-19 infection are strictly monitored for coagulation profiles (Moady et al., 2022). Those who have COVID-19 typically develop circulating antibodies that affect PTT; hence, oral anticoagulation is also administered in these individuals. Patients with high D-dimer levels and low fibrinogen levels are often treated further.

## 6. The Outcomes of COVID-19 Patients with LVAD

Many variables affect whether patients with LVADs with COVID-19 infection survive or pass away. A few factors are the activation of cytokines, abnormal immune system, and the patient demographics (elderly with numerous comorbidities), which decide the patient outcome (Fig. 4). Acute SARS-CoV-2 infections have been reported in LVAD implant patients at the same time as COVID-19 pandemic. A cohort study involving 28 patients with LVADs with confirmed COVID-19 was conducted (Iyengar et al., 2022). Out of these, 24 patients received inpatient care, whereas four patients received outpatient care. HeartWareTM, HVADTM, HeartMate3 (HM3), & HeartMate2 (HM2) (Ravichandran et al., 2021) were used in these patients. Five patients required mechanical breathing, and the death rate was around 32%. In a second registry of 80 patients, eight people succumbed during protracted hospital stays, and the other 60% needed hospitalization (Mallhi et al., 2017). This study has reported that hematological and hemodynamical alterations have an immediate effect on the patient's well-being. These changes in physiological parameters certainly affect the functioning of assist cardiac devices, which further decide the fatality rate. Therefore, appropriate improvement in cardiac assist devices is required to compensate for the changes in hemodynamical parameters as per the requirement during COVID-19 disease (Zakrzewski et al., 2021).

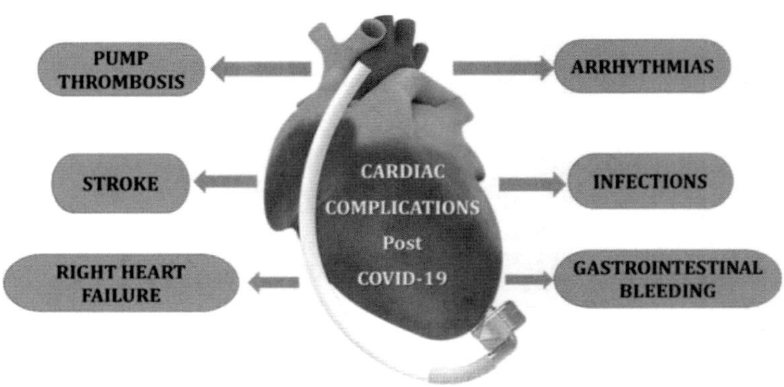


**Figure 4.** Cardiac complications observed during/post-COVID 19.

## 7. Pump Thrombosis Management after COVID-19

The cardiac device output is impacted by a number of device-related parameters, including outflow graft subluxation, extrinsic compression, low pump speed, and local heat created by the pump. In order to prevent pump thrombosis, patient-related variables such as high blood pressure, infections, congestive heart failure, and hypercoagulable states are required to be treated. Clinical management-related problems such as subtherapeutic and inadequate anticoagulation are also required to be regulated to prevent unwanted complexities.

## 8. Thrombus Surface Effect on Hydrodynamics in LVAD

Due to increased flow resistance, thrombotic surfaces in LVAD parts may restrict the flow route (Malone et al., 2023). This significantly reduces the device's pumping capacity and efficiency. Clinically, it may result in reduced pump output which can limit the left ventricle's ability to empty itself. It may result in experiencing the symptoms of dyspnea, weakness, or dizziness (Quick and Walker, 2004) of heart failure in LVAD patients. Eventually, the patient may experience symptoms like dyspnea, weakness, or dizziness before going into cardiogenic shock.

## 9. LVAD Input and Outflow Graft Cannula Placement

Anchored techniques have been proven helpful in reducing the growth of thrombi (Jeschke et al., 2004). For pump fixation, research employed epicardial, transmural, and transmural with or without back stitch techniques. The epicardial stitching of the sewing ring was the most critical result of pump stability, with a decreased risk of bleeding and thrombus formation. Transmural surgery without a back stitch has formed thrombi that can flow through the cannula into the blood (Hanke et al., 2027). A less invasive procedure was used, including an anterior lateral thoracotomy and an upper hemi-sternotomy (Sidhu et al., 2020). LVAD inflow cannula configurations are illustrated from the left lateral view (Fig. 5).

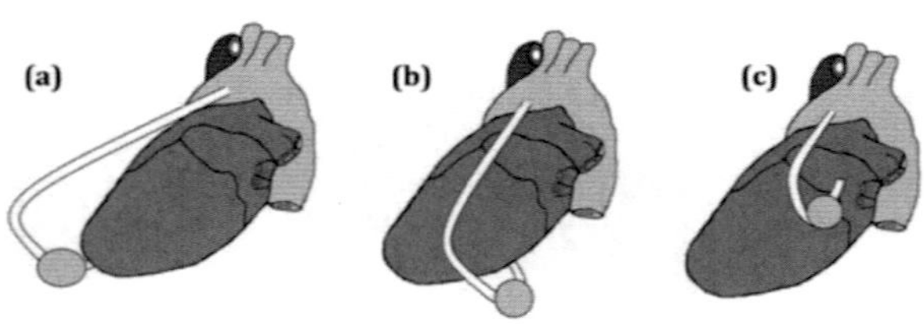


| Fig. | Inflow Cannula | Insertion site for LVADs |
|---|---|---|
| (a) | LV Inferior Wall | Severe dilatation, LV apical aneurysm, narrow chested patients |
| (b) | Left Atrium | Hypertrophic cardiomyopathy or unviable wall |
| (C) | LV Apex | Standard practice and preferred site |

**Figure 5.** LVAD inflow cannula arrangements to reduce thrombi (adapted from Malone et al., 2023).

## 10. Coating to Reduce Blood Coagulopathy for Mechanical Circulatory Cardiac Support Devices

There are studies that have used surface texturing to promote blood compatibility to decrease thrombus development and inflammatory reactions in medical/cardiac assist devices (Fig. 6). The two basic surface modification methods that can be used are surface passivation of material surfaces and bioactive surface treatments and coatings (Mandracci et al., 2016). By creating a barrier at the device-blood interface by changing the surface chemistry (such as hydrophilicity) or material physical structure, the inherent thrombogenicity of the material surface can be reduced (e.g., topography). To pharmacologically inhibit the coagulation reaction directly, a bioactive surface strategy involving long-term immobilization with an active agent or local drug delivery can also be done.

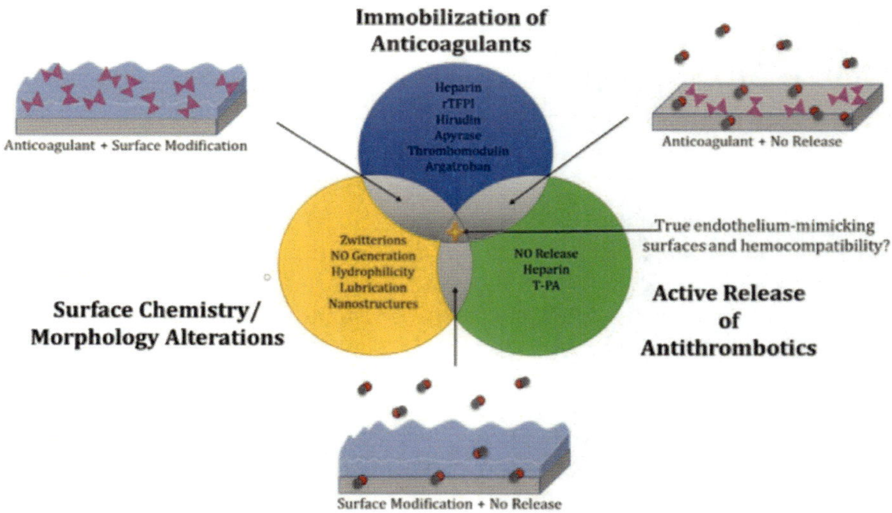


**Figure 6.** Combination strategies for the development of antithrombotic surface.

## 11. Futuristic Approach to Combat COVID-19-related Thrombosis

A centrifugal intrapericardial-flow route pump such as HeartMate 3 (HM3) LVAD (Abbott, Abbott Park, IL, USA) (Mehra et al., 2021) with the fully magnetically levitated frictionless rotor and a fixed intrinsic pulse generator has emerged as an advanced LVAD system. An inflow cannula, pump housing, control electronics, outflow graft and bend relief, and percutaneous driveline are among the LVAD components. This design has definitely replaced the HMII LVAD of the past (Kirklin et al., 2014). It is exposed to reduced blood shear stress with intervals in the blood flow channel, which are approximately ten to twenty times larger than those of other devices. The rotor also rotates and levitates with no friction, which reduces pump thrombosis. The standstill in the pump is also dispersed by a synthetic pulse produced by higher rotor speed variations, which also provide larger blood flow pathways.

Compared to the HMII LVAD, this device's *in vivo* testing exhibited negligible hemolysis (Berk et al., 2019) and blood cell shearing. Early studies with multicenter validation and clinical confirmation of these advantages have been conducted. Von Willebrand antigen circulating high molecular weight multimers were seen to shear less frequently and exhibit less hemolysis. In HM3 and HMII, there is an inverse relationship between the pressure head across the pump and the flow through the pump. Compared to HMII, the pressure differential between centrifugal route pumps in HM3 can operate over a significantly more comprehensive range of flows.

The possibility of Thrombus formation is present throughout the entire LVAD system (Reider et al., 2017). The left ventricle's inflow cannula, the pump's inner housing, and the outflow graft that connects it to the aorta can all develop thrombi. The differences in these components' quantitative features that a thrombus might produce can be exploited to determine exactly where the thrombus formed. For instance, modern LVADs were made only to use the energy required to spin the rotor at a specific pace. The device inquiry arrangements can be developed to detect thrombus on the rotor, which demands more energy to maintain the momentum. LVADs with magnetic levitation will offer these circumstances and reduce the likelihood of thrombosis caused by COVID-19 (Mancini and Colombo, 2015).

## 12. Conclusions

This chapter presents the association between COVID-19 and hypercoagulation in blood. In addition, the increased level of cytokines and other immunological factors significantly affect the health of myocardial tissue. Extraordinarily heightened inflammatory response in COVID-19 results in thrombo-inflammation. Thrombosis is closely associated with cardiac assistive devices' complications, leading to malfunctioning. Future innovations and approaches shall focus on developing cardiac assist devices that have reduced thrombosis during functioning. Surface modification of cardiac devices can be one of the strategies to reduce thrombosis. Cardiac pump parameters and design shall be optimized to reduce pump thrombosis. This will help minimize the failure of the cardiac device, thus preventing patient mortality associated with such complications.

## References

Ahmad, S., Manzoor, S., Siddiqui, S., Mariappan, N., Zafar, I., Ahmad, A. et al. (2022, August). Epigenetic underpinnings of inflammation: Connecting the dots between pulmonary diseases, lung cancer and COVID-19. pp. 384–398. *In*: Seminars in Cancer Biology (Vol. 83). Academic Press.

Berk, Z.B., Zhang, J., Chen, Z., Tran, D., Griffith, B.P. and Wu, Z.J. (2019). Evaluation of *in vitro* hemolysis and platelet activation of a newly developed maglev LVAD and two clinically used LVADs with human blood. Artificial Organs 43(9): 870–879.

Brown, M.G. and Angela, G.C. (2022). Leaders launch a new global health partnership. Cardiology 7: 17.

Cid-Menendez, A. and Dar, O. (2020). Mechanical circulatory support use in advanced heart failure. J. Cardio Cardiovasu Med. 4: 013.

Connors, J.M. and Levy, J.H. (2020). COVID-19 and its implications for thrombosis and anticoagulation. Blood 135(23): 2033–2040.

Dandel, M. and Hetzer, R. (2018). Temporary assist device support for the right ventricle: pre-implant and post-implant challenges. Heart Failure Reviews 23: 157–171.

Fang, J.C. and O'Gara, P.T. (2015). 13 History and Physical Examination: An Evidence-Based Approach.

Frick, William H., Ryan D. Mallory, Maya Guglin, Eve Anderson, Erin L. Lushin, Rey P. Vivo et al. (2020). Unusual case of pump thrombosis in LVAD patient with COVID-19—Diagnostic Challenges.

Haddad, F., Doyle, R., Murphy, D.J. and Hunt, S.A. (2008). Right ventricular function in cardiovascular disease, part II: pathophysiology, clinical importance, and management of right ventricular failure. Circulation 117(13): 1717–1731.

Han, J. and Trumble, D.R. (2019). Cardiac assist devices: early concepts, current technologies, and future innovations. Bioengineering (Basel). 2019 Feb 15; 6(1): 18. doi: 10.3390/bioengineering6010018. PMID: 30781387; PMCID: PMC6466092.

Hanke, J.S., Krabatsch, T., Rojas, S.V., Deniz, E., Ismail, I., Martens, A. et al. (2017). *In vitro* evaluation of inflow cannula fixation techniques in left ventricular assist device surgery. Artificial Organs 41(3): 272–275.

Henry, T.D. and Kereiakes, D.J. (2022). The direct and indirect effects of the COVID-19 pandemic on cardiovascular disease throughout the world. European Heart Journal.

Hosseinipour, M., Gupta, R., Bonnell, M. and Elahinia, M. (2017). Rotary mechanical circulatory support systems. Journal of Rehabilitation and Assistive Technologies Engineering 4: 2055668317725994. https://doi.org/10.1177/2055668317725994.

Iba, T., Levy, J.H., Levi, M. and Thachil, J. (2020). Coagulopathy in COVID-19. Journal of Thrombosis and Haemostasis 18(9): 2103–2109.

Imazio, M., Klingel, K., Kindermann, I., Brucato, A., De Rosa, F.G., Adler, Y. et al. (2020). COVID-19 pandemic and troponin: indirect myocardial injury, myocardial inflammation or myocarditis? Heart 106(15): 1127–1131.

Iyengar, A., Cohen, W., Han, J., Helmers, M., Kelly, J.J., Patrick, W. et al. (2022). Effects of body mass index on presentation and outcomes of COVID-19 among heart transplant and left ventricular assist device patients: a multi-institutional study. Asaio Journal 69(1): 43–49.

Jeschke, M.G., Rose, C., Angele, P., Füchtmeier, B., Nerlich, M.N. and Bolder, U. (2004). Development of new reconstructive techniques: use of Integra in combination with fibrin glue and negative-pressure therapy for reconstruction of acute and chronic wounds. Plastic and Reconstructive Surgery 113(2): 525–530.

Kaptoge, S., Pennells, L., De Bacquer, D., Cooney, M.T., Kavousi, M., Stevens, G. et al. (2019). World Health Organization cardiovascular disease risk charts: revised models to estimate risk in 21 global regions. The Lancet Global Health 7(10): e1332–e1345.

Kaur, S., Bansal, R., Kollimuttathuillam, S., Gowda, A.M., Singh, B., Mehta, D. et al. (2021). The looming storm: blood and cytokines in COVID-19. Blood Reviews 46: 100743.

Kichloo, A., Dettloff, K., Aljadah, M., Albosta, M., Jamal, S., Singh, J. et al. (2020). COVID-19 and hypercoagulability: a review. Clinical and Applied Thrombosis/Hemostasis 26: 1076029620962853.

Kirklin, J.K., Naftel, D.C., Kormos, R.L., Pagani, F.D., Myers, S.L., Stevenson, L.W. et al. (2014). Interagency Registry for Mechanically Assisted Circulatory Support (INTERMACS) analysis of pump thrombosis in the HeartMate II left ventricular assist device. The Journal of Heart and Lung Transplantation 33(1): 12–22.

Mallhi, T.H., Khan, A.H., Sarriff, A., Adnan, A.S. and Khan, Y.H. (2017). Determinants of mortality and prolonged hospital stay among dengue patients attending tertiary care hospital: a cross-sectional retrospective analysis. BMJ Open 7(7): e016805.

Malone, G., Abdelsayed, G., Bligh, F., Al Qattan, F., Syed, S., Varatharajullu, P. et al. (2023). Advancements in left ventricular assist devices to prevent pump thrombosis and blood coagulopathy. Journal of Anatomy 242(1): 29–49.

Mancini, D. and Colombo, P.C. (2015). Left ventricular assist devices: a rapidly evolving alternative to transplant. Journal of the American College of Cardiology 65(23): 2542–2555.

Mandracci, P., Mussano, F., Rivolo, P. and Carossa, S. (2016). Surface treatments and functional coatings for biocompatibility improvement and bacterial adhesion reduction in dental implantology. Coatings 6(1): 7.

Mehra, M.R., Cleveland Jr, J.C., Uriel, N., Cowger, J.A., Hall, S., Horstmanshof, D. et al. (2021). Primary results of long-term outcomes in the MOMENTUM 3 pivotal trial and continued access protocol study phase: a study of 2200 HeartMate 3 left ventricular assist device implants. European Journal of Heart Failure 23(8): 1392–1400.

Moady, G., Ben Gal, T. and Atar, S. (2022). Management of patients with left ventricular assist device during the COVID-19 pandemic. Medicina 58(1): 116.

Morlacchi, P. and Nelson, R.R. (2011). How medical practice evolves: Learning to treat failing hearts with an implantable device. Research Policy 40(4): 511–525.

Nosé, Y. (1998). Design and development strategy for the rotary blood pump. Artificial Organs 22(6): 438–446.

Piperata, A., Bottio, T. and Gerosa, G. (2020). COVID-19 infection in left ventricular assist device patients. Journal of Cardiac Surgery 35(11): 3231–3234. https://doi.org/10.1111/jocs.14969.

Prinzing, A., Herold, U., Berkefeld, A., Krane, M., Lange, R. and Voss, B. (2016). Left ventricular assist devices-current state and perspectives. J. Thorac Dis. 2016 Aug; 8(8): E660–6. doi: 10.21037/jtd.2016.07.13. PMID: 27621895; PMCID: PMC4999658.

Quick, G. and Walker, J.S. (2004). Woman with exertional dyspnea, weakness, and dizziness. Consultant 44(4): 633–637.

Ranard, L.S., Fried, J.A., Abdalla, M., Anstey, D.E., Givens, R.C., Kumaraiah, D. et al. (2020). Approach to acute cardiovascular complications in COVID-19 infection. Circulation: Heart Failure 13(7): e007220.

Ravichandran, A.K., Cowger, J.A., Feller, E.D., Mahr, C., Hiesinger, W., Klein, L. et al. (2021). An analysis of driveline infections with left ventricular assist devices utilizing carbothane versus pellethane driveline sheaths. The Journal of Heart and Lung Transplantation 40(4): S434–S435.

Reider, C., Moon, J., Ramesh, V., Montes, R., Campos, J., Herold, B. et al. (2017). Intraventricular thrombus formation in the LVAD-assisted heart studied in a mock circulatory loop. Meccanica 52: 515–528.

Reyes, M.V. (2020). The disproportional impact of COVID-19 on African Americans. Health and Human Rights 22(2): 299.

Saeed, S. and Rajani, R. (2021). The cardiovascular complications in COVID-19: Focus on acute cardiac injury. Pakistan Journal of Medical Sciences 37(3): 908.

Samuels, L., Tuluca, A., Samuels, M., Robbins, T., Bonita, R. and Bozorgnia, B. (2021). COVID-19 associated left ventricular assist device pump thrombosis with hemolysis and pigment nephropathy: case report. Ann Cardiovasc Surg. 4(1): 1027.

Sidhu, K., Lam, P.H. and Mehra, M.R. (2020). Evolving trends in mechanical circulatory support: Clinical development of a fully magnetically levitated durable ventricular assist device. Trends Cardiovasc Med. 2020 May; 30(4): 223–229. doi: 10.1016/j.tcm.2019.05.013. Epub 2019 Jun 4. PMID: 31201005.

Slaughter, M.S., Pagani, F.D., Rogers, J.G., Miller, L.W., Sun, B., Russell, S.D. et al. (2010). Clinical management of continuous-flow left ventricular assist devices in advanced heart failure. The Journal of Heart and Lung Transplantation 29(4): S1–S39.

Wong, Y.P., Khong, T.Y. and Tan, G.C. (2021). The effects of COVID-19 on placenta and pregnancy: what do we know so far?. Diagnostics 11(1): 94.

Zakrzewski, J., Coyle, L., Aicher, T., Chickerillo, K., Gallagher, C., Kuper, K. et al. (2021). Impact of COVID-19 on patients supported with a left ventricular assist device. Asaio Journal 67(11): 1189.

# Chapter 12

# New Trends in Bio-Fluid Flow Modelling

*Akshoy Ranjan Paul*[1,*] and *Suvash C. Saha*[2]

## 1. Introduction

The study of fluid dynamics has long been a cornerstone in understanding various physical phenomena, from the motion of celestial bodies to the aerodynamics of airplanes. However, with the rapid advancements in computational power and the ever-growing interdisciplinary nature of scientific research, the focus has now shifted towards a much more intricate and fascinating realm: bio-fluid flow modeling.

Bio-fluid flow modeling deals with the intricate and dynamic movement of fluids within living organisms, encompassing a wide range of applications in biology, medicine, and bioengineering. This emerging field has witnessed unprecedented growth in recent years, fueling an extensive exploration of new trends that revolutionize our understanding of biological systems and their behavior at the fluidic level.

Bio-fluid flow modeling is essential for understanding the mechanics and behavior of biological systems. Recent advancements in computational methods, imaging techniques, and experimental approaches have led to exciting new trends in this field. Integrating Computational Fluid Dynamics (CFD) with high-performance computing enables more accurate simulations and visualization of fluid flow, facilitating the study of cardiovascular diseases and respiratory disorders. Another emerging trend is the incorporation of multi-scale and multi-physics approaches,

[1] Department of Applied Mechanics, Motilal Nehru National Institute of Technology Allahabad, Prayagraj-211004, Uttar Pradesh, India.

[2] School of Mechanical and Mechatronic Engineering, University of Technology Sydney (UTS), 15 Broadway, Ultimo, NSW-2007, Australia.

* Corresponding author: arpaul@mnnit.ac.in

DOI: 10.1201/9781003287636-12

considering the complex interactions between different physical phenomena in biological systems. This requires developing computational frameworks and numerical techniques to capture the interplay between different scales and physics, enhancing our understanding of bio-fluid dynamics. Advanced imaging technologies, such as MRI, CT, and optical imaging, have further expanded possibilities for experimental bio-fluid flow modeling. These techniques provide detailed data on flow patterns and fluid-structure interactions, validating computational models and enabling the development of personalized medicine approaches. New technologies have provided large data sets for biological flow modeling, enabling advanced techniques like Artificial Intelligence (AI), Machine Learning (ML), Deep Learning (DL) and predictive analytics. High-Performance Computing (HPC) enhances accuracy and efficiency by utilizing parallel computing and cloud computing. By leveraging these trends, researchers aim to optimize treatments and interventions and improve overall well-being and quality of life. Some of these latest, evolving and upcoming modeling trends in the biological flows are, therefore, discussed in the following sections.

## 2. Physiologically Based Pharmacokinetic (PBPK) Modeling

Physiologically based pharmacokinetic (PBPK) modeling is a mathematical technique to predict the concentration of drugs in the body over time and also to assess the pharmacokinetic behavior of drugs in humans and other species. The concept of PBPK was first described by Teorell in 1937. It is based on physiological features of the body such as body size, organ and tissue mass, organ perfusion and metabolic rates. PBPK models can be used to describe drug Absorption, Distribution, Metabolism, and Excretion (ADME). They can also be used to explore the factors that influence drug pharmacokinetics, such as age, sex, and disease state. PBPK models can be used to simulate drug exposure to various scenarios, such as dose, route of administration and changing physiological parameters, in order to predict drug exposure in different populations and to optimize the efficacy and safety of drug therapy. Insufficient drug concentration levels can lead to ineffective treatment, while excessively high concentrations of drugs can cause side effects or toxicity. PBPK, a potent tool, relies on advanced computational power to handle the intricate and demanding mathematical equations required for accurate quantitative estimations and predictions of pharmacokinetics (PK). By considering physiological and anatomical characteristics, as well as the physical and chemical properties of a drug, PBPK offers a mechanistic approach to studying and forecasting drug PK.

PBPK models rely on assumptions regarding the rates of various processes, and there are instances where these rates may be uncertain. To address this uncertainty, sensitivity analyses can be conducted to assess the impact of such unknowns. Additionally, it's worth noting that PBPK models typically represent an average individual with a specific disease, lacking the ability to account for inter-individual and unexplained variabilities (unlike population PK modeling). However, this limitation can be mitigated by performing sensitivity analyses that incorporate a range of values for crucial characteristics, allowing for a more comprehensive understanding of the model's outcomes.

The fundamental concept underlying PBPK is that changes or variations in an individual's physiology, which can be mathematically described, can influence the pharmacokinetic (PK) properties (ADME) of a compound and consequently affect the desired therapeutic outcome. Through PBPK modeling and simulation, we can explore the effects of these changes in a virtual environment before administering drugs to patient populations. PBPK is a robust tool that can be continually modified and updated as new knowledge about the compound and physiology is acquired. By following 'predict-learn-confirm' cycles, these models not only help explain clinical observations but more importantly, through simulations, can predict untested clinical outcomes. PBPK is strategically employed by drug developers to anticipate and evaluate the impact of previously unrecognized physiological factors that directly influence drug PK. These factors encompass intrinsic elements (such as pregnancy, genetics, disease, organ dysfunction, and race) and extrinsic elements (like diet, smoking, alcohol use, and other medications). The power of PBPK lies in its ability to examine the influence of these critical parameters in drug development. Ultimately, PBPK enables researchers and physicians to gain a better understanding of how dose selection should be adjusted in various diseases, life stages, and physiological events that can alter the PK of compounds.

Yoo and Ito (2017) employed a hybrid PBPK-CFD model that incorporated a Computationally Simulated Person (CSP). The virtual CSP consisted of two interconnected models—a virtual manikin and a virtual airway. These models were directly linked by connecting the nostril and oral surface openings. The CSP accurately replicated the geometry of the human body and the respiratory tract, encompassing the nasal and oral cavities as well as the bronchial tubes. This CSP incorporated thermoregulation and respiratory tract models to assess indoor environmental quality. The researchers utilized the CSP to develop a comprehensive numerical simulation procedure for analyzing airflow, temperature, and contaminant transport through Computational Fluid Dynamics (CFD). By utilizing the coupled PBPK-CFD-CSP analysis, they investigated inhalation exposure to formaldehyde emitted from indoor floor materials under steady-state conditions. The study quantitatively examined respiratory tissue doses and the distribution of inhaled contaminants.

Das et al. (2020) assessed the seasonal risk associated with carcinogenic metals (specifically arsenic, cadmium, and lead) present in particulate matter (PM2.5) on various organs and tissues, particularly the lungs. They employed an integrated model comprising the Human Respiratory Tract (HRT) and physiologically-based pharmacokinetic (PBPK) models to calculate the deposition of these metals in different organs. The calculations revealed that lead primarily accumulates in tissues such as bone, muscle, and blood, indicating significant deposition. Cadmium deposition predominantly occurs in the lungs, while arsenic deposition is observed in the lungs, muscles, and skin. Notably, the majority of the studied metals were found to be deposited in the lungs, with 45 of arsenic and 70% of cadmium being present in their bio-dissolved contents.

Recently, Ladumor and Unadkat (2022) created an Oral Inhalation (OI)-PBPK model with the aim of predicting the regional deposition in the respiratory system, as well as the concentrations of Orally Inhaled Drugs (OIDs) in both systemic and specific tissue regions. This is particularly valuable for assessing drug concentrations at the site of action, which is challenging to measure directly in humans. The developed OI-PBPK model has the potential to optimize OID formulations, delivery systems, and dosing schedules and facilitate the evaluation of bioequivalence in generic OID products. Moreover, this model can be interconnected with other advanced technologies such as organs-on-chips, pharmacodynamic models, and quantitative systems pharmacology models. This integration allows for the prediction and assessment of the safety and efficacy of OID, enabling a comprehensive understanding of the drug's effects.

Microfluidics is a technology used to precisely manipulate small volumes of fluids within miniature structures, often on a sub-micron scale, also referred to as lab-on-a-chip. The merging of microfluidics and cell culture has resulted in Organ-on-a-Chip (OoC) technology, which has the potential to replicate intricate human physiology in a laboratory setting. By combining physiologically based pharmacokinetic (PBPK) modeling with OoC, researchers believe they can develop a powerful tool to comprehend the absorption, distribution, metabolism, and excretion (ADME) processes of drugs crucial for new drug development, potentially replacing the need for animal testing (Yang et al., 2022). The integration of PBPK modeling on OoC is expected to become a crucial component in areas such as new drug development, personalized medicine, and other relevant fields.

## 3. Computational Fluid-Particle Dynamics (CPFD)

Computational Fluid Dynamics (CFD) is a powerful tool for simulating the flow of fluids in biomedical systems. CFD provides detailed information about the flow of fluids in complex geometries, allowing for the accurate prediction of a variety of flow parameters, such as pressure, velocity, temperature, and viscosity. The application of CFD to biomedical flow modeling has been gaining traction in recent years as it offers a great deal of control over the modeling process and allows for the simulation of complex scenarios. CFD can be used to model the flow of blood and other fluids in the body, such as in the cardiovascular, respiratory, digestive, renal, and urinary systems as described in the different chapters of this book. It can also be used to study the effects of external forces on the body, such as gravity, wind, and waves. In addition, CFD can be used to model the flow of fluids in medical devices, such as stents, catheters, and artificial hearts. CFD is also being used to investigate the effects of drug delivery systems on the body and to predict the effects of new drug treatments.

The power of CFD has been used when coupled with multiphase flow models. Computational Fluid-Particle Dynamics (CFPD) combines the principles of CFD with modeling techniques for studying the behavior of particles in fluid flows. These modeling techniques, namely Euler-Euler and Euler-Lagrangian methods, enable the analysis of multiphase flows and the dynamics of particles within them, as schematically shown in Fig. 1 (Tu et al., 2012). CFPD models provide a detailed description of fluid and particle behaviors by solving conservation laws with constitutive equations, offering high-resolution local data. These validated models, which are non-invasive, cost-effective, and accurate, serve as a complement

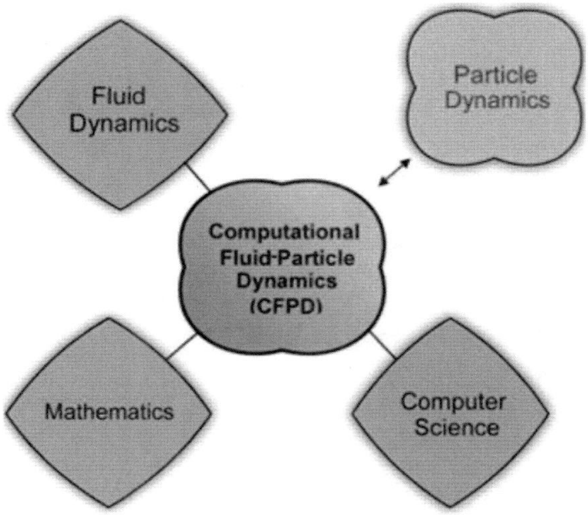


**Figure 1.** Structure of Computational fluid particle dynamics (CFPD).

to *in-vitro* and *in-vivo* studies on unconventional lung aerosol dynamics. They enable simulations of challenging fluid-particle flows that are difficult to replicate experimentally, providing new insights for medical device enhancements, targeted drug delivery, lung therapeutics, disease diagnostics, and health risk assessments (Feng et al., 2016). CFPD models have potential contributions in several areas that include (i) understanding the mechanisms behind the transport of inhaled aerosols, (ii) virtualizing corrective surgeries to improve abnormal airway function, (iii) assisting toxicologists in evaluating the health effects of inhaled particulate matter, (iv) designing innovative carriers for delivering drugs to the lungs, and (v) establishing guidelines for limiting occupational health hazards (like workplace exposure to gasoline fumes, new nanoparticles and micron fibers, etc.).

## 4. Multiple Path Particle Dosimetry (MPPD)

Dosimetry involves assessing the internal dose of a substance in individuals or populations and establishing a relationship between external exposure and biological effects. Computer models are valuable tools in dosimetry, reducing the reliance on extensive experimental studies. Various lung dosimetry models have been developed to estimate the deposition of particles with different sizes, shapes, and densities in specific regions of the respiratory tract. Additionally, models have been created to determine the *in-vivo* deposition and retention of inhaled particles, including aerosols and nanoparticles. Different options are available for estimating respiratory tissue dosimetry after inhalation exposure, each with its own strengths and weaknesses. The Environmental Protection Agency (EPA) utilizes two models, namely, the Regional Deposited Dose (RDD) model and the Multiple Path Particle Dosimetry (MPPD) model, to estimate aerosol deposition in the respiratory tracts of laboratory animals and humans.

The MPPD model is a product of Applied Research Associates, Inc.'s development. The MPPD model comprises ten distinct stochastic multiple-path models, enabling the estimation of particle deposition and clearance of inhaled particles of uniform or varying sizes ranging from ultrafine (1 nm) to coarse (100 μm) in various respiratory zones of humans (ARA, 2023). This computational tool (the MPPD model) provides insights into how particles of different sizes, ranging from ultrafine to coarse, deposit and clear in the respiratory system. It employs both single-path and multiple-path methods to track airflow and calculate aerosol deposition in the lungs. The single-path method calculates deposition along a representative path in each airway generation, while the multiple-path method considers deposition in all lung airways, providing detailed information about specific lobes and airways. The deposition calculations within each airway consider diffusion, sedimentation, and impaction efficiencies of particles. The model also incorporates empirical equations to account for aerosol filtration by the nose and mouth. Furthermore, the MPPD model considers the clearance of deposited particles from the lung, accounting for factors such as mucociliary transportation, phagocytosis by pulmonary alveolar macrophages, dissolution, particle concentrations, breathing patterns, airway regions, and generation number.

The MPPD model distinguishes between constant and variable exposure scenarios when considering exposure to Particulate Matter (PM). In the constant scenario, specific parameters such as breathing frequency and tidal volume remain fixed for a given PM concentration. Conversely, in the variable scenario, these parameters can vary based on hourly or activity patterns. For this particular study, seasonal average values of PM1, PM2.5, and PM10 are utilized to represent constant exposure. The model takes into account provided values for upper respiratory tract volume, functional residual capacity, breathing frequency, and tidal volume. Among the available options, the upright body orientation is chosen in this study. The model assumes that all particles enter the respiratory tract through the nose, although other breathing conditions, such as oral, oronasal, and endotracheal, are also available. By inputting the necessary data, the model is employed to assess deposition and clearance doses (Manojkumar et al., 2019).

The MPPD model has been integrated with the PBPK model to assess the bio-distribution of inhaled particles. However, the MPPD, ICRP (International Commission on Radiological Protection), and NCRP (National Council on Radiation Protection and Measurement) models are limited to calculating deposition and clearance under constant exposure conditions. To address this limitation, the Exposure Dose Model (ExDoM) was developed to estimate deposition and clearance under variable continuous exposure conditions. Additionally, the integrated Exposure and Dose Modeling and Analysis System (EDMAS) was created to overcome the inherent shortcomings of the MPPD, ICRP, and NCRP models. Unlike ExDoM, EDMAS can account for time-dependent changes in particle size and composition caused by nucleation, condensation, coagulation, and gas-phase chemical reactions. The EDMAS model has been evaluated and demonstrated good agreement with experimental data and results from other models. However, one drawback of this mechanistic model is that it requires numerous physiological parameters.

## 5. Multiscale Modeling

Multiscale modeling is an approach that combines different levels of modeling in order to simulate complex biological systems. It can be used to study the interaction between different biological processes at different length scales, such as the interaction between biochemical and physiological processes. Multiscale modeling is a powerful tool for biomedical flow modeling. It is used to investigate the interactions between different biological components at different scales. Multiscale modeling is used to simulate the interactions between multiple biological components, such as cells, tissues, and organs, in order to better understand the complex dynamics of biological systems. Multiscale modeling can be used to study a wide range of biological processes, including cell motility, biochemical pathways, and tissue remodeling. Multiscale modeling can also be used to develop predictive models for drug delivery, tissue engineering, and tissue regeneration. By combining data from different scales, multiscale modeling provides an improved understanding of the interactions between different components in the biological system, enabling researchers to develop more effective treatments and therapies.

The field of multi-scale modeling of the respiratory system has seen the emergence of hybrid approaches. These approaches include combining a one-dimensional volume-filling airway skeleton with a four-dimensional moving CFD model that extends to the initial generations of bronchi (Yin et al., 2013). Another approach involves incorporating a Stochastic Individual Path (SIP) approximation for all bronchiolar airways and acinar moving wall models into the main lobar bronchi of an upper and large airway CFD model (Longest et al., 2016). There is also the option of using a Whole-Lung Airway Model (WLAM) composed of adjustable triple bifurcation units attached to the main lobar bronchi of an upper airway CFD model (Kolanjiyil and Kleinstreuer, 2017). These approaches offer varying levels of computational efficiency by simplifying airway anatomy and boundary conditions for the deep lung. However, as the demand for patient-specific models or models that consider the effects of pulmonary disease on aerosol transport and deposition increases, there will be a need for additional multi-scale approaches that bridge the gap between site-specific and whole-airway modeling.

In order to achieve this objective, Kuprat et al. (2016) combined essential components of CFPD models for large conducting airways with the widely utilized MPPD model. This integration allowed for the incorporation of patient-specific anatomy and physiology, serving as the foundation for simulating the transport and deposition of aerosols throughout the entire respiratory system during the complete breathing cycle. Initial simulations demonstrated the effectiveness of this approach in studying nanoparticle transport across different scales. Building upon this work, Kuprat et al. (2021) further expanded the approach to encompass a range of aerosol sizes and developed efficient bi-directional coupling procedures between three-dimensional CFPD models and personalized one-dimensional Navier-Stokes airflow and particle transport using MPPD modeling for the distal region of the lungs. Similar to previous multi-scale models, the computational costs of this approach primarily arise from the three-dimensional domain, which is necessary for site-specific aerosol deposition analysis.

## 6. Systems Biology, Synthetic Biology and Systems Bioengineering

Systems biology is an interdisciplinary field that seeks to understand the complex behaviors of biological systems through the integration of multiple data sources, such as genomics and proteomics. It is a powerful tool to study the interactions between different parts of a system and to understand how changes in one part of the system affect the other parts. According to Ideker et al. (2001), "Systems biology does not investigate individual genes or proteins one at a time, as has been the highly successful mode of biology for the past 30 years. Rather, it investigates the behavior and relationships of all of the elements in a particular biological system while it is functioning." System biology offers a holistic approach to biomedical research by combining physical, computational, and experimental sciences to unravel the intricacies of complex biological systems. In biomedical flow modeling, systems biology is used to build detailed models of the different components of a biological system, such as the metabolic and signaling pathways and the interactions between them. These models can then be used to investigate the effects of changes in one part of the system on the others, as well as to predict the behavior of the system as a whole. By combining systems biology with traditional flow modeling techniques, researchers can gain a deeper understanding of the processes that take place in a biological system, which can lead to the development of new treatments and therapies.

Biology diverges from fields like engineering, lacking definitive governing equations such as Maxwell's or Navier-Stokes'. In biological systems, assumptions and mathematical frameworks should be subject to discussion and debate, considering the interdisciplinary nature of biology that encompasses chemical, electrical, and mechanical aspects. While certain theories may apply effectively in specific contexts, they may fall short in others when applied to biomolecules, cells, and tissues. This necessitates a 'trust-but-verify' approach, where models and experiments in systems biology serve as important crosschecks. Non-engineering approaches in systems biology embody the scientific method, emphasizing hypothesis generation, testing, and revision (Fig. 2A). Although the general paradigm of engineering-based problem-solving aligns with systems biology, specific implementation details differ. Systems biology is driven by inquiries rather than product development. However, engineering design principles minimize aimless exploration and prioritize tangible outcomes (Fig. 2B).

The intersection of biology and engineering can be approached from different perspectives. One such perspective is the bottom-up approach of synthetic biology, where biological systems are created using specific components and principles, similar to building circuits (Fig. 3A). On the other hand, systems biology encompasses both top-down and middle-out approaches, aiming to understand and model complex biological systems by analyzing their higher-level behaviors and incorporating detailed knowledge of their components and interactions (Fig. 3B). The essence of systems biology lies in reverse engineering, focusing on feedback, adaptation, and dynamics. Mathematical models play a crucial role as deductive tools, testing existing understanding and alternative hypotheses through falsifiable predictions. Failures in model-experiment comparisons are not dead-ends but rather indications that new regulatory mechanisms are yet to be defined. Synthetic biology, on the other hand,

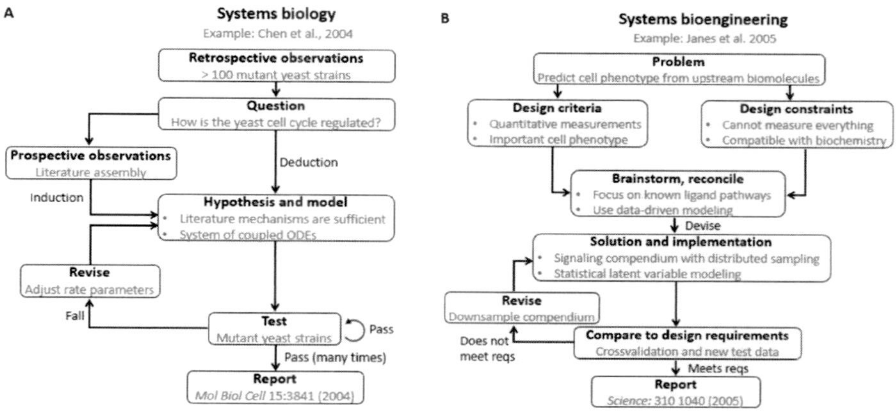


**Figure 2.** Flow chart for systems biology and systems bioengineering [Janesa et al., 2017].

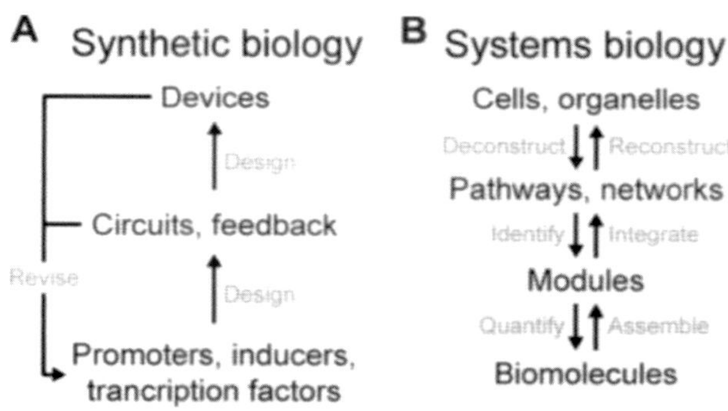


**Figure 3.** Different intersections of biology and engineering. (A) The bottom-up approach of synthetic biology. (B) Top-down and middle-out approaches of systems biology.

demands analytical flexibility, given the dynamic nature of biological regulatory networks. Mutation and adaptation to changing environments introduce infinite numbers of unknowns, requiring systems biology to grapple with uncertainties. As a result, the synergy between systems biology and synthetic biology is expected to establish a new set of engineering-inspired principles for the life sciences.

Systems biology has proposed various models for biological systems that rely on the complex kinetic reactions of different components. However, these models face practical limitations due to a lack of sufficient kinetic information. Currently, deterministic chemical kinetics serve as the primary framework for modeling biological systems. Yet, applying this framework to model biological systems is challenging due to the scarcity of information regarding chemical kinetics. Additionally, the slow transport of reactants can significantly impact reaction rates.

In biological systems, particularly enzymatic reactions, the reaction speed is much faster than the transport of reactants. Therefore, it is essential to consider the transport behavior of components in biological system models. To address these limitations, a computational fluid dynamics (CFD) approach is introduced, incorporating the transport of components and the kinetics of biological reactions in biological system modeling.

## 7. Big Data, Artificial Intelligence (AI), Machine Learning (ML), Deep Learning (DL)

In the realm of biological flow modeling, researchers have traditionally focused on specific length and time scales, leading to separate investigations in different fields. However, with the advent of high-performance computing architectures, multiscale simulation approaches have emerged, bridging these distinct research areas. Physics-based descriptions are crucial for understanding the behavior of bio-fluids within living organisms, as microscale properties often impact overall behavior. The study of such phenomena across various lengths and time scales generates large datasets. Consequently, mathematical modeling, experiments, and computer simulations have been traditional scientific approaches, now supported by the fourth paradigm and advancements in computer science, including High-Performance Computing (HPC), Artificial Intelligence (AI), Machine Learning (ML), DeepLearning (DL) and data mining methods. The following graph (Fig. 4) indicates the exponential growth of research publications in the last five years that used these advanced techniques in the areas of biological modeling.

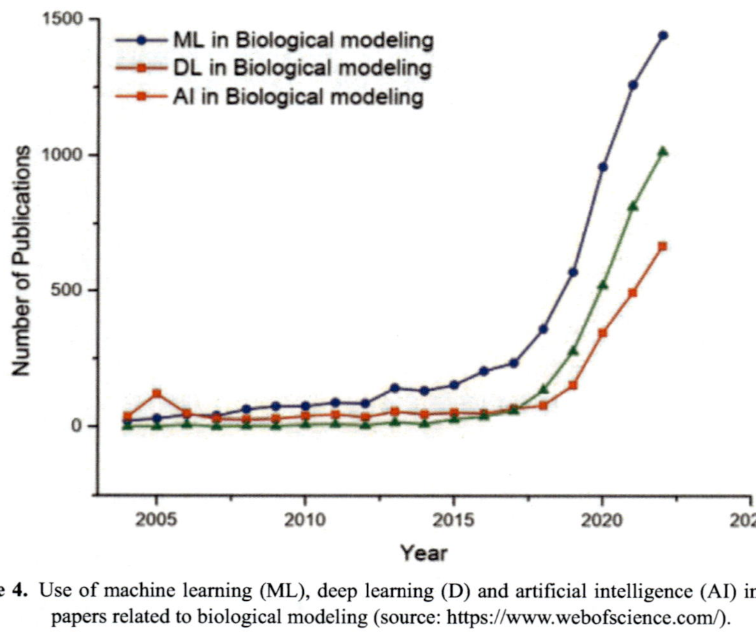


**Figure 4.** Use of machine learning (ML), deep learning (D) and artificial intelligence (AI) in research papers related to biological modeling (source: https://www.webofscience.com/).

## *7.1 Big Data*

According to a report published by The Royal Society (2017), 90% of the world's data has been produced in the past 5 years. Systems bioengineering involves the integration of data generation, predictive modeling, and quantitative analysis. This approach differs from the computational and informatic efforts in biomedical Big Data, which aim to enable the reuse of large, accessible data repositories through novel algorithms. Systems bioengineers engage in the early stages of data generation, allowing them to shape the study design from the beginning and assess progress toward the study's goals as data is collected. On the other hand, Big Data science focuses on the long-term storage and interoperability of observations for broader applications beyond the original intentions. Both stages are crucial in ensuring the enduring value of experimental investments. The National Science Foundation (USA) recognizes the leverage of molecular-level Big Data for predicting biological phenotypes as one of its 10 Big Ideas (NSF, 2023). Moving forward, systems bioengineering has the potential to benefit from data science while also contributing to its advancement.

## *7.2 Artificial Intelligence (AI)*

In the past few years, the combination of medical data collection and analysis with advancements in artificial intelligence has opened up exciting possibilities in the field of biomedicine. Machine learning (ML) and Deep learning (DL), in particular, have emerged as a highly promising tool for extracting valuable insights from medical data and achieving various objectives. It offers the potential to discover new alternative solutions and address complex biomedical challenges in a more automated, efficient, and timely manner. The following sub-sections discuss the applications of machine learning and deep learning in biological modeling and the challenges associated with them.

## *7.3 Machine Learning (ML)*

Machine learning (ML) is a type of artificial intelligence that can be used to analyze Big Data (i.e., large datasets) and to identify patterns in the data. Machine learning, as a subset of artificial intelligence (AI), utilizes statistical techniques to analyze data, construct algorithms, and make predictions. It is typically divided into two categories: supervised learning and unsupervised learning. In supervised learning, the focus is on making predictions for labeled data by leveraging the knowledge derived from training data that maps inputs to corresponding outputs. On the other hand, unsupervised learning involves exploring and identifying patterns within unlabeled data before making any predictions or classifications.

ML can be used to develop predictive models of drug action and to identify new drug targets. ML can be used to improve the accuracy of biomedical flow modeling. This can be achieved by using machine learning algorithms to identify patterns and features in flow data. These features can then be used to identify underlying physical phenomena, such as turbulence, and to improve the accuracy of flow predictions. Additionally, ML can be used to develop models that are able to predict the behavior of fluids in biomedical applications. This can be used to improve the accuracy of diagnosis, treatment, and prevention of diseases. Machine learning provides accurate and cost-effective predictions, outperforming traditional numerical simulations in terms of computational efficiency. It captures relevant data behavior, eliminates irrelevant features, and offers explainable techniques. In contrast, simulations are limited by high computational costs, restricting their application to small systems and short durations. Machine learning bridges this gap, enabling results that are comparable to experimental findings (Sofos et al., 2022).

Combining multiscale modeling and machine learning is a powerful approach to understanding complex phenomena and generating predictive models. While multiscale modeling alone struggles to efficiently integrate diverse datasets, the integration of machine learning overcomes this limitation. This synergistic combination enables the incorporation of underlying physics, addresses ill-posed problems, and explores extensive design spaces. By integrating machine learning and multiscale modeling, we can gain new insights into disease mechanisms, discover novel targets and treatment approaches, and make informed decisions for human health (Alber et al., 2019). However, challenges remain in combining these two approaches, such as creating robust mechanistic models with limited data, managing ill-posed problems, effectively exploring large design spaces to identify correlations, accurately predicting system dynamics for causal analysis, and understanding the limitations of both machine learning and multiscale modeling.

Figure 5 demonstrates how multiscale modeling can leverage machine learning (ML) to create surrogate models that bridge the gap between different scales (Baker et al., 2018). The behavior of an atom within a molecule, liquid, or solid is influenced by the forces it experiences. By accurately learning the relationship between these forces and the atomic chemical environment using big-data techniques, simulations can be accelerated. ML has been employed from the 'bottom' to extract classical

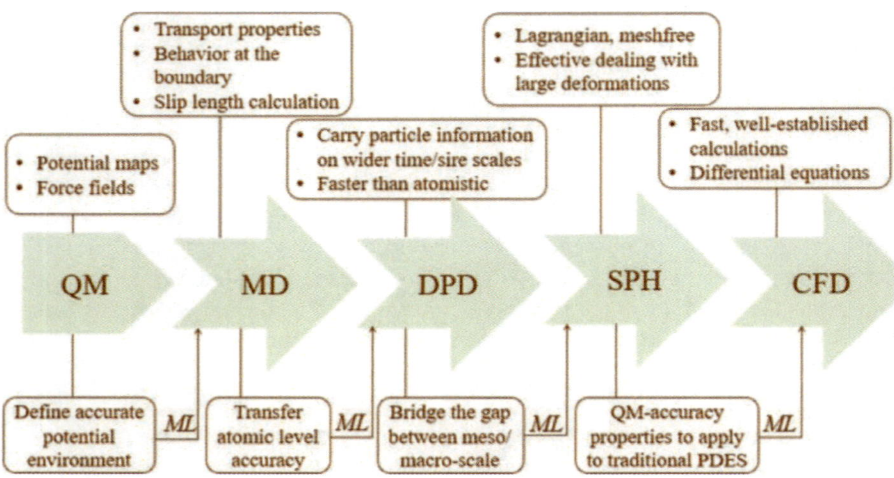


**Figure 5.** ML-enabled multiscale modeling [Sofos et al., 2022].

Potential Energy Surfaces (PES) from Quantum Mechanical (QM) calculations. This enables efficient Molecular Dynamics (MD) simulations that account for quantum effects. Learning the dynamics across scales, specifically the deterministic macroscale and the stochastic microscale, can significantly reduce computational time and effort (Holland et al., 2015). Techniques have been proposed to minimize the number of simulations at lower scales by incorporating the Dissipative Particle Dynamics (DPD) mesoscale method. Additionally, the integration of experimental data into the Smoothed-Particle Hydrodynamics (SPH) mesoscale particle simulation method using ML offers a comprehensive framework for studying the rheological properties of complex fluids (Bai et al., 2021).

## *7.4 Deep Learning (DL)*

The rapid progress of biological and medical technologies has resulted in the accumulation of vast quantities of biological and physiological data, including medical images, which provide valuable insights into human health and disease. Deep learning algorithms, which are based on artificial neural networks, have emerged as powerful tools for extracting complex patterns and features from this data. By utilizing multi-layered deep neural networks, deep learning can interpret various types of data, such as images, sounds, and texts. Although deep learning has predominantly been applied to tasks like image recognition, video analysis, and natural language processing, it also holds potential in the field of life sciences.

In the domain of medical imaging, deep learning has revolutionized the way diseases are diagnosed and assessed. Unlike traditional machine learning methods that relied on manually designed features, deep learning's feature extraction capabilities have surpassed expectations, making it applicable in various medical applications, including brain tumor segmentation (Pereira et al., 2016) Cerebral microbleeds from MR images (Dou et al., 2016), Coronary artery calcium scoring (Wolterink et al., 2016), Mammographic lesions detection (Kooi et al., 2017) and hemorrhages detection in fundus (Van Grinsven et al., 2016). Deep learning's ability to extract data-driven features has also made it increasingly popular in genomic sequence analysis. However, challenges persist when applying deep learning to the biomedical field. One challenge is the requirement for large labeled datasets, which can be time-consuming and necessitate expert annotations. Additionally, there are privacy concerns related to the collection and protection of medical image data. Moreover, biomedical data often suffer from class imbalance, where the data from normal classes outweighs that from other classes. Addressing these challenges requires advancements in technology to handle the demands of high-resolution inputs, training speed, and memory requirements associated with analyzing medical images.

Precision medicine research aims to utilize all available biological, biomedical, and health data for active learning. The wealth of health data generated from wearable sensors, smartphones, and medical devices provides unprecedented opportunities. Deep learning serves as a promising tool for interpreting this data, enabling tasks such as disease prediction, prevention, diagnosis, prognosis, and therapy. Deep learning has emerged as a transformative technology in the biomedical field, finding applications in medical imaging, genomic sequence analysis, and precision medicine. Despite challenges, the potential of deep learning to extract meaningful insights from complex biological and medical data offers great promise for advancing our understanding of human health and enhancing healthcare outcomes such as forecasting epidemics/pandemics, preventing illnesses, and making informed clinical judgments (Cao et al., 2018).

## 8. Conclusions

The field of biological engineering is undergoing significant changes, expanding beyond its traditional focus areas to encompass cellular and molecular levels of healthcare. While it initially emerged from the work of biologists, engineers now play a crucial role in developing laboratory instrumentation and bioinformatic tools. Advancements in cellular engineering and molecular biology have opened up new possibilities for treating diseases at the tissue, cellular, and genetic levels, going beyond traditional pharmaceutical approaches.

Systems biology, an interdisciplinary approach that utilizes mathematical models, has emerged as a powerful tool in studying biological systems. It aids in the identification of new drug targets and the analysis of drug effects on complex biological networks. The integration of systems biology and systems engineering recognizes the importance of engineering principles in advancing these fields. Systems biology involves quantitatively modeling biological systems, offering predictive models for design purposes. On the other hand, systems engineering focuses on creating biological circuits using well-specified components, drawing parallels with the microelectronics industry. These fields also explore the development of functional surfaces and chemical cells for targeted cellular interactions, presenting opportunities for the development of novel therapies.

Biomedical engineers are poised to play a critical role in this paradigm shift, necessitating the introduction of new modules and courses in their training programs. The field of bio-fluid flow modeling is experiencing rapid advancements driven by computational methods, experimental techniques, and imaging technologies. High-performance computing, multi-scale/multi-physics modeling, artificial intelligence, machine learning, deep neural networks, and advanced imaging technologies have opened up new avenues for understanding and predicting fluid dynamics within biological systems. These developments have profound implications for healthcare, enabling the design of more effective diagnostics, treatment strategies, and personalized medicine approaches.

By capitalizing on these emerging trends, researchers are poised to unravel the complexities of bio-fluid flow, ultimately contributing to improved well-being and enhanced quality of life for individuals worldwide. The integration of engineering principles, computational methods, and advanced imaging techniques holds great promise in advancing our understanding of biological systems and revolutionizing healthcare practices.

## References

Alber, M., Buganza Tepole, A., Cannon, W.R. et al. (2019). Integrating machine learning and multiscale modeling—perspectives, challenges, and opportunities in the biological, biomedical, and behavioral sciences. npj Digit. Med. 2: 115. https://doi.org/10.1038/s41746-019-0193-y.

ARA. Multiple-Path Particle Dosimetry Model (MPPD v 2.11). Available online: https://www.ara.com/products/multiple-pathparticle-dosimetry-model-mppd-v-211 (accessed on 3 July 2023).

Bai, J., Zhou, Y., Rathnayaka, C.M., Zhan, H., Sauret, E. and Gu, Y. (2021). A data-driven smoothed particle hydrodynamics method for fluids. Eng. Anal. Bound. Elem. 132: 12–32.

Baker, R.E., Pen˜a, J-M., Jayamohan, J. and Je´rusalem, A. (2018). Mechanistic models versus machine learning, a fight worth fighting for the biological community? Biol. Lett. 14: 20170660. http://dx.doi.org/10.1098/rsbl.2017.0660.

Cao, C., Liu, F., Tan, H. Song, D., Shu, W., Li, W. et al. (2018). Deep learning and its applications in biomedicine. Genomics Proteomics Bioinformatics 16: 17–32. https://doi.org/10.1016/j.gpb.2017.07.003.

Das, A., Habib, G., Perumal, V. and Kumar, A. (2020). Estimating seasonal variations of realistic exposure doses and risks to organs due to ambient particulate matter-bound metals of Delhi. Chemosphere 260: 127451. https://doi.org/10.1016/j.chemosphere.2020.127451.

Dou, Q., Chen, H., Yu, L., Zhao, L., Qin, J., Wang, D. et al. (2016). Automatic detection of cerebral microbleeds from MR images via 3D convolutional neural networks. IEEE Trans Med Imaging 35: 1182–95.

Feng, Y., Xu, Z. and Haghnegahdar, A. (2016). Computational fluid-particle dynamics modeling for unconventional inhaled aerosols in human respiratory systems. Ch. 3. *In*: Volkov, K. (ed.). Aerosols - Science and Case Studies. IntechOpen Limited, London. http://dx.doi.org/10.5772/65361.

Holland, D.M., Lockerby, D.A., Borg, M.K., Nicholls, W.D. and Reese, J.M. (2015). Molecular dynamics pre-simulations for nanoscale computational fluid dynamics. Microfluid. Nanofluid. 18: 461–474.

Ideker, T., Galitski, T. and Hood, L. (2001). A new approach to decoding life: Systems Biology. Annu. Rev. Genomics Hum. Genet. 2: 343.

Janesa, K.A., Chandranb, P.L., Fordc, R.M., Lazzarac, M.J., Papina, J.A., Peircea, S.M. et al. (2017). An engineering design approach to systems biology. Integr Biol (Camb). 2017 July 17; 9(7): 574–583. doi:10.1039/c7ib00014f.

Kolanjiyil, A. and Kleinstreuer, C. (2017). Computational analysis of aerosol-dynamics in a human whole-lung airway model. Journal of Aerosol Science 114: 310–316.

Kooi, T., Litjens, G., van Ginneken, B., Gubern-Me´rida, A., Sa´ nchez, C.I., Mann, R. et al. (2017). Large scale deep learning for computer aided detection of mammographic lesions. Med. Image Anal. 35: 303–12.

Kuprat, A.P., Kabilan, S., Price, O., Asgharian, B. and Corley, R.A. (2016). Multiscale modeling of aerosol deposition in the complete respiratory tracts of animals and humans. Toxicologist 150: 286.

Kuprat, A.P., Jalali, M., Jan, T., Corley, R.A., Asgharian, B., Price, O. et al. (2021). Efficient bi-directional coupling of 3D computational fluid-particle dynamics and 1D Multiple Path Particle Dosimetry lung models for multiscale modeling of aerosol dosimetry. Journal of Aerosol Science 151(105647). https://doi.org/10.1016/j.jaerosci.2020.105647.

Ladumor, M.K. and Unadkat, J.D. (2022). Predicting regional respiratory tissue and systemic concentrations of orally inhaled drugs through a Novel PBPK Model. Drug Metab Dispos 50: 519–528. dx.doi.org/10.1124/dmd.121.000789.

Longest, P.W., Tian, G., Khajeh-Hosseini-Dalasm, N. and Hindle, M. (2016). Validating whole-airway CFD predictions of DPI aerosol deposition at multiple flow rates. Journal of Aerosol Medicine and Pulmonary Drug Delivery 29: 461–481.

Manojkumar, N., Srimuruganandam, B. and Shiva Nagendra, S.M. (2019). Application of multiple-path particle dosimetry model for quantifying age specified deposition of particulate matter in human airway. Ecotoxicology and Environmental Safety 168: 241–248. https://doi.org/10.1016/j.ecoenv.2018.10.091.

National Science Foundation (NSF). (2023). 10 Big Ideas for Future NSF Investments. https://www.nsf.gov/about/congress/reports/nsf_big_ideas.pdf [accessed on 7 July 2023].

Pereira, S., Pinto, A., Alves, V. and Silva, C.A. (2016). Brain tumor segmentation using convolutional neural networks in MRI images. IEEE Trans Med. Imaging 35: 1240–51.

Sofos, F., Stavrogiannis, C., Exarchou-Kouveli, K.K., Akabua, D., Charilas, G. and Karakasidis, T.E. (2022). Current trends in fluid research in the era of artificial intelligence: a review. Fluids 7: 116. https://doi.org/10.3390/fluids7030116.

Teorell, T. (1937). Studies on the diffusion effect upon ionic distribution: II. Experiments on ionic accumulation. J. Gen. Physiol. 21:107–122. doi: 10.1085/jgp.21.1.107.

The Royal Society. (2017). Machine learning: the power and promise of computers that learn by example. Report no. DES4702. ISBN: 978-1-78252-259-1. Issued April 2017.

Tu, J., Inthavong, K. and Ahmadi, G. (2023). Computational Fluid and Particle Dynamics (CFPD): An introduction. Computational Fluid and Particle Dynamics in the Human Respiratory System 18: 1–18. doi: 10.1007/978-94-007-4488-2_1.

Van Grinsven, M.J.J.P., van Ginneken, B., Hoyng, C.B., Theelen, T. and Sanchez, C.I. (2016). Fast convolutional neural network training using selective data sampling: application to hemorrhage detection in color fundus images. IEEE Trans Med Imaging 35: 1273–84.

Wolterink, J.M., Leiner, T., de Vos, B.D., van Hamersvelt, R.W., Viergever, M.A. and Isˇgum I. (2016). Automatic coronary artery calcium scoring in cardiac CT angiography using paired convolutional neural networks. Med. Image Anal. 34: 123–36.

Yang, Y., Chen, Y., Wang, L., Xu, S., Fang, G., Guo, X. et al. (2022). PBPK modeling on organs-on-chips: an overview of recent advancements. Front Bioeng. Biotechnol. 14(10): 900481. doi: 10.3389/fbioe.2022.900481.

Yin, Y., Choi, J., Hoffman, E.A., Tawhai, M.H. and Lin, C.L. (2013). A multiscale MDCT image-based breathing lung model with time-varying regional ventilation. J. Comp Phys. 244: 168–192.

Yoo, S-J. and Ito, K. (2018). Numerical prediction of tissue dosimetry in respiratory tract using computer simulated person integrated with physiologically based pharmacokinetic–computational fluid dynamics hybrid analysis. Indoor and Built Environment 27(7): 877–889. doi:10.1177/1420326X17694475.

# Index


## A

Airway Segmentation 10, 19, 22
Airway Structure Modeling 1, 3
Aneurysm 87–89, 91–95, 97, 98
Aortic aneurysm 87, 88, 91, 92, 94, 98
aqueous humor 142, 147, 150, 151
arterial aneurysm 100, 102, 107, 110–115
Artificial Intelligence (AI) 232, 240, 241, 244
Atherosclerosis 93

## B

Bio-fluid flow modelling 231
Biomechanics of CSF flow 125
Blood flow 101–103
Brain ventricles and CSF flow 118, 120, 121, 123, 124, 131

## C

Cardiac Assist Device 215, 225, 227, 228
Cardiovascular Disease 214
Cerebrospinal fluid (CSF) 118–135
Cervical subarachnoid space 127, 128
Computational Fluid Dynamics (CFD) 2, 30, 31, 33, 51, 65, 118, 121, 135, 158, 163–165, 170–172, 175, 178, 181–185, 231, 233, 234, 237, 240
Computational Fluid-Particle Dynamics (CFPD) 234, 235, 237
COPD 50, 53, 70
COVID-19 215, 216, 220–228
CSF flow dynamics 122, 127
CSF production and circulation 123

## D

Deep Learning (DL) 2, 18, 22, 232, 240, 241, 243, 244
direct drug delivery technology 83
Discrete Element Method 67
DPI 53, 54, 65–68, 70, 71
drug absorption 74–80
drug delivery 50–54, 58, 65–68, 70, 71, 74, 75, 78, 81, 83, 84, 189–199, 202–205, 207, 208
drug formulations 74, 75
dry eye syndrome 147

## E

eye health 142, 154

## F

Finite Element Analysis (FEA) 158, 164
finite volume method 102
first pass metabolism 74, 83
flow field 31, 36, 37, 40, 47
Fluid-Structure Interaction (FSI) 164

## G

Gastrointestinal Flow 168
Geometry Modeling 4, 5, 7
glaucoma 142, 143, 147, 150, 151, 154

## H

Health 1
Heart Failure 214–217, 221, 223, 226
High-Performance Computing (HPC) 231, 232, 240, 244
Human Airway 3
Human Respiratory Tract 51, 60, 69
Hydrodynamic Lubrication 174

## I

*In vitro* CSF models 129, 131
Inhaler 50–54, 57, 66–68, 70, 71

Interstitial Fluid 162, 163, 165, 174
intraocular pressure 141, 142, 150

**L**

Laminar Flow 161, 172
Left Ventricle Assist Device (LVAD) 215–217, 219, 221, 223–228
Lymphatic Flow 158, 159, 161, 163

**M**

Machine Learning (ML) 232, 240–244
magnetic field 102–104, 107, 114–116
Magnetic resonance imaging (MRI) 120, 122, 124–131, 134, 135
Microfluidics 234
Mucus Flow 158, 177, 178
Multiple Path Particle Dosimetry (MPPD) 235–237
Multi-scale modelling 237

**N**

nanocarriers 190
Nanoparticles 189–193, 195–197, 199, 200, 202, 205, 207
nanotechnology 189, 204
nasal mucosa 74, 78, 79, 81–83
Nasal route 74, 75, 77–79, 81–84
Neural Networks 3, 18–22
Numerical Methods 64, 66, 68, 69
Numerical Modeling of CSF 129, 131, 134

**O**

ocular fluid dynamics 150, 154
Ocular fluids 150, 154
ophthalmic care 153
Oral Inhaler 51
Organ-on-a-Chip (OoC) 234

**P**

particle deposition 30, 31, 35, 36, 44–47
patient compliance 82, 83
Physiological flow 87, 92, 94
Physiologically Based Pharmacokinetic (PBPK) modelling 232–234
pMDI 52, 53, 66, 70
Pulmonary drug delivery 50, 51, 66, 68, 70, 71

**R**

Respiratory Disease 2, 12
Respiratory Tract 12, 18
Reverse-Engineering 4

**S**

Shear Stress 170, 184
Splanchnic Circulation 165–168
stenosis airway 30, 31, 40, 41, 43
Synovial Fluid 172–176
Synthetic Biology 238, 239
Systems Bioengineering 238, 239, 241
Systems Biology 238, 239, 244

**T**

targeted drug delivery 83
tear film 143, 145, 149, 150
Thrombosis 218–222, 224, 226–228
Tumor 51, 56
Turbulence flow 62, 65

**U**

upper airway 29–32, 34, 41
Urine Flow 158, 182–185

**V**

Viscoelastic fluid 92, 98
Viscosity 158, 160, 171, 175, 177, 178, 182
vitreous humor 146

**W**

wall shear stress 101, 102, 107, 110, 111, 115, 116
Womersley Number 161

Made in the USA
Las Vegas, NV
16 September 2026

bf06237c-1daa-4962-a557-2e4a96ab1b94R03